# SUPERACIDS AND ACIDIC MELTS AS INORGANIC CHEMICAL REACTION MEDIA

# SUPERACIDS AND ACIDIC MELTS AS INORGANIC CHEMICAL REACTION MEDIA

## Thomas A. O'Donnell

Thomas A. O'Donnell
School of Chemistry
The University of Melbourne
Parkville, Victoria, 3052
Australia

**Library of Congress Cataloging-in-Publication Data**

O'Donnell, Thomas A.
   Superacids and acidic melts as inorganic chemical reaction media /
Thomas A. O'Donnell.
      p.      cm.
   Includes bibliographical references and index.
   ISBN 1-56081-035-1.—ISBN 3-527-28092-8
   1. Ionization.   2. Solution (Chemistry)   3. Chemistry, Inorganic.
   4. Superacids.   5. Molten salts.   I. Title.
   QD561.034   1992
   546'.26423—dc20                                                    92-15542
                                                                          CIP

**British Library Cataloging in Publication Data**

A catalogue record for this book is available from the British Library.

Printed in the United States of America
ISBN 1-56081-035-1 VCH Publishers
ISBN 3-527-28092-8 VCH Verlagsgesellschaft

Printing History:
10 9 8 7 6 5 4 3 2 1

Published jointly by

| | | |
|---|---|---|
| VCH Publishers, Inc. | VCH Verlagsgesellschaft mbH | VCH Publishers (UK) Ltd. |
| 220 East 23rd Street | P.O. Box 10 11 61 | 8 Wellington Court |
| New York, New York 10010 | D-6490 Weinheim | Cambridge CB1 1HZ |
| | Federal Republic of Germany | United Kingdom |

# Preface

To date there has been much more investigation of organic chemical reactions in superacidic media than is the case for inorganic systems. Chemists generally regard as superacids those solvent systems comparable in strength with 100% sulfuric acid or of greater strength. The only current book dedicated to superacid chemistry is that by George Olah and his colleagues, cited as reference 13 in chapter 1 of this book. There is a general introduction to superacidic systems in that book, but virtually all of the rest of it is devoted to organic chemical reactions in superacids. Their treatment of inorganic systems is very brief.

Ronald Gillespie pioneered the application of rigorous physicochemical investigation to solutions of sulfuric acid and fluorosulfuric acid to demonstrate the nature of cations of the halogen and chalcogen elements in these superacids, and he and his colleagues measured Hammett Acidity Functions in these two acids and in hydrogen fluoride. Without those acidity measurements, much of the systematization of generation and stabilization of unusual cations which is presented in this book could not have been undertaken.

The general aim of this book is to demonstrate that solute speciation depends on the acidity or basicity of nonaqueous media, as for water. Cations are stable in a superacid or molten salt of enhanced acidity, anions are stable when the particular medium—superacid or melt—is made basic and binary compounds formed between transition element cations and the bases of the solvents are insoluble in neutral media; that is, binary fluorides, bisulfates, and fluorosulfates are insoluble in the neutral superacids or melts, just as transition metal "hydroxides" are insoluble in neutral water.

While there is nothing untoward about the existence of cations in "normal" oxidation states in superacids and acidic melts, a special feature of superacid chemistry is that cations can be generated in fractional or very low oxidation states. Cations like $I_2^+$, $Se_4^{2+}$, $Bi_5^{3+}$, $Ti^{2+}$, and $Sm^{2+}$ will be discussed in chapters 4, 5, and 7. Their stability toward base-induced disproportionation is also a major theme.

It will be demonstrated that the general nature of cationic speciation is essentially the same in all nonaqueous solvents, regardless of the chemical nature or the temperature domain of the solvent. Further, it will be shown that the level of acidity or basicity of the solution medium is the principal, and frequently the sole determinant of the particular nature of unusual ions generated and stabilized in solution.

Many of the major books on nonaqueous chemistry tend to compartmentalize the discussions of solute-solvent interaction by presenting separate chapters on individual nonaqueous solvents written by different authors. Frequently chapters tend to be recitals of the physical properties of individual solvents, not always related to the solution chemistry, and of solubilities, solvolysis, redox, and acid-base properties of a wide range of solutes. There is rarely any apparent attempt to provide a uniform approach or to deal with related solutes in different solvents. Molten salt chemistry is often presented as though it were significantly different from inorganic solution chemistry in more conventional nonaqueous solvents.

Where this book differs from others is that each of several of the later chapters will deal with a particular group of inorganic cations—polyatomic or monatomic—and, *within each chapter,* the generation and stability of that group of cations will be considered in all four of the superacids HF, $HSO_3F$, $CF_3SO_3H$, and $H_2SO_4$ *and* in chloroaluminate melts and, where possible, will be related to the behavior of those cations in aqueous solution.

Chapter 1 uses very brief accounts of the self-ionization and acid-base equilibria of individual members of a wide range of solvents, which are liquids at or near room temperature, starting with water and then proceeding through more basic and then more acidic protonic solvents, then to nonprotonic solvents, and finally to molten salts in an attempt to simplify and generalize fundamental acid-base concepts.

It is suggested in chapter 1 that certain definitions of acid-base behavior, for example, the Lux-Flood and the solvo-acid definitions, are unnecessary and can be incorporated into a general treatment of acid-base behavior within the Lewis system. Throughout the book acid-base reactions are treated solely within the Lewis system. Of course, it is recognized that some elements of Bronsted acidity and basicity must enter into discussions of equilibria in protonic superacids because of the increase or decrease of concentration of the solvated proton which occurs when a Lewis acid or a Lewis base is added.

Chapter 2 is devoted to a description of the physicochemical techniques, based largely on conductance and cryoscopy, which have been used to elucidate the nature of cationic species generated in superacid media. It also describes the spectrophotometric, potentiometric, and other experimental approaches which have been used to determine acidity levels in protonic superacids and indicates why Hammett Acidity Functions are chosen to delineate acid-base levels in the descriptive solution chemistry in chapters 4–7.

There is detailed discussion in chapter 3 of individual Lewis acids of the four protonic superacids $H_2SO_4$, $CF_3SO_3H$, $HSO_3F$, and HF because these are the four acids in which, to the extent that the information is available, generation and stability of the cations described in chapters 4–7 are discussed.

Chloroaluminates represent fairly simple melt systems for chemical investigations and are the only melts discussed in this book. There appear to be virtually no reports of identification of cationic species, other than those derived from the media themselves, in any acidic molten salts except those based on $AlCl_3$. As stated toward the end of chapter 1, acid-base behavior in molten salts based on oxoanions, e.g., nitrates and sulfates, appears to be similar to that in chloroaluminates, but those molten salt systems are themselves very much more complex. The author believes that similar principles for cationic and anionic generation and stabilization to those established in this book for protonic superacids and chloroaluminates would apply in molten sulfates and nitrates and in media that are even more complex, for example, geological magma and industrial slags.

In each of the chapters 4–7 of this book an individual group of cations will be presented—polyatomic cations of nonmetals and of metals in chapters 4 and 5 and monoatomic transition metal cations in "normal" oxidation states in chapter 6 and in unusually low oxidation states in chapter 7. As stated above, the generation and stability of each group of cations will be described for all four of the superacids considered in this book, in chloroaluminate melts and, where possible, in aqueous solutions.

Because of the present state of development of inorganic solution chemistry in highly acidic media, it is not possible to give approximately even weighting to the chemistry in each of the four protonic superacids and in chloroaluminate melts. The major contribution to the chemistry of homopolyatomic cations of nonmetals (chapter 4) has been Gillespie's work on halogens and chalcogens in $HSO_3F$ while the studies of homopolyatomic cations of metals (chapter 5) have been conducted largely in molten salts by Corbett and others. Most of the definitive work on characterization of solvated cations of transition metals has been carried out in HF at Melbourne, although there is a significant body of work in acidic molten salts; but most molten salt investigations of transition metal compounds have involved anionic species in basic melts and lie outside the scope of this book.

The reported experimental data on preparation and stability of cationic species presented in chapters 4–7 are drawn together in chapter 9 to produce a set of principles which govern the acidity dependence of cation stabilization and, conversely, the fact that cations in unusually low oxidation states undergo base-induced disproportionation reactions. This framework is then used in chapter 10 to allow some reinterpretation and systematization of reactions previously reported in the chemical literature and to suggest some general synthetic strategies for future cation syntheses in nonaqueous media.

Realization that stabilization of unusual polyatomic cations in highly acidic media was governed by a very small set of general principles led to what is essentially a corollary in a converse sense—polyatomic anions such as Zintl ions are generated in strongly basic media or in the absence of acidic or electrophilic species. Chapter 8 is offered to illustrate this latter generalization. It is not intended to be a comprehensive or exhaustive treatment of a literature which is itself limited and fragmented. There has been almost no deliberate experimental control of basicity in attempts to form specific polyatomic anions.

Many investigations of synthesis of polyatomic cations and anions in acidic and basic media or under "naked" conditions over the last two to three decades have resulted in elucidation of some fascinating structures of new ions—indeed, much work was directed toward crystallographic investigation and attempts to describe bonding in these structures. However, the general thrust of this book is to illustrate and systematize the acid-base dependence of generation and stability of ions *in solution*, so there is virtually no diagrammatic representation of structures. Occasionally they are dealt with descriptively when they provide confirmation or elucidation of the nature of species in solution.

Some compounds having low thermal stability in solution or in the solid state are discussed in appropriate chapters, but transient species are not reported. For example, $Cl_3^+$ is given as the only chlorine cation, existing in a stable solid compound at $-78°C$. $Cl_2^+$, identified in the gas phase from its electronic band spectrum under conditions of low-pressure discharge, is not discussed.

Although many references to individual papers in the chemical literature have been necessary, particularly for more recent work, reviews and chapters have been used for the convenience of readers in situations where those reviews are comprehensive and authoritative and have not become dated.

It is a pleasure to record my appreciation for so much of the research reported in this book that has been carried out by my postgraduate research students, particularly those recent ones who worked on superacid systems and whose yet-unpublished work is reported in several chapters. In particular I wish to acknowledge the imaginative and meticulous work done by Russell Cockman and John Besida on characterization of metal and nonmetal cations in HF. Their research led to subsequent work within my group by later students on characterization of species in $H_2SO_4$, $HSO_3F$, and $CF_3SO_3H$ and also led to the reinterpretation of much earlier work in acidic melts.

I wish also to express my gratitude to many research colleagues for stimulating discussions on superacid and melt systems. In particular I owe much to discussions and correspondence with Professor Ronald Gillespie of McMaster University and Professor Charles Hussey of the University of Mississippi.

Miss Wendy Edwards coped cheerfully and competently with my badly handwritten drafts and converted them to typescripts which I could offer with confidence to the publishers.

Finally, I wish to pay a grateful tribute to my wife, Pat, who provided immense encouragement during the compilation of this work and who willingly sacrificed much of what might have been leisure time in my "retirement."

T.A. O'D

*Melbourne*
*February 1992*

# Contents

# CHAPTER 1

# Self-ionization and Acid-base Equilibria in Protonic and Nonprotonic Liquids

Much inorganic chemistry can be considered and discussed in acid-base terms in a wide range of liquid media, some of which may themselves be very complex. A considerable degree of simplification is introduced when the liquid medium is itself a single liquid, but the same general approach to acid-base reactions and equilibria that is presented here for single liquids is believed to apply for systems based on mixed solvents or in very complex media such as industrial slags or geological melts.

The purpose of this introductory chapter is to discuss in very general terms self-ionization and acid-base equilibria in a wide range of single liquids. Some, such as water, sulfuric acid and bromine trifluoride, are liquids at ambient temperature. Others such as alkali metal chloroaluminates are solids at ambient temperatures and their solution equilibria must be studied at appropriate elevated temperatures. In most, if not all, of previously published books describing nonaqueous solvent systems, high-temperature melts have been discussed in chapters separate from those describing room-temperature liquids. This can lead to a view, arrived at consciously or unconsciously, that melts are very different in their general solute-solvent interactions from room-temperature liquids. In this book solute behavior in acidic and basic melts will be treated in the same chapters as acidic and basic ambient temperature liquids, protonic or nonprotonic. Different chapters will be used to discuss the chemistry of different classes of solutes in a range of solvents. The temperature domain of a particular liquid or solvent has relatively little effect on the way in which its acid-base properties affect speciation or stability of solutes.

Many textbook authors and editors have classified liquids acting as solvents on the basis of their being acidic or basic. This arbitrary classification seems to rest on an acceptance of water—the most familiar solvent of inorganic chemical systems—as being "neutral." Obviously water or any other solvent can itself be acidic, neutral, or basic and an acidified aqueous solution may well be more acidic than a solution in anhydrous acetic acid, for example, which as a solvent within this classification might be deemed to be "acidic." Similarly, it is necessary to define acidic, neutral, and basic solutions in liquid ammonia chemistry.

A more valid differentiating classification depends on whether acid-base equilibria in the solvent depend, or do not do so, on proton transfer in the self-ionization for the liquid. A typical protonic (or protic) liquid HA (e.g., $H_2O$, $NH_3$, or HF) is one described by the self-ionization (or autoprotolysis) described by the equation

$$HA \rightleftharpoons H^+ + A^- \qquad\qquad 1.1$$

The bare proton has no separate existence in any liquid and both it and the anion will be solvated to greater or less extent, so the self-ionization equilibrium can be written

$$(m + n)\ HA \rightleftharpoons H^+_{(HA)_m} + A^-_{(HA)_n} \qquad\qquad 1.2$$

where m molecules of HA are coordinated to $H^+$ and n molecules to $A^-$ to form the solvated ions. Alternatively, and somewhat more simply, the equilibrium can be represented as

$$HA \rightleftharpoons H^+_{solv} + A^-_{solv} \qquad\qquad 1.3$$

In the absolutely pure liquid, a condition probably never achieved under experimental conditions, the concentration of $H^+_{solv.}$ equals that of $A^-_{solv.}$; i.e., the system is neutral. Any compound that causes an increase of $H^+_{solv.}$ concentration over that of $A^-_{solv.}$ is an acid while, of course, a base acts in the opposite direction.

For protonic solvents which are not superacids, it is usual but not essential in all cases to use the Bronsted definition of an acid; i.e., it is a compound HX capable of donating a proton fully or partially to a base, in this case the solvent HA:

$$HX + HA \rightarrow H_2A^+ + X^- \qquad\qquad 1.4$$

A Bronsted base Y is capable of accepting a proton from an acid which again can be the solvent HA:

$$Y + HA \rightarrow HY^+ + A^- \qquad\qquad 1.5$$

thereby increasing the concentration of $A^-_{solv.}$ over that of $H^+_{solv.}$ in the solution equilibrium (i.e., rendering the solvent basic).

A more general definition of acids and bases that can be applied more widely than the Bronsted definitions, but which includes the Bronsted concept of acids and bases, was put forward by G. N. Lewis. The reaction between gaseous $NH_3$ and gaseous $BF_3$ to form the 1:1 solid adduct $NH_3.BF_3$ provides an example of Lewis acidity and basicity. $NH_3$ in providing a pair of electrons to constitute a covalent bond in forming the adduct is acting as a Lewis base and $BF_3$ in accepting a pair electrons to form the bond is acting as a Lewis acid. There is no proton transfer in this acid-base reaction even though there are hydrogen atoms *in the base*. An even more clear-cut distinction between Lewis and Bronsted definitions is in a reaction which is typical of many used later in this book:

$$BF_3 + F^- \rightarrow BF_4^- \qquad\qquad 1.6$$

The Lewis acid-base definitions apply regardless of whether the reaction occurs as a gas-solid one, for example, between gaseous $BF_3$ and solid KF, or occurs in an interacting solvent, such as HF, or in a noninteracting one, such as $CH_3CN$.

As stated earlier, there are no logical grounds for regarding water as a neutral solvent, as many textbook authors do, implicitly or explicity. Obviously, an aqueous solution, like that in any other solvent, can be acidic, neutral, or basic. However, water is the most familiar and the most studied of solvents and therefore can be used as a reference solvent in discussions of acidity and basicity in other systems. To greater or less extent the discussion of self-ionization in water and of interaction of acids and bases with it as a solvent applies to other solvents to be presented in this chapter.

Discussions of each solvent will be limited to acid-base interactions in the solvent, presenting simply a descriptive account of a limited range of Bronsted and Lewis acids and bases, where appropriate for each solvent. Detailed and quantitative treatment, where available, will be given in chapter 3 after experimental procedures for determining the nature of speciation and acid-base strengths of solutes have been described in chapter 2.

There will be no attempt to present detailed physical properties, comprehensive solubility relationships, or solvolysis reactions for organic or inorganic solutes, nor will there be any discussion of redox behavior of solutes. This information is available in great detail in book series such as J. J. Lagowski's *The Chemistry of Nonaqueous Solvents.**

## 1.1 Protonic Solvents

### 1.1.1 Water

The most fundamental, and primitive, equation for the self-ionization of water is

$$H_2O \rightleftharpoons H^+ + OH^- \qquad\qquad 1.7$$

---

*For the convenience of readers and as far as possible, references to solvents, other than for the four protonic superacids dealt with in detail in this chapter, will be drawn from the chapters in Lagowski's series of volumes published over a period of years by Academic Press.

As a recognition of the fact that the bare proton has no meaningful existence in water or any other solvent, the self-ionization is frequently represented as

$$2H_2O \rightleftharpoons H_3O^+ + OH^- \qquad\qquad 1.8$$

The oxonium ion $H_3O^+$ has been characterized by spectroscopic and crystallographic structural analysis in some crystalline species (e.g., $H_3O^+SbF_6^-$, $(H_3O^+)_2IrF_6^{2-}$ and $H_3O^+TiF_5^-$) and it exists as such in solution in superacids, where the basic molecular entity $H_2O$ is being protonated, but there is no real evidence that it exists in normal aqueous solutions. Species like $H_5O_2^+$ probably exist in very concentrated acidic aqueous solutions, but in normal concentration ranges $H_9O_5^+$ is the most likely entity, with some further loose binding of water molecules probably occurring. Additionally, equation 1.8 is oversimplified in that the $OH^-$ ion will also be coordinated to water molecules in aqueous solution. Species like $O_2H_3^-$ are known in the solid state.

A very general form of the self-ionization equation for water is

$$H_2O \rightleftharpoons H_{aq}^+ + OH_{aq}^- \qquad\qquad 1.9$$

where recognition is made of solvation of the ions by $H_2O$ molecules without any attempt to define the extent of coordination.

It would be consistent with other self-ionization equations presented in this chapter to write

$$H_2O \rightleftharpoons H_{solv}^+ + OH_{solv}^- \qquad\qquad 1.10$$

For this self-ionization at 25°C, the ionic product $K_w$, $[H^+][OH^-]$, is $1.0 \times 10^{-14}$ mol$^2$ L$^{-2}$ at 25°C, i.e., $pK_w$ is 14, leading to a pH of 7 for water that is exactly neutral. Therefore acidic solutes that cause an increase in the $H^+$ concentration, $[H^+]$, lower the pH of the solution and, in similar fashion bases raise the pH.

Typical Bronsted acids in water are HCl and $CH_3COOH$. The former is effectively completely ionized in aqueous solution:

$$HCl + H_2O \rightarrow H_3O^+ + Cl^- \qquad\qquad 1.11^*$$

The strongest acid that can exist in aqueous solution is $H_3O^+$, so HCl, a stronger acid, is completely deprotonated. $CH_3COOH$ is a much weaker acid than $H_3O^+$ and is involved in an equilibrium that lies well to the left:

$$CH_3COOH + H_2O \rightleftharpoons H_3O^+ + CH_3COO^- \qquad\qquad 1.12$$

Bronsted acids need not be molecular species, like HCl or $CH_3COOH$; they may be cationic or anionic, reacting with water to form $H_3O^+$ in equilibria such as

---

*Effective solvation of ions is not represented in equations 1.11–1.18.

$$NH_4^+ + H_2O \rightleftharpoons H_3O^+ + NH_3 \hspace{4cm} 1.13$$

and

$$HSO_4^- + H_2O \rightleftharpoons H_3O^+ + SO_4^{2-} \hspace{3cm} 1.14$$

The hydroxide ion $OH^-$ is the strongest base of the aqueous acid-base system and compounds such as NaOH and $Ca(OH)_2$ act as sources of the base $OH^-$ on dissolution in water. An entity, such as the oxide ion $O^{2-}$, which is a stronger base than $OH^-$, will react to completion with water to form $OH^-$ when solid $Na_2O$ is added:

$$O^{2-} + H_2O \rightarrow 2OH^- \hspace{4.5cm} 1.15$$

Typical weak Bronsted bases, uncharged or charged, are $NH_3$ and $CH_3OO^-$, accepting protons from the solvent:

$$NH_3 + H_2O \rightleftharpoons NH_4^+ + OH^- \hspace{3.5cm} 1.16$$

$$CH_3COO^- + H_2O \rightleftharpoons CH_3COH + OH^- \hspace{2.5cm} 1.17$$

Almost invariably, Bronsted acid-base concepts are used in discussing aqueous systems because of the ease of carrying out calculations which are based on the concentration of acid or base, the constant value of $pK_w$, and experimentally derived values of $pK_a$ and $pK_b$, where $K_a$ and $K_b$ are the equilibrium constants for weak acids and weak bases interacting with water as in the equilibria given as equations 1.12–1.14, 1.16, and 1.17.

Lewis acid-base concepts, although of more general applicability over a wide range of inorganic chemical systems, cannot yet be applied quantitatively in the way that the Bronsted approach can be used. However, they should be considered briefly for aqueous systems. The hydroxide ion, with several pairs of electrons available to form bonds, is the base of the system and acts as such with protons, with nonmetals in high oxidation states, and with transition metal cations in hydroxocomplexes. A weak Lewis acid of the aqueous solvent system is boric acid, $H_3BO_3$, which has been shown to decrease the pH, not by proton donation to $H_2O$ to increase the $H_3O^+$ concentration, but by the following reaction:

$$B(OH)_3 + OH^- \rightleftharpoons B(OH)_4^- \hspace{3.5cm} 1.18$$

thereby reducing the $OH^-$ concentration and shifting the self-ionization equilibrium in favor of increase in the $H_3O^+$ concentration.

## 1.1.2 Solvents More Basic Than Water

Basic solvents do not have much direct relevance to the principal areas of subject matter of this book, namely, inorganic systems in superacid media. However, they are dealt with briefly at this point for completeness of treatment of a range of solvents more basic or more acidic than water. Another reason is that in chapter 8 it will be shown, again briefly, that there is a corollary to the general proposition developed extensively in this book, namely, that unusual inorganic cationic species are generated and stabilized in very highly acidic media, that collorary being that unusual anionic species are frequently generated and are stable in highly basic media.

### 1.1.2.1 Ammonia

For about a century there has been extensive study of liquid ammonia as a solvent, the major preoccupation being with determination of the solubilities and solvolysis reactions of a very wide range of organic and inorganic compounds. Much of this work has been reviewed by Lagowski and Moczygemba,[1] who quote a value for the self-ionization constant in the low-boiling liquid ammonia as $1.9 \times 10^{-33}$ mol$^2$ L$^{-2}$ at $-50°C$ with an estimated value of $1.6 \times 10^{-30}$ at 25°C. The self-ionization is usually written as

$$2NH_3 \rightleftharpoons NH_4^+ + NH_2^- \qquad\qquad 1.19$$

More fundamentally it might be written as

$$NH_3 \rightleftharpoons H^+ + NH_2^- \qquad\qquad 1.20$$

In equation 1.19 it is recognized that the bare proton cannot exist in ammonia. The solvated proton can be isolated from $NH_3$ solutions in compounds containing the $NH_4^+$ cation considerably more easily than in the case of the isolation of the oxonium cation $H_3O^+$ (section 1.1.1). $NH_4^+$ is the most acidic species that can exist in liquid ammonia and $NH_2^-$ the most basic and is the formal Lewis base of the solvent system.

Addition of $NH_4NO_3$, for example, increases the acidity and $NaNH_2$ makes the system more basic. Acids such as HCl, which are very strong in water, are fully deprotonated and acids like $CH_3COOH$, relatively weak in water, exhibit the same level of acidity in the strongly basic $NH_3$ as HCl and other "strong" acids. Imide $NH^{2-}$ and azide $N^{3-}$ are stronger bases than $NH_2^-$. Azide is protonated by $NH_3$ to $NH^{2-}$ and to $NH_2^-$ in turn. There appears to have been little study of Lewis acids, as such, in $NH_3$. The nitrates of Zn, Al, and Mg are reported to be solvolysed by solutions of $NaNH_2$ in $NH_3$ to $Zn(NH_2)_2$, $Al(NH_2)_3$, and $Mg(NH_2)_2$ (see ref. 1), and $BCl_3$ yields $B(NH_2)_3$. With additional $NaNH_2$ the metal amides yield $Zn(NH_2)_4^{2-}$, $Al(NH_2)_4^-$, and $Mg(NH_2)_4^{2-}$ (see ref. 1), indicating that they are amide-ion acceptors in strongly basic $NH_3$. However, there does not appear to have been any investigation of the intrinsic Lewis acid strength of these amides, that is, of the extent to which the amides themselves would enhance the $NH_4^+$ concentration of neutral ammonia by combining with $NH_2^-$.

One of the best-known solution properties of liquid $NH_3$ is its ability to react with alkali metals to produce, at low concentrations, solutions containing the alkali metal cations and solvated electrons.

### 1.1.2.2 Hydrazine

Hydrazine is considerably less basic than ammonia and its self-ionization can be represented as

$$2N_2H_4 \rightleftharpoons N_2H_5^+ + N_2H_3^- \qquad 1.21$$

The self-ionization constant is given as about $10^{-13}$ mol$^2$ L$^{-2}$.[2] Acids stronger than $N_2H_5^+$ react to give this, the hydrazinium cation. These include acids very weak in water, for example, benzoic acid.

### 1.1.2.3 Amines

Aliphatic amines, such as methylamine, ethylamine, and n-butylamine, have basicities similar to that of ammonia and, like ammonia, they solvate the electron.[3] Ethylenediamine is more basic than ammonia, phenol being shown by potentiometric titration to be quite a strong acid in it.[3] Ethylenediamine is introduced here because, as will be shown in chapter 8, it has been used fairly widely recently as the medium for the generation of polyatomic anions of several main group elements.

## 1.1.3 Solvents More Acidic Than Water

Anhydrous acetic acid is not classified as a superacid. R. J. Gillespie has used this term to describe any acid of strength comparable with or more acidic than 100% sulfuric acid, but acetic acid is presented briefly here as a bridge in acid strength between water and anhydrous sulfuric acid, the weakest in acidity of the superacids. Although many phenomenological observations have been made on a very wide range of solvents, such as determination of solubilities of a vast number of organic and inorganic compounds and characterization of reactions of such compounds, the greatest extent and depth of physicochemical study of inorganic chemical systems interacting with individual solvents has been for the superacids sulfuric acid, trifluoromethylsulfuric acid, fluorosulfuric acid, and hydrogen fluoride. After acetic acid, their acid-base chemistry will be presented in that order—the order of increasing acidity of the systems.

### 1.1.3.1 Anhydrous Acetic Acid

The self-ionization of acetic acid can be written reasonably realistically as

$$2CH_3COOH \rightleftharpoons CH_3COOH_2^+ + CH_3COO^- \qquad 1.22$$

Protonation of the carbonyl group occurs so the solvated proton in the pure solvent can be written as $CH_3COOH_2^+$ or, better, as $CH_3.C(OH)_2^+$. The ionic product is given as $10^{-14.5}$ mol$^2$ L$^{-2}$ at 25°C, so the extent of self-ionization is similar to that for water, but the dielectric constant of 6.3 is very low by comparison with water (78.3 at 25°C). As for other protonic acid-base systems, the protonated cation, in this case $CH_3.C(OH)_2^+$, is the most acidic species that can exist in the solvent system and $CH_3COO^-$ the strongest base.

As a consequence of the low dielectric constant of the medium, specific and molar conductances of ammonium and alkali metal acetates in acetic acid are much lower than the corresponding 1:1 electrolytes in water; ion-pairing occurs to a very large extent.[4] The chloranil electrode has been used to obtain relative strengths of bases in $CH_3COOH$ by titration against $HClO_4$ solutions.[4]

$HClO_4$ is effectively completed deprotonated in $CH_3COOH$, but extensive ion-pairing in this solvent of low polarity to form the associated species $CH_3.C(OH)_2^+.ClO_4^-$ makes it appear to be a weak acid. A $CH_3COOH$ solution $10^{-2}$ M in $HClO_4$ is reported to have a pH of about 3.4.[5]

A glass electrode, shown to have a Nernstian response in anhydrous acetic acid, when coupled with the reference electrode $Hg$—$Hg_2(CH_3COO)_2$—$CH_3COO^-$, was used to titrate the conventional strong acids of the aqueous systems against pyridinium acetate, a soluble source of the base of the system, acetate.[6] Experimentally determined $pK_a$ values for $HClO_4$, $HBr$, $H_2SO_4$, and $HCl$ were 5.3, 6.8, 7.4 (for the first dissociation step $H_2SO_4 \rightleftharpoons H^+ + HSO_4^-$), and 8.8, respectively.

Many anhydrous metal chlorides, such as $ZnCl_2$, when dissolved in anhydrous $CH_3COOH$, give precipitates such as $Zn(CH_3COO)_2$ which are amphoteric, dissolving as excess acetate is added to form $[Zn(CH_3COO)_4]^{2-}$. However, as is the case with many nonaqueous solvent systems, there does not seem to have been any serious investigation of the Lewis acidity of these binary acetates.

## 1.1.3.2 Sulfuric Acid

Much of the early drive for reliable physicochemical measurements and observations in anhydrous sulfuric acid came from the use of this solvent in synthetic and mechanistic organic chemistry. For example, Gillespie, working with Ingold and other colleagues, identified the nitronium ion in the $HNO_3$—$H_2SO_4$ mixture used for nitrating aromatic compounds, as will be shown in section 2.3.

Pure sulfuric acid is extremely viscous, boils with decomposition at about 300°C, and freezes at 10.371°C. Its dielectric constant is greater than that of water and the pure liquid is extremely acidic. Its Hammett Acidity Function, $H_0$, described as a measure of relative acidities of superacids in chapter 2, is $-11.93$. For comparison an aqueous solution containing 20 weight percent of $H_2SO_4$ has a value for $H_0$ of $-1.06$, nearly 11 powers of 10 less acidic. These physical properties and the self-ionization equilibrium data given below have been reported by Gillespie and Robinson.[7]

The specific conductance of $1.044 \times 10^{-2} \, \Omega^{-1} \, cm^{-1}$ is very high at 25°C, compared with values for most other protonic liquids. In fact, it has been calculated that, at 25°C, the concentration of all ionic species in pure $H_2SO_4$ is 0.049 M. This indicates that ionization processes in liquid sulfuric acid are extensive. They are also complex. For the typical ionization that occurs for all protonic solvents—

$$2H_2SO_4 \rightleftharpoons H_3SO_4^+ + HSO_4^- \qquad\qquad 1.23$$

the equilibrium constant at 25° is given as $2.7 \times 10^{-4}$.

A second, but significant, ionization process is

$$2H_2SO_4 \rightleftharpoons H_3O^+ + HS_2O_7^- \qquad\qquad 1.24$$

for which the equilibrium constant is $5.1 \times 10^{-5}$ at 25°C. The two most likely impurities in $H_2SO_4$, resulting from adventitious addition during preparation or handling or from dissociation of the 100% liquid, are $H_2O$ and $SO_3$. Each of these causes enormous enhancement of the extent of ionization in the solvent. Addition of $SO_3$ to $H_2SO_4$ can be considered as leading to formation of $HS_2O_7^-$ and an equilibrium constant of $1.4 \times 10^{-2}$ is given for the equation frequently written in Bronsted acidity terms as

$$H_2S_2O_7 + H_2SO_4 \rightleftharpoons H_2SO_4^+ + HS_2O_7^- \qquad\qquad 1.25$$

In the Conclusion of this chapter (section 1.3) it will be stated that this equilibrium can be viewed as a direct interaction between the formal Lewis acid $(SO_3)$ and Lewis base $(HSO_4^-)$ of the solvent system. The avidity of $H_2SO_4$ for $H_2O$—a likely impurity even from the surface of glass equipment—is shown by the equilibrium constant being quoted as 1 for the equation

$$H_2O + H_2SO_4 \rightleftharpoons H_3O^+ + HSO_4^- \qquad\qquad 1.26$$

It is apparent from the above that simple conductivity measurements are not reliable in determining the purity of $H_2SO_4$ or in monitoring the progress or extent of reactions in the solvent. The extensive equilibria would be disturbed by the presence of impurities or on addition of reactive solutes. However, as will be shown in chapter 2, measurement of the freezing point can be used as a very good criterion for the purity of 100% $H_2SO_4$, and more specialized conductance measurements give reliable data in monitoring reactions in solution.

Again, as is common with all protonic solvents, the most acidic species that can exist in pure $H_2SO_4$ is the solvated proton $H_3SO_4^+$ and the strongest base is the anion of the system, $HSO_4^-$, in this case.

**Bronsted Acids.** Because of the high acidity—the very low proton-acceptor properties—of anhydrous $H_2SO_4$, very few Bronsted acids have been identified in this solvent, and those that have been studied are weak. The value of $K_a$ for $HClO_4$ is about $10^{-4}$ while the values for $HSO_3F$ and $H_2S_2O_7$ are given as $2.3 \times 10^{-3}$ and $1.4 \times 10^{-2}$.

*Bronsted Bases.*  The corollary to the limited availability of solutes which can donate protons to $H_2SO_4$ to any significant extent is that there will be a vast number of Bronsted bases in this solvent. Thus most compounds with coordinately unsaturated O, N, S, and P atoms, that is, those with electron pairs available for bonding to $H^+$, will be protonated. Unless dehydrated or otherwise solvolysed by $H_2SO_4$, alcohols, aldehydes, ketones, amines, and carboxylic acids will be protonated, usually completely. The base of the solvent system $HSO_4^-$ will be produced effectively quantitatively. Many bases that are almost immeasurably weak in water are strongly protonated in $H_2SO_4$. Nitrated aromatic compounds, although weak, have observable base strength in $H_2SO_4$ as shown by the values of basicity constants determined conductometrically and listed in Table 1.1. The low base strengths of these and of several chemically related compounds are such that they can be used as indicators in highly acidic media in the spectrophotometric determination of Hammett Acidity Functions (chapter 2).

Water presents a special case because it is a likely impurity or contaminant in $H_2SO_4$ solutions under investigation. $HSO_4^-$, produced in amounts equivalent to the protonated water $H_3O^+$, will render the $H_2SO_4$ more basic than might be expected. Many oxoanions and oxoacids are solvolysed, for example, $HNO_3$ to $NO_2^+$ and HCl and HF to the halosulfuric acids, but $H_3PO_4$ presents a special case, being protonated to $H_4PO_4^+$, that is, acting as a strong base in $H_2SO_4$, with all four oxygens protonated as $P(OH)_4^+$.

*Lewis Acids.*  The most obvious Lewis acid of the solvent system is $SO_3$, which bonds to $HSO_4^-$ to form the anion $[O_3S—O—SO_2(OH)]^-$. Decrease in concentration of $HSO_4^-$ causes displacement of equilibrium in the self-ionization equation 1.23 and consequent increase in concentration of $H_3SO_4^+$, that is, increase in acidity of the system.

Many authors, including Gillespie, prefer to describe the action of $SO_3$ in increasing the acidity of $H_2SO_4$ as an interaction of $SO_3$ with $H_2SO_4$ to form $H_2S_2O_7$ and then to consider $H_2S_2O_7$ as a Bronsted acid in the solvent. This approach will be discussed in section 3.1.1.

Boric acid, $H_3BO_3$, was found to increase the conductance of pure $H_2SO_4$ markedly. This was interpreted as solvolysis of $B(OH)_3$ to $B(OSO_3H)_3$, the three molecules of $H_2O$ released being protonated to $H_3O^+$ with consequent formation of 3 $HSO_4^-$ ions. Using the conductance and cryscopic techniques

**Table 1.1.**  Basicity Constants of Weak Bases in Sulfuric Acid[7]

| Base | $K_b$ |
|------|------:|
| *p*-Nitrotoluene | $9.5 \times 10^{-2}$ |
| *o*-Nitrotoluene | $6.7 \times 10^{-2}$ |
| *m*-Nitrotoluene | $2.3 \times 10^{-2}$ |
| Nitrobenzene | $1.0 \times 10^{-2}$ |
| *p*-chloronitrobenzene | $4 \times 10^{-3}$ |

described in chapter 2, it was shown that $B(OSO_3H)_3$ acts as a strong Lewis acid reacting with $HSO_4^-$ to form $[B(OSO_3H)_4]^-$ with consequent increase of $H_3SO_4^+$ concentration in the self-ionization reactions for $H_2SO_4$. Again, many authors postulate the formation and existence of a Bronsted acid $HB(OSO_3H)_4$, but there seems to be some artificiality about this proposal.

*Lewis Bases.* The base of the solvent system is the $HSO_4^-$ anion. So addition of any soluble, ionic bisulfate, for example, where the cation might be $NH_4^+$ or an alkali metal cation, will quantitatively enhance the basicity of the solvent systems.

The sulfate ion is a stronger base than $HSO_4^-$ and reacts with the solvent according to the equation

$$SO_4^{2-} + H_2SO_4 \rightarrow 2HSO_4^- \qquad\qquad 1.27$$

This reaction resembles the reaction of the very strong base $O^{2-}$ with $H_2O$ (equation 1.15).

## 1.1.3.3 Trifluoromethylsulfuric ("Triflic") Acid

Of the four superacids dealt with in detail in this book, triflic acid is the next most acidic after sulfuric acid. Its Hammett Acidity Function is now believed to be about $-14.3$. A vast amount of work, typified in a review by Howells and McCown,[8] has been directed toward reactions and syntheses of organic compounds in this acid, but the investigation of inorganic chemical systems has been very limited, being restricted to Hammett function measurements of Lewis acids[9,10] and of bases[10] in the solvent and to some study of the generation and stability of a few and transition metal nonmetal cations in acidified triflic acid.[10]

The self-ionization can be written as

$$2CF_3SO_3H \rightleftharpoons CF_3SO_3H_2^+ + CF_3SO_3^- \qquad\qquad 1.28$$

As for $H_2SO_4$ Bronsted acids will be few and very weak, while the very high acidity of the solvent will mean that there is a vast number of Bronsted bases. The important ones within the context of this book, other than the potential impurity water, are the nitro- and halonitro- benzenes and toluenes that are used as indicators in the spectrophotometric determination of Hammett Acidity Functions in this solvent.

Alkali metal triflates are convenient sources of the Lewis base of the system, $CF_3SO_3^-$, which is the entity produced in solution when Bronsted bases are protonated. Relatively few acids have been studied that are genuinely Lewis acids of the solvent system, that is, which are themselves binary triflates. $B(OSO_2CF_3)_3$ has been used as a strong acid in Hammett function measurements[9,10] and also in generation of cationic species in triflic acid.[10] It undoubtedly interacts with the base of the solvent system to form $[B(OSO_2CF_3)_4]^-$.

Other Lewis acids such as $SbF_5$[10,11] and $PF_5$[11] have been investigated spectrophotometrically for acid strength in triflic acid or have been used to increase the acidity of triflic acid in generation of inorganic cations in the acid.[10] These fluorides are not Lewis acids *of the solvent system* in the way that $B(OSO_2CF_3)_3$ is. They probably form mixed anionic species such as $[SbF_5(OSO_2CF_3)]^-$ and possibly more complex ones, in the same fashion as mixed anions are formed between pentafluorides and fluorosulfates, as discussed in the section immediately below.

Hammett Acidity Function measurements have been reported for the perfluoroalkylsulfuric acids $C_nF_{2n+1}SO_3H$ for which n = 1, 2, 4, and 6. These acids become somewhat less acidic with increase in molecular weight, values of $H_0$ being given as $-14.1 \pm 0.1$, $-14.0 \pm 0.1$, $-13.2 \pm 0.1$, and $-12.3 \pm 0.3$ for the acids for which n = 1, 2, 4, and 6, respectively.[12] A more recent determination[10] of $H_0$ for $CF_3SO_3H$, presented in chapter 2, gives a value of $-14.33$ for $CF_3SO_3H$, but the order of acid strengths for $C_nF_{2n+1}SO_3H$ is of interest here. The $H_0$ values reported for the range of acids $C_nF_{2n+1}SO_3H$ are likely to be somewhat less negative than the true values because of a defect in methodology used by the researchers. They measured $H_0$ values for added Lewis acid but not for added Lewis base—a necessary requirement, as shown in section 2.4.1.

## 1.1.3.4 Fluorosulfuric Acid

This solvent, formally the member of the series given earlier for which $n = 0$, is, as expected from the general trend, more acidic than triflic acid. As will be shown in chapter 2, its Hammett Acidity Function value is reported as $-15.07$. Within the limits of experimental error, this is essentially the same as the value reported for anhydrous HF, namely, $-15.1$. However, the addition of similar Lewis acids causes much more enhancement of acidity in HF than for $HSO_3F$—hence the order of presentation of the two superacids in this general introductory chapter.

Organic chemists have exploited the acidic properties of this convenient solvent in the preparation and identification of carbonium species and in study of acid catalysis of organic reactions, particularly when the acidity has been enhanced by the addition of Lewis acids such as $SbF_5$. The book by Olah and colleagues[13] outlines much of this work. Fluorosulfuric acid has a viscosity similar to that of water and, unlike HF, can be handled in glass apparatus. Its boiling point of 162.7°C is much lower than that of $H_2SO_4$, so it can be distilled away reasonably conveniently from reaction mixtures. Its low freezing point ($-88.98$°C) has enabled NMR spectra to be recorded at temperatures low enough for proton transfer to be sufficiently slowed to allow signals to be obtained from the acidic proton of the conjugate acids of a large number of weak bases. These physical properties provided Gillespie with an excellent medium for the generation by controlled oxidation, and subsequent characterization, of a large number of nonmetal cations in this solvent of very low basicity.

For the self-ionization equilibrium,

$$2HSO_3F \rightleftharpoons H_2SO_3F^+ + SO_3F^- \qquad \qquad 1.29$$

the ionic product is about $4 \times 10^{-8}$ mol$^2$ kg$^{-2}$.[14] There is much less ionization in the solvent than for $H_2SO_4$ but much more than for HF, the next superacid to be discussed.

***Bronsted Acids and Bases.*** The acidity of the solvent is such that there is no real evidence of any solute acting as a Bronsted acid; there is in fact some evidence for slight protonation of $H_2SO_4$ by $HSO_3F$.[14] As was the case for the two preceding superacids, protonation of even very weak bases will occur, to a greater extent than in $H_2SO_4$, because of the differing acidities of the two solvents. Table 1.1 gave values for the basicity constants, $K_b$, of $1.0 \times 10^{-2}$ and $4 \times 10^{-3}$ for nitrobenzene and $p$-nitrochlorobenzene in $H_2SO_4$. In $HSO_3F$ nitrobenzene is fully protonated and $K_b$ for the chlorocompound is $7.6 \times 10^{-1}$.[14]

***Lewis Acids and Bases.*** In pursuing a program of synthesis of inorganic compounds in $HSO_3F$, Aubke and co-workers have prepared the only compounds which can be classified as strong Lewis acids of the solvent system, that is, compounds in which each of the ligands on a metal is the base of the system, $SO_3F^-$. They have prepared several fluorosulfates and fluorosulfatocomplexes of noble metals[15] and have demonstrated that the conductances of $Au(OSO_2F)_3$ and $Pt(OSO_2F)_4$ in $HSO_3F$ are almost as great as that of $SbF_2(OSO_2F)_3$, which has been shown by Hammett function measurements to be a very strong Lewis acid in $HSO_3F$. The high conductances are interpreted as resulting from the formation of $[Au(OSO_2F)_4]^-$ and $[Pt(OSO_2F)_6]^{2-}$ in solution with resulting increase in $H_2SO_3F^+$ concentration.[16] However, no acidity function measurements have been made on solutions of the Au(III) or Pt(IV) compounds. More recently, Aubke and Cicha[17] have synthesized $Ta(OSO_2F)_5$ and demonstrated by Hammett function measurements that it is closely comparable in Lewis acid strength with $SbF_2(SO_3F)_3$, described later as the previously known strongest Lewis acid of the $HSO_3F$—$SbF_5$—$SO_3$ system.

Most acidity measurements or studies of reactions in $HSO_3F$ of enhanced acidity have involved use of $SbF_5$ or of $SbF_5$—$SO_3$ mixtures as the Lewis acids. Gillespie[14] showed by NMR spectroscopy that the base of the solvent reacts with $SbF_5$ to form an octahedral anion,

$$SbF_5 + SO_3F^- \rightleftharpoons [F_5SbOSO_2F]^- \qquad \qquad 1.30$$

thereby reducing the $SO_3F^-$ concentration in the self-ionization equilibrium (1.29) and enhancing the $H_2SO_3F^+$ concentration. The NMR spectrum could

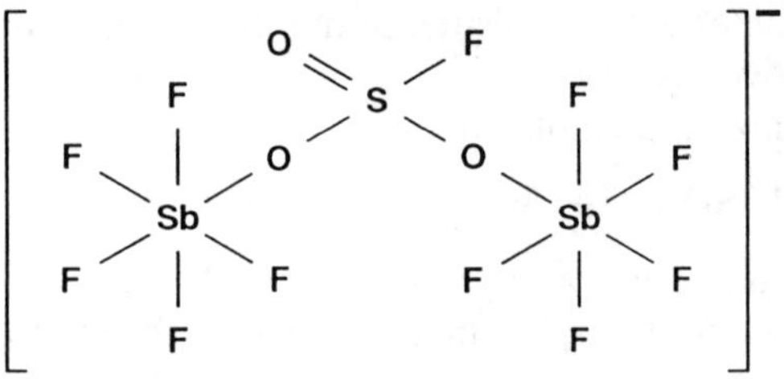

**Figure 1.1.** The $[SO_3F(SbF_5)_2]^-$ Anion.

be explained only by involving the formation of the fluorosulfatobridged species in Figure 1.1 in addition to $[F_5SbOSO_2F]^-$.

Higher conductances, that is, greater acidities, were achieved by adding $SO_3$ to the $HSO_3F$—$SbF_5$ solutions until the $SO_3$:$SbF_5$ ratio was approximately 3:1.[14] It was concluded that $SO_3$ was inserted progressively into ligand positions:

$$SO_3 + SbF_5 \rightleftharpoons F_4Sb(OSO_2F) \qquad\qquad 1.31a$$
$$2SO_3 + SbF_5 \rightleftharpoons F_3Sb(OSO_2F)_2 \qquad\qquad 1.31b$$
$$3SO_3 + SbF_5 \rightleftharpoons F_2Sb(OSO_2F)_3 \qquad\qquad 1.31c$$

$F_2Sb(OSO_2F)_3$ has been described as the strongest formal Lewis acid of the $HSO_3F$—$SbF_5$—$SO_3$ system, accepting a base from the solvent to form $[F_2Sb(OSO_3F)_4]^-$ and causing very great enhancement of the $H_2SO_3F^+$ concentration.

Gillespie and co-workers[18] used conductances of binary fluorides in $HSO_3F$ to establish an order of Lewis acidity: $SbF_5 > BiF_5 > AsF_5 \approx TiF_4 > NbF_5 \approx PF_5$ and have used $SO_3$ to enhance the Lewis acidity of $SbF_5$ and of $AsF_5$.

The Lewis base of the solvent system is $SO_3F^-$, added most conveniently as an alkali metal fluorosulfate.

### 1.1.3.5 Hydrogen Fluoride*

As stated earlier and as will be described in chapter 2, the Hammett Acidity Function for HF has been measured as $-15.1$, a value which is indistinguishable within the limits of experimental error from the value of $-15.07$ given for $HSO_3F$. However, when the strong Lewis acid $SbF_5$ is studied in both HF and $HSO_3F$, HF provides the most acidic solutions. $H_0$ values for HF—$SbF_5$ solutions are about $-21$, whereas, even when the acidity of $SbF_5$ in $HSO_3F$ is enhanced by the addition of $SO_3$, $H_0$ values are about $-19$.

---

*In referring to hydrogen fluoride solutions, the convention used will be that adopted by most inorganic chemists, certainly by fluorine chemists. Aqueous solutions, in which HF is the solute, are referred to as hydrofluoric acid. Anhydrous hydrogen fluoride as the pure compound or as a solvent will be referred to as "hydrogen fluoride."

HF—$SbF_5$ solutions are the most acidic protonic solvent systems known currently. It will be shown in chapter 2 that this generalization is valid, regardless of the methods used to determine relative acidities of superacidic media based on HF and $HSO_3F$.

Hydrogen fluoride is a very volatile liquid at room temperature, its boiling point being 19.51°C and its melting point −83.55°C. It has a high dielectric constant, 83.6 at 0°C and 175 at −73°C, and is therefore a good solvent for compatible ionic solutes. Atmospheric moisture acts as a strong base and is readily taken into solution. Coupled with the volatility, this chemical property of HF means that it is best handled in vacuum systems. However, it is very corrosive and reacts with glass and other materials used in conventional vacuum systems. In this sense, studies of solutions in HF present greater experimental difficulties than those in the superacids discussed earlier that, with their solutions, can be studied in conventional glass equipment which may require only minor modification. However, suitable materials are now available for construction of vacuum systems, reaction vessels, and spectroscopic and electrochemical cells for study of HF solutions. Selected metals, polymers (such as Teflon and KelF) and "synthetic sapphire," more correctly called crystalline alumina, can be used. Appropriate choice of materials and details of construction of fluoride-resistant cells, reactors, and vacuum lines have been discussed[19,20] and the use and commercial availability of "synthetic sapphire" reaction tubes have been reported recently.[21]

In its simplest form the self-ionization of HF can be written

$$HF \rightleftharpoons H^+ + F^- \tag{1.32}$$

It is customary to recognize, at least, the first solvation of the proton to give $H_2F^+$. There is now spectroscopic and structural evidence[22] to show that, for the HF—$SbF_5$ system, $H_2F^+$ is the dominant cation only in very concentrated $SbF_5$ solutions (greater than 40 mol%) but that solvated $H_3F_2^+$ is the cationic species in more dilute solutions. Largely because solids such as $KHF_2$ are isolated from HF solutions, many authors write the anion in the self-ionization equilibrium as $HF_2^-$:

$$3HF \rightleftharpoons H_2F^+ + HF_2^- \tag{1.33}$$

However, evaporation of a solution of CsF in HF gives a residue of $CsH_2F_3$ and all of the anions $F(HF)_n^-$ for $n = 1-4$ have been characterized structurally.[23] So there seems to be no unique and specific self-ionization equation that can be written. The general form that recognizes no specific degree of solvation of the proton or of the fluoride ion is probably the best:

$$HF \rightleftharpoons H_{solv}^+ + F_{solv}^- \tag{1.34}$$

The low specific conductance of "pure" HF and calculations[21] based on measured Hammett Acidity Functions for HF solutions of known acidity and basicity suggest a value of $10^{-12.3}$ mol$^2$ kg$^{-2}$ for the ionic product for HF.

A special feature of HF as a solvent for synthetic work is its very long usable potential range of about 4.5 V between anodic and cathodic release of $F_2$ and $H_2$. This makes HF a good solvent for the generation and study of very strong oxidants and very strong reductants. $Ag^{2+}$, $NiF_6^{2-}$, $MnF_6^{2-}$, $Ti^{2+}$, and $Sm^{2+}$ have all been obtained as stable solutions in HF.

***Bronsted Acids.***    As was the case for the fluorosulfuric acid system, little if any Bronsted acidity will be expected in HF because of the great acidity of the solvent, that is, its very low basicity and proton acceptor capability. Stein and Appleman[24] made a Raman spectroscopic study of protonation of $ClO_4^-$ and of $BrO_4^-$ in HF and showed that the anions were partially protonated to the undissociated conjugate acids. However, the protonation would cause a large enhancement of the concentration of the base $F^-$ in the resulting solutions over that in the "pure" HF, as will be discussed in detail in chapter 2. Hammett functions would probably be of the order of $-10$. It would be difficult to correlate the relative protonation of the perhalates in basic HF with an ability of the parent perhalic acids to protonate pure HF at its Hammett Function of $-15.1$.

***Bronsted Bases.***    As for the other superacids, there will be a vast number of compounds showing total or extensive proton acceptance from HF. Bases, often immeasurably weak in water, will be very strong in HF and, for the reasons given in chapter 2, will cause a greater increase in solution basicity in HF for the addition of a particular concentration of solute than will occur for the other superacids.

***Lewis Acids.***    The most-studied Lewis acid of the HF solvent system is $SbF_5$, a fluoride acceptor according to the equation

$$SbF_5 + F^- \rightarrow SbF_6^- \qquad\qquad 1.35$$

This reaction goes virtually to completion at low concentrations of $SbF_5$ and causes a massive displacement in the HF self-ionization. The low value of the ionic product for this self-ionization, compared with those for $H_2SO_4$ and $HSO_3F$, means that acidity increases more rapidly in HF for a given addition of $SbF_5$ than for the other two superacids. As the concentration of $SbF_5$ in HF is increased to about 5 M and beyond, polymeric anions of the type $Sb_2F_{11}^-$, $Sb_3F_{16}^-$, and $Sb_nF_{5n+1}^-$ become progressively more significant.

The next-strongest acid of the HF solvent system is $AsF_5$, which, even at low concentrations, is present as the molecular entity in equilibrium with $AsF_6^-$ and $As_2F_{11}^-$. The transition metal pentafluorides $TaF_5$ and $NbF_5$ form dimeric anions, with $TaF_5$ somewhat weaker than $AsF_5$ but considerably stronger than $NbF_5$. $PF_5$ appears to be vanishingly weak as a Lewis acid in HF. $BF_3$ is intermediate in strength between $TaF_5$ and $NbF_5$ and there has been some study of the acidity of transition metal oxide tetrafluorides

(MOF$_4$) in HF. Quantitative measurements of relative strengths of all of these Lewis acids in HF and the nature of the species formed when they accept a fluoride ion will be discussed in detail in chapter 3.

*Lewis Bases.*    The fluoride ion is the Lewis base of the HF solvent system and inorganic fluorides can act as sources of the base, depending on their solubility. Alkali metal and ammonium fluorides are convenient sources of the base. As Table 1.2 shows, they are significantly more soluble than alkaline earth fluorides that are, in turn, much more soluble than transition metal difluorides.[25]

The xenon fluorides offer an interesting example of base strength. XeF$_2$ and XeF$_4$ are essentially molecular in the solid state and provide very little fluoride ion when dissolved in HF. XeF$_6$, the structure of which is based on tetrameric or hexameric rings of XeF$_5^+$ units with bridging F$^-$ ions, provides high concentrations of F$^-$ on dissolution in HF. ClF$_3$ and BrF$_3$, strong fluoride donors in solid adducts such as ClF$_2^+$.SbF$_6^-$ and BrF$_2^+$.SbF$_6^-$, are weak sources of F$^-$ in HF. The chalcogen tetrafluorides, which also form solid adducts with fluorides that are Lewis acids, are weak fluoride donors to HF, decreasing in strength in the order SF$_4$ > SeF$_4$ > TeF$_4$.

There has been relatively little study of anhydrous hydrogen halides other than HF as solvents. Their low boiling points (e.g., $-84.1°C$ for HCl) lead to great experimental difficulties in manipulation of their solutions. Because of their low dielectric constants ($< 10$) they are poor solvents for ionic compounds and extensive, if not complete, ion-pairing occurs in solution. Most quantitative investigations have been restricted to conductivity measurements indicating 1:1 interaction for Lewis acids such as BCl$_3$ with sources of base such as (CH$_3$)$_4$N.HCl$_2$, which have suitable solubility in the solvent of low dielectric strength. Those studies indicate very small degrees of self-ionization which can be represented as

$$3HX \rightleftharpoons H_2X^+ + HX_2^-$$    1.36

where X = Cl, Br, or I.

**Table 1.2.** Solubilities of Metal Fluorides in Hydrogen Fluoride at Ambient Temperature (mol kg$^{-1}$)[25]

| LiF | NaF | CsF | AgF | TlF |
|---|---|---|---|---|
| 3.96 | 7.17 | 11.8 | 6.55 | 26.0 |
| MgF$_2$ | CaF$_2$ | SrF$_2$ | BaF$_2$ | |
| 0.004 | 0.10 | 1.18 | 0.03 | |
| FeF$_2$ | CoF$_2$ | NiF$_2$ | CuF$_2$ | |
| 0.00064 | 0.0037 | 0.0038 | 0.001 | |

## 1.2 Nonprotonic Solvents

By comparison with water and with many other protonic solvents, especially the superacids, there has been very limited quantitative physicochemical investigation of nonprotonic solvents, although for many of these solvent systems solubilities of a wide range of organic and inorganic solutes have been determined and solute-solvent interactions have been studied descriptively. An exception to this generalization would be the area of high-temperature and room-temperature chloroaluminate "melts" (see later), where there is now a considerable body of electrochemical and spectroscopic work. Of special interest in the presentation of this book would be information on relative strengths of acids and bases in the various nonprotonic media. Very little information of this sort seems to exist. Unlike the situation for water and for the acidic protonic solvents, there are no acidity scales which allow comparison of acidity and basicity for nonprotonic solvents. For many systems, the proposed self-ionization equilibria are largely inferential, the proposals being based on observed reactions in the solvents.

A very large range of nonprotonic solvents has been studied with varying degrees of rigor. Some classes of such solvents will be surveyed briefly in this section of chapter 1, and a limited number of examples will be offered in each class. The major object in presenting this material is to provide a setting for discussion of the acid-base behavior of chloroaluminates so that behavior in such media can, in turn, be compared with acid-base–controlled chemical behavior in protonic superacids. In later chapters in this book generation and stability of cations of metallic and of nonmetallic elements will be discussed in terms of acidity and basicity levels in superacids and in chloroaluminates.

## 1.2.1 Molecular Nonmetal Oxides

Sulfur dioxide was one of the earliest of the nonaqueous solvent systems to be investigated. Studies date back to the turn of the century, as do those with liquid ammonia. Solubilities, as well as solvolysis, redox, and metathetical reactions were reported in some considerable detail. The metathetical reaction between $Cs_2SO_3$ and $CsCl$ to give insoluble $CsCl$ was interpreted as a neutralization reaction, based on a proposed self-ionization equilibrium involving the cation (acid) $SO^{2+}$ and the anion (base) $SO_3^{2-}$. Subsequent radiochemical investigations involving use of the isotopes $^{18}O$ and $^{35}S$ have thrown considerable doubt on this interpretation of the self-ionization that accounts for the small conductance of pure liquid $SO_2$. It is now believed that the "neutralization" reaction may be very complex, involving $SOCl^+$ as an intermediate. Certainly it would be difficult for $SO_2$ with a dielectric constant less than 20 to maintain a self-ionization equilibrium based on doubly charged cations and anions.

Dinitrogen tetroxide, $N_2O_4$, has a small but measurable conductance and a very low dielectric constant at room temperature. It is proposed that it self-ionizes as

$$N_2O_4 \rightleftharpoons NO^+ + NO_3^- \qquad\qquad 1.37$$

NOCl, a potential source of $NO^+$, increases the acidity of $N_2O_4$, making it a better solvent for a metal such as zinc. The cation is described as undergoing a neutralization reaction with $NO_3^-$ when NOCl in $N_2O_4$ reacts with $AgNO_3$ to give a precipitate of AgCl. The "solvates," $Cu(NO_3)_2.N_2O_4$ and $Fe(NO_3)_3.N_2O_4$, generated by reaction of the metals with $N_2O_4$, can be formulated as $NO[Cu(NO_3)_3]$ and $NO[Fe(NO_3)_4]$, complexes based on the acid and base species of the proposed solvent self-ionization equilibrium.

## 1.2.2 Molecular Halides and Oxohalides

Widely varying degrees of work have been done on a range of molecular halides and oxohalides as examples of self-ionizing nonprotonic solvents. The work has been reviewed fairly recently[26], and the systems will be mentioned only very briefly here. Somewhat more emphasis will be given later to one class of molecular halides, the interhalogen compounds.

Conductances and other properties of the chlorides, bromides, and iodides of antimony(III) and of arsenic(III) and of a range of solutes in these compounds have been interpreted in terms of a generalized self-ionization:

$$2AX_3 \rightleftharpoons AX_2^+ + AX_4^- \qquad\qquad 1.38$$

Halide donors, such as alkali metal, nitrosyl, or tetraalkyl ammonium halides lead to enhanced conductance of the trihalide solvent and to the formation of adducts containing anions such as $AX_4^-$ and are considered bases. Chloride acceptors are classed as Lewis acids giving residual adducts such as $SbX_6^-$, $SnX_6^{2-}$, and $AlX_4^-$.

Of all the halogen compounds presented in this section, $POCl_3$ has received the most attention. Many chlorides, similar to those mentioned immediately above, which are acids or sources of the base chloride for the $POCl_3$ system, have been studied, leading to a postulated self-ionization:

$$POCl_3 \rightleftharpoons POCl_2^+ + Cl^- \qquad\qquad 1.39$$
$$\text{or} \quad 2POCl_3 \rightleftharpoons POCl_2^+ + POCl_4^- \qquad\qquad 1.40$$

For some systems, at least, this simple self-ionization has been challenged, as discussed by Paul and Singh.[26]

Several other oxide halides have been studied using a similar experimental approach to that adopted for $POCl_3$. These include NOCl, $SOCl_2$, $SOBr_2$, $SO_2Cl_2$, and $SeOCl_2$. The self-ionization equilibria have usually been interpreted as involving the chloride ion, appropriately solvated, and the resulting cation derived from the parent oxide chloride by loss of chloride.

## 1.2.3 Interhalogen Compounds

Despite its very great reactivity, bromine trifluoride is the member of the interhalogens which has received the greatest amount of reliable physico-chemical study, most of it from the pioneering work of Eméleus and colleagues reported in 1949. The Cambridge group measured a relatively high specific conductance for $BrF_3$ compared with those of $IF_5$, which was very low, and $ClF_3$ having an intermediate value. This and the following work were reviewed in 1973.[27]

It was observed that the addition of compounds that were converted to KF and $SbF_5$ by $BrF_3$ led to an enhancement of the conductance of $BrF_3$ and, on removal of the solvent, the involatile residues were identified analytically as $KBrF_4$ and $SbBrF_8$. Subsequently these compounds were shown crystallographically to be $K^+BrF_4^-$ and $BrF_2^+SbF_6^-$. Solutions of each of these compounds in $BrF_3$ gave a 1:1 minimum in a conductimetric "titration"; that is, a neutralization reaction had occurred.

These observations led Eméleus and colleagues to propose a self-ionization:

$$2BrF_3 \rightleftharpoons BrF_2^+ + BrF_4^- \qquad\qquad 1.41$$

An ionic product $[BrF_2^+][BrF_4^-]$ of about $4 \times 10^{-4}$ has been estimated for this equilibrium.

It is worth comment that the proposed speciation in this self-ionization provides an example of the dangers of postulating that the ionic entities identified as residual solids after acid-base reactions in a solvent are necessarily those involved in the fundamental self-ionization equilibrium. It seems easier to write the $BrF_3$ self-equilibrium as

$$BrF_3 \rightleftharpoons BrF_{2\,solv}^+ + F_{solv}^- \qquad\qquad 1.42$$

Then the ionic entities can be defined simply within the Lewis acid–Lewis base framework; that is, $BrF_2^+$ is an electron-pair acceptor and $F^-$ an electron-pair donor; the ion $BrF_4^-$ is recognized as being particularly stable in isolated solids and probably in solution is a labile entity. It is then not necessary to invoke acid-base definitions, such as solvoacids and solvobases, in order to describe neutralization reactions of solutions of the solids $KBrF_4$ and $BrF_2.SbF_6$ when dissolved in $BrF_3$. The neutralization reaction becomes simply the reverse of equation 1.42. This point will be discussed in a more general context in the concluding part of this chapter.

Acceptance of equation 1.42 as the most fundamental way of representing the self-ionization of $BrF_3$ makes it easy to see that KF, AgF and similar compounds are sources of the Lewis base $F^-$ for the systems and that, to greater or less extent, $SbF_5$, $AsF_5$, $SnF_4$, and so on, will accept $F^-$ in $BrF_3$ to act as Lewis acids.

## 1.2.4 Molten Ionic Halides and Chloroaluminates

There is a very large body of research based on study of solutes in molten binary halides, most of it using alkali metal chlorides and fluorides or eutectic mixtures of individual halides. Dissolution of a wide range of elements and compounds in molten chlorides has been reviewed by Kerridge.[28] Gruen and McBeth[29] summarized much transition metal chloride solution behavior in molten chlorides and Young and White[30] have described the behavior of some *d*- and *f*-transition metal fluorides in alkali metal fluorides and their eutectics. Not much of this work in chlorides and fluorides has direct relevance to the material to be presented later in this book. Molten alkali metal chlorides and fluorides are, of their nature, basic solution media containing the ions $Cl^-$ and $F^-$. In the main, we shall be more interested in solution behavior in acidic media and the way in which that behavior changes as acidic media pass through neutrality to being basic. Solutes in basic chloride and fluoride media will be of interest only to the extent that they provide information within the acidity-basicity context.

While the fluoroaluminate cryolite ($Na_3AlF_6$) is the subject of a vast amount of technological interest, lying, as it does, at the heart of the aluminium smelting industry, chloroaluminate systems have received far more extensive fundamental research investigation insofar as study of solute speciation is concerned. As solution media, the chloroaluminates provide a very convenient range of acid-base properties, depending on the stoichiometry of $AlCl_3$—MCl mixtures, where MCl is an alkali metal chloride—or, in more recent work, it may be a chloride with an organic cation, such as 1-butyl-pyridinium chloride or 1-methyl-3-ethylimidazolium chloride. $AlCl_3$—NaCl mixtures suffer from an experimental disadvantage in comparison with those based on organic cations in that their melt range is roughly 250–500°C depending on their $AlCl_3$ content, whereas members of the other class of chloroaluminates are liquids at or very close to room temperature and, in principle, spectroscopic and electrochemical investigations are easier to carry out. However, as will be seen later, the range of attainable acidity is more limited in room-temperature chloroaluminates than in the higher-temperature melts.

For chloroaluminates systems, $AlCl_3$ is the Lewis acid of the system, $Cl^-$ is the Lewis base, and in the 1:1 reaction to form $AlCl_4^-$, as shown, neutralization occurs:

$$AlCl_3 + Cl^- \rightleftharpoons AlCl_4^- \qquad\qquad 1.43$$

Extensive investigation, based on conductance measurements, Raman spectroscopy, and potentiometric use of a chloride-sensing electrode, has provided much information on the anionic speciation in chloroaluminate systems. $AlCl_4^-$ and $Cl^-$ are the only species in basic systems, but $Al_2Cl_7^-$ and $Al_3Cl_{10}^-$ become progressively more significant as additions of $AlCl_3$ are made to $AlCl_4^-$.

There has been virtually no work done on progressive adjustment, as such, of pCl in chloroaluminate systems. For high-temperature melts, solute speciation has been observed in many cases over the whole melt stoichiometry range from pure acidic $AlCl_3$ through neutral $AlCl_4^-$ to basic $Cl^-$ when the $AlCl_3$:$Cl^-$ ratio in individual experiments has been varied.

Room-temperature melts differ from high-temperature melts in two significant aspects. Mixtures of $AlCl_3$ and the pyridinium or other chloride are not molten near room temperature when the ratio acid:base is greater than 2:1. This places a limit on the level of acidity for study of solutes in this type of solvent, whereas, in high-temperature melts, solute speciation can be studied (at about 230°C) under the most acidic conditions, that is, in pure $AlCl_3$. The second difference is that in pure $AlCl_3$ the concentration of ionic species derived from the solvent alone is vanishingly small, and at a ratio of 2:1 for $AlCl_3$:$Cl^-$ in high-temperature melts, $AlCl_3$, $Al_2Cl_7^-$, and $AlCl_4^-$ have been shown to be present in the high-temperature melt in approximately equal concentrations. However, the 2:1 acid-base mixture for room-temperature melts is believed to be almost exclusively ionic, containing little or no molecular $AlCl_3$. Spectroscopy of cationic species (chapters 4–7) indicates very similar solvent-solute interactions in both classes of chloroaluminate, suggesting that the differences in temperature domain and melt ionic speciation are not particularly significant in this context. Indeed, as will be suggested briefly toward the end of this chapter, differences in temperature domain on solute-solvent interaction do not appear to have major effects on solute speciation even for quite dissimilar solvents, for example, for melts compared with superacids.

## 1.2.5 Oxoanionic Melts

Many melts involving oxoanions are of great industrial importance, for example, carbonates, sulfates, and silicates. Nitrates have been subjected to most research investigation and much of this work has been reviewed by Kerridge.[28] Ionization processes and therefore acid-base equilibria in nitrates are not well established.

Carbonates have been put forward as a classic example of one of the methods of defining nonprotonic acids and bases as proposed by Lux and Flood and used in this context by many textbook authors. The definition involves examples like CaO reacting with $CO_2$ to form $CaCO_3$ and considers oxide transfer as the essential criterion for acidity and basicity within this framework. $CO_2$ in accepting $O^{2-}$ is defined as an acid with CaO, providing $O^{2-}$ as the base. As discussed later, it seems easier to consider $O^{2-}$ as an electron-pair donor (a Lewis base) and $CO_2$ as a Lewis acid.

Molten silicates, immensely important in metallurgical and other industrial slags and in geological magma, can also be considered within a simple Lewis acid–Lewis base context, the essential neutralization reaction being

$$SiO_2 \text{ (Lewis acid)} + 2O^{2-} \text{ (Lewis base)} \rightleftharpoons SiO_4^{4-} \qquad 1.44$$

Anions with $SiO_2:O^{2-}$ ratios higher than 1:2 would then be intermediate in acidity between acidic $SiO_2$ (1:0) and neutral $SiO_4^{4-}$ (1:2). Thus the ratio for $(SiO_3^{2-})_n$ is 1:1; that for $(Si_4O_{11}^{6-})_n$ is 4:3 and that for $(Si_2O_5^{-})_n$ is 2:1.

## 1.3 Conclusion

The very brief review of ionizing media given in this chapter rests on descriptive differences between a selection of solvents and on a qualitative description of acid-base equilibria in those solvents. It has been designed to set a background to the main content of this book, which is a discussion, quantitative where possible, of the nature of nonmetal and metal cationic speciation in superacids and chloroaluminate melts and the dependence of that speciation on varying levels of acidity and basicity of the media. This introductory review brings to the surface certain generalizations that appear to be worth presenting here at the conclusion of the chapter.

There appear to be too many unnecessary definitions of acids and bases. For example, the solvosystem is very popular with authors of textbooks for undergraduates. An acid is defined within this system as the cationic species involved in the self-ionization equation as written conventionally, and the anionic species is defined as the base. The self-ionization of $NH_3$ (equation 1.19) is frequently cited as an example, with $NH_4^+$ the acid of the solvent system and $NH_2^-$ the base. It seems far easier to describe $NH_4^+$ as a Bronsted acid, just as $H_3O^+$ is described as a Bronsted acid in water, even if the simpler Lewis acid-base neutralization process (equation 1.20) is not acceptable. Many authors would see the need to use the solvosystem definition as considerably greater when dealing with nonprotonic solvents where Bronsted acidity does not apply. For example, $BrF_2^+$ (equation 1.41) can be perceived easily as a Lewis acid but $BrF_4^-$ does not come within the definition of a Lewis base. However, if $BrF_4^-$ is recognized as solvated $F^-$ and the source of the entity $(F^-)$ which is transferred in acid-base reactions in $BrF_3$, then the Lewis acid-base description (equation 1.42) is quite adequate.

Another unnecessary compartmentalization of acid-base reactions appears to be the Lux-Flood description of an oxide donor, for example, CaO, as a base when it reacts with $SiO_2$, the oxide acceptor, to form $CaSiO_3$. Again, it seems far easier to think of CaO as the source of the Lewis base $O^{2-}$, which then reacts with the Lewis acid $SiO_2$, as in equation 1.44 and in the examples given immediately after that equation.

There are very many situations in nonaqueous chemistry where a Lewis acid like $SbF_5$ reacts with the Lewis base $F^-$ to form $SbF_6^-$ (equation 1.35), whether that is a direct reaction with solid NaF, the source of $F^-$, or with HF to cause displacement of the self-ionization of the Bronsted acid HF to enhance the concentration of $H_{solv}^+$ (equation 1.34). There is no evidence for, nor is there any need to specify, the existence of a Bronsted acid $HSbF_6$—

protonation of a terminal fluorine ligand of $SbF_6^-$ seems most unlikely. It would be quite artificial to consider the formation, on reaction of HF with the appropriate Lewis acids, of a very strong acid $HSbF_6$, a Bronsted acid $HAsF_6$ of intermediate strength, and very weak Bronsted acids such as $HBF_4$ or $HPF_6$, and yet some authors[31] use species such as these in discussions of enhanced acidity.

It does seem somewhat more reasonable to postulate the protonation of an oxygen of a ligand in acids of the fluorosulfuric and sulfuric acid systems. However, it also seems unnecessary to specify the acidic entities as being in these solvents in the Bronsted acid form, for example, $HSbF_5(OSO_2F)$ and $HB(HSO_4)_4$.[32] It seems simpler to propose direct Lewis acid–Lewis base interaction of $SbF_5$ with $SO_3F^-$ anions and of $B(HSO_4)_3$ with $HSO_4^-$ to form the octahedral entity $[SbF_5(OSO_2F)]^-$ and tetrahedral $[B(HSO_4)_4]^-$ than to postulate some intermolecular reaction to form $H[SbF_5OSO_2F]$ and $H[B(HSO_4)_4]$ which then act as Bronsted acids. Formation of $[SbF_5(OSO_2F)]^-$ and of $[B(HSO_4)_4]^-$ causes displacement of the self-ionization in the parent acids $HSO_3F$ (equation 1.29) and $H_2SO_4$ (equation 1.23). Of course, depending on the basicity of entities such as $[F_5SbOSO_2F]^-$ and $[B(HSO_4)_4]^-$, they may be protonated to some extent in $HSO_3F$—$SbF_5$ and $H_2SO_4$—$B(HSO_4)_3$ mixtures. $H_2S_2O_7$ is somewhat more acceptable as a Bronsted acid in its own right because it has been reported that it can be isolated as such, but, particularly in dilute solutions of $SO_3$ in $H_2SO_4$, it seems simpler to consider the increase in acidity of $H_2SO_4$ on addition of $SO_3$ as resulting from direct stoichiometric interaction of the Lewis acid $SO_3$ with the base $HSO_4^-$ to form $HS_2O_7^-$ and the displacement of the self-ionization reactions of $H_2SO_4$. Depending on the composition of the $H_2SO_4$—$SO_3$ system, $HS_2O_7^-$ could be protonated to greater or less extent. The desirability of treating this system in Lewis acid–Lewis base terms will be developed more fully in section 3.1.1.

A major aim in the presentation of this introductory chapter has been to try to offer a unified presentation of protonic and nonprotonic solvents in terms of Bronsted and Lewis acidity and basicity and, where appropriate, the interaction of these two acid-base systems.

Another aim, to be developed in much more detail later in the book, is to demonstrate that many inorganic chemical reactions which lead to the formation of relatively unusual species such as polyatomic halogen and chalcogen cations in unusually low oxidation states occur in similar fashion in different media regardless of differences in chemical nature or temperature domain of the media. It will be shown that these species can be generated in each of the superacids at appropriate levels of acidity. They are also formed in acidic chloroaluminates at temperatures considerably above room temperature. For all the media, whether superacids or melts, increase in basicity leads to very similar disproportionation patterns for many cationic solutes. This generalization will be developed in detail in chapters 4, 5, and 7.

It is to be hoped that, if a generalized approach can be developed for solute-solvent interaction in a wide range of solvents under different physical conditions, much unnecessary and artificial compartmentalization can be removed from textbook treatments of nonaqueous systems, where different solvents are frequently presented in virtual isolation and, most frequently, in a context in which implicitly or explicitly, molten salt chemistry is perceived as a separate area of nonaqueous chemistry. For simplicity chloroaluminates will be the only molten salt systems presented in this book, but it is also to be hoped that recognition of the commonality of large areas of inorganic chemistry in very diverse solvents may lead to a simplified acid-base approach to complex melts of great importance, such as industrial slags or geological magma.

After the discussion in subsequent chapters of the species that do exist in a wide variety of solvents, depending on whether each solvent is acidic, neutral, or basic, it will be shown in chapter 9 that, in a general sense, solute speciation is very similar in all solvents, regardless of whether the solvent is water, a basic or an acidic protonic solvent, or a nonprotonic solvent which is liquid at room temperature or one which is a high-temperature molten salt.

# References

1. J.J. Lagowski, G.A. Moczygemba in *The Chemistry of Nonaqueous Solvents,* Vol. II. J.J. Lagowski, ed. Academic Press, 1967, p. 319.

2. D. Bauer, P. Gaillochet in Ref. 1, Vol. VA, 1978, p. 251.

3. G. Charlot, B. Trémillon, *Chemical Reactions in Solvents and Melts.* Pergamon Press, 1969, pp. 254, 256.

4. A.I. Popov in *The Chemistry of Nonaqueous Solvents,* Vol. III. J.J. Lagowski, ed. Academic Press, 1970, p. 241.

5. B. Trémillon, *Chemistry in Non-aqueous Solvents.* Reidel Publishing Company, 1974, pp. 112, 113.

6. L. Le Port, Comm. Energie At (France), Rappt. 2904, 1966, cited in Ref. 5.

7. R.J. Gillespie, E.A. Robinson in *Advances in Inorganic Chemistry and Radiochemistry,* Vol. 1. H.J. Emeléus, A.G. Sharpe, eds. Academic Press, 1959, p. 385.

8. R.D. Howells, J.D. McCown, *Chem. Rev., 77,* 69 (1977).

9. A. Engelbrecht, E. Tshager, *Z. Anorg. Allg. Chem., 433,* 19 (1977).

10. R. Adrien, Ph.D. Thesis, University of Melbourne (1992).

11. J. Liang, Ph.D. Thesis, McMaster University (1976).

12. J. Grondin, R. Sagnes, A. Commeyras, *Bull. Soc. Chim de France, No. 11–12,* 1779 (1976).

13. G.A. Olah, G.K. Surya Prakash, J. Sommer, *Superacids.* Wiley-International, 1985.

14. R.J. Gillespie, *Accts. Chem. Res., 1,* 202 (1968).

15. K.C. Lee, F. Aubke, *Inorg. Chem., 18,* 389 (1979).

16. K.C. Lee, F. Aubke, *Inorg. Chem., 23,* 2124 (1984).

17. W. Cicha, Ph.D. Thesis, University of British Columbia (1989).

18. R.J. Gillespie, K. Ouchi, G.P. Pez, *Inorg. Chem.*, *8*, 63 (1969).

19. T.A. O'Donnell, in *Comprehensive Inorganic Chemistry*, Vol. 2. J.C. Bailar, H.J. Eme-léus, R.S. Nyholm, A.F. Trotman-Dickenson, eds. Pergamon Press, 1973, pp. 1015–1019.

20. J.H. Canterford, T.A. O'Donnell, in *Technique of Inorganic Chemistry*, Vol. VII. H.B. Jonassen, A. Weissberger, eds. Wiley-Interscience, 1968, pp. 273–306.

21. J. Besida, T.A. O'Donnell, *Inorg. Chem.*, *28*, 1669 (1989).

22. (a) I. Gennick, K.M. Harmon, M.P. Potvin, *Inorg. Chem.*, *16*, 2033 (1977). (b) D. Mootz, D. Boenigk, *Z. Anorg. Allg. Chem.*, *544*, 159 (1987).

23. (a) B. Bonnet, G. Mascherpa, *Inorg. Chem.*, *19*, 785 (1980). (b) D. Mootz, K. Bartmann, *Angew. Chem., Int. Ed. Engl.*, *27*, 391 (1988).

24. L. Stein, E.H. Appleman, *Inorg. Chem.*, *22*, 3017 (1983).

25. A.W. Jache, G.H. Cady, *J. Phys. Chem.*, *56*, 1106 (1952).

26. R.C. Paul, G. Singh in *The Chemistry of Nonaqueous Solvents*, Vol. VB. J.J. Lagowski, ed., Academic Press, 1978, pp. 197–268.

27. Ref. 19, pp. 1054–1062.

28. D.H. Kerridge in *The Chemistry of Non-aqueous Solvents*. Vol. VB. J.J. Lagowski, ed., Academic Press, 1978, pp. 270–329.

29. D.M. Gruen, R.L. McBeth, *Pure Appl. Chem.*, *6*, 23 (1963).

30. J.P. Young, J.C. White, *Anal. Chem.*, *32*, 799 (1960).

31. An example of this usage is in the highly respected textbook by F.A. Cotton and G. Wilkinson, *Advanced Inorganic Chemistry*, 5th ed. Wiley-Interscience, 1988, pp. 108–109.

32. See R.J. Gillespie, T.E. Peel in *Advances in Physical Organic Chemistry*, Vol. 9. V. Gold, ed., Academic Press, London, 1972, pp. 1–24.

# CHAPTER 2

# Experimental Determination of Solute Speciation and of Acidity and Basicity in Highly Acidic Media

For many nonaqueous solvents, physicochemical investigation of speciation of solutes has been limited to measurement of conductances of solutions. While this has provided much qualitative information on relative acidities and basicities and on solvolysis reactions, it is difficult to obtain quantitative data from these measurements. This is particularly the case if the solvent has an extensive and complex self-ionization, as with anhydrous sulfuric acid. "Background" conductivity from the solvent and shift in equilibria for the ions of the solvent make rigorous investigation based on conductance measurements of dissolution processes for solutes difficult or impossible. Simple conductance measurements are also very difficult to interpret reliably if the dielectric constant of the medium is low because of the large extent of ion-pairing that occurs. Anhydrous acetic acid, with its dielectric constant of only 6.3, presents a special case, and even in liquid ammonia ($\varepsilon = 22$), ion-pairing has been shown to be extensive.

Many investigators have constructed very elegant apparatuses to perform conductimetric titrations or have made conductance measurements after progressive additions of basic solutions to acidic solutions. However, the only information this approach provides is the stoichiometry of the neutralization reactions. For example, titration of $NH_4Cl$ in $NH_3$ against $KNH_2$ in $NH_3$ shows a 1:1 reaction. Similarly, in the hydrogen halides $BX_3$ in $HX$ reacts on a 1:1 basis with $R_4NX$ in $HX$. $BrF_2^+SbF_6^-$ in $BrF_3$ neutralizes $Ag^+BrF_4^-$ in $BrF_3$ in similar fashion.

Potentiometric titrations have been used to obtain data on relative strengths of acids and bases in a very small number of nonaqueous solvents. Popov[1] and Trémillon[2] have reviewed work in anhydrous acetic acid extending over a long period. The mineral acids $HClO_4$, $HBr$, $H_2SO_4$, $HCl$, and $HNO_3$ have been titrated against pyridinum acetate—the acetate ion being the strong base of the system—using chloranil and glass electrodes as indicator electrodes. The acids have been shown to decrease in Bronsted acid strength in the order given above, but the value of pKa for $HClO_4$ is only

4.85 (very close to the value of $pK_a$ for $CH_3COOH$ in water) because extensive ion-pairing results in a very low concentration of free $H_{solv}^+$ in solution. Similarly, bases like pyridine, ammonia, and acetates have very low values of $K_b$, typically 6.1–6.7.

The major potentiometry carried out on the superacids, which are the central systems for this book, has been on measurement of acid strengths in anhydrous HF. One investigation was based on the use of the hydrogen electrode as the indicator in potentiometric titrations and the other was based on the use of the chloranil electrode. These will be discussed below, and the results, designed to establish an order of Lewis acid strengths in HF, will be compared with determinations of Hammett Acidity Functions in HF, measurements which depend on spectrophotometric determination of concentrations of appropriate indicators in acid and base form in the solvent.

Simple direct conductance measurements have been used with Lewis acids and bases in all of the superacids to establish orders of acid or base strengths on a nonquantitative basis. Additionally, direct conductance measurements have given indications of the extent of solvent-solute interaction for solutes other than acids or bases without giving definitive information on the nature of the species formed by solvolysis or synthesis. Firmer information on solute speciation has been obtained by combining cryoscopy with selective conductance measurements that monitor only the concentration of the base of the solvent which is produced in any solvolysis reaction. The most comprehensive and quantitative information on strengths of acids and bases in all four of the major superacids studied to date has come from spectrophotometric determination of Hammett Acidity Functions.

The way in which cryoscopy, selective conductance measurements, and Hammett Function measurements have been applied to superacids will be described in the remainder of this chapter. For chloroaluminates, the only molten salt system to be described in detail in this book, there is no equivalent of the Hammett Acidity Function—the measurement of which depends on protonation by superacids of extremely weak bases acting as indicators. Electrochemical methods are widely used to determine the solvent species in these largely ionic liquids and solute speciation is determined by electronic and Raman spectroscopy and voltammetry.

## 2.1  Cryoscopy in Superacids

In his pioneering work on the application of a complementary set of physicochemical experimental methods to superacids, to anhydrous sulfuric acid in the first instance, R. J. Gillespie overcame the limitations of the application of simple, direct conductance measurements in monitoring the purity of the solvent and in attempting to characterize the species resulting in solution from solute-solvent interaction. He found the well-established technique cryoscopy to be a powerful tool in determining the number of particles existing or generated in a superacid. Subsequently he extended and refined the use of conductance measurements to indicate the number found in solution of the anions which are the base of the solvent when a solute is added. By combining these two experimental approaches in complementary fashion, he was able to deduce not only the number of particles in solution but also their nature. He then used spectroscopic and crystallographic methods to confirm these deductions. This combination of experimental approaches is exemplified in a later section of this chapter.

Cryoscopy is the physicochemical technique in which the extent of depression of the freezing point of a solvent can be related to the molal concentration of the solute *and* to the number of mols of particles per mol of solute formed when the solute is dissolved in the solvent. If a solute dissolves without association, dissociation, or solvolytic interaction, there is formed 1 mol of solute particles per mol of solute. In sufficiently dilute solution, the experimentally measured difference between the freezing point of the solution and that of the solvent ($\Delta T_f$) increases essentially linearly with concentration of the solute, expressed as mol fraction. The accepted convention in this situation is that the slope of the curve for the plot of $\Delta T_f$ vs. concentration of a solute that neither dissociates nor associates is that characteristic of a solute for which $v = 1$, where $v$ is called the cryoscopic coefficient. If the solute were to dissociate to give 2 mol of particles per mol of solute, the depression of the freezing point at any particular concentration would be twice that given in the example above; that is, the slope of the plot of $\Delta T_f$ vs. concentration would be twice that for which $v = 1$, and such a solute is said to have a value of 2 for $v$. If a solute dissolves or reacts solvolytically to give $n$ particles, regardless of whether $n$ is greater or less than unity, and therefore to cause $n$ times the freezing point depression, the value of $v$ is $n$. These generalities hold regardless of the chemical nature of the particles and regardless of whether the particles formed in solution are charged or uncharged, that is, are ions or molecules.

In reporting on a project that was part of Ingold's extensive program of investigation of nitration of aromatic hydrocarbons in $H_2SO_4$, Gillespie described an all-glass cryoscope in which he could measure the freezing point of supposedly pure $H_2SO_4$, make quantitative additions of acid and base ($SO_3$ and $H_2O$) to it without exposing the acid to atmospheric moisture, and measure the new freezing points.[3] Adventitious moisture must be excluded from the system because of the avidity with which $H_2SO_4$ absorbs water and the resultant effect that this has on the self-ionization equilibria (equation 1.26). The shift in equilibria increases the basicity of the solvent, $H_2O$ being protonated, causing an increase in the concentration of $HSO_4^-$.

Figure 2.1 shows the experimentally determined freezing points of $H_2SO_4$ to which $SO_3$ and $H_2O$ have been added separately in incremental amounts.[4] A well-defined maximum of 10.371°C is observed at 100% $H_2SO_4$, the depression of freezing point on addition of $SO_3$ or $H_2O$ being due to the increased concentrations of resulting ionic species over the already large number of ions present from the true self-ionization of pure $H_2SO_4$.

(See section 1.1.3.2.) So despite this large "background" ionization, measurement of freezing point provides an excellent criterion for the purity of the solvent—much more reliable than direct conductance measurements would be.

The principal advantage of applying cryoscopy to study of superacid solutions is that it gives sound preliminary information on solute-solvent interactions. Figure 2.2 summarizes determinations of freezing points of several solutions for which different solutes have been added to pure $H_2SO_4$.[4] Addition of sulfuryl chloride, $SO_2Cl_2$, to $H_2SO_4$ causes no enhancement of the conductance—the solute is not ionized in solution. It dissolves simply as molecular $SO_2Cl_2$. Curve A shows a small but steady decrease in freezing point of the $H_2SO_4$ solution as $SO_2Cl_2$ is added. This is a demonstration of the situation for which the cryscopic coefficient $\nu$ equals unity.

For the solute $NaHSO_4$ (curve B), for which conductance measurements show marked enhancement, the freezing-point depression at any particular concentration is twice that for $SO_2Cl_2$ at the same concentration. It is easy to deduce that $NaHSO_4$ dissociates into the solvated ions $Na^+$ and $HSO_4^-$, that is, $\nu = 2$. For benzophenone (curve C) it is observed that $\nu = 2$, so protonation of the solute by the superacidic solvent is a reasonable postulate. It had been thought by earlier workers studying highly acidic media that $HNO_3$ would be protonated by $H_2SO_4$ to $H_2NO_3^+$. If this were to be the case, the value of $\nu$ for $HNO_3$ solutions in $H_2SO_4$ would be 2. However, curve D shows that $\nu = 4$. This apparent difficulty is easily resolved in the following section and the interpretation of solute-solvent interaction for $HNO_3$ as a solute can then be applied to the case of the solute $KNO_3$ for which $\nu = 6$ (curve E).

Subsequent to the work on the $H_2SO_4$ solvent system, Gillespie and co-workers[5] demonstrated a marked maximum in the freezing-point curve for

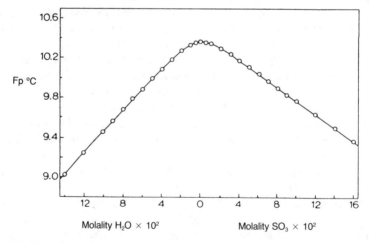

**Figure 2.1.** Freezing Points for the $H_2O$—$H_2SO_4$—$SO_3$ system.

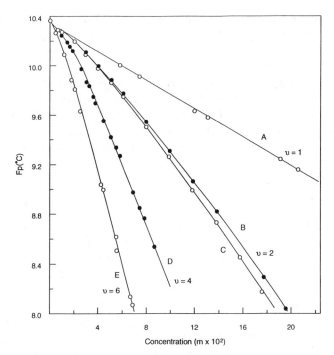

**Figure 2.2.** Depression of Freezing Point in $H_2SO_4$.
A, $SOCl_2$; B, $NaHSO_4$; C, $(C_6H_5)_2CO$; D, $HNO_3$; E, $KNO_3$.

the HF—$SO_3$ system at the 1:1 ratio, that is, for pure $HSO_3F$ which is pre-pared by reacting equimolar amounts of HF and $SO_3$. They measured ex-perimental values of unity for $v$ for the compounds $S_2O_6F_2$, $S_2O_5F_2$, and $CH_3SO_3F$ which are essentially nonelectrolytes when dissolved in $HSO_3F$. They observed values of 2 for $v$ for the dissociating solutes $MSO_3F$ (where M = Li, K, Rb, and Cs) and for benzoic acid that is fully protonated by $HSO_3F$. Having demonstrated the reliability of cryoscopy in $HSO_3F$, Gilles-pie later used this technique as part of his approach to the identification of cations of the halogens and chalcogens that he generated by oxidation of the elements in $HSO_3F$, as will be presented later in this chapter.

In 1970, Gillespie and Humphreys[6] reported a cryoscope similar in general design to the one used for $H_2SO_4$ and $HSO_3F$, but constructed from KelF, Teflon, and platinum, so that they could make cryoscopic measurements in anhydrous HF. They showed $HSO_3F$, $XeF_2$, $SO_2$, $SF_6$, and $S_2O_5F_2$ to be non-electrolytes, while the alkali metal fluorides and $NH_4F$ gave the expected value of $v = 2$. Subsequently Gillespie and colleagues used HF cryoscopy as part of a study of the Lewis acidity of the pentafluorides $SbF_5$, $AsF_5$, and $PF_5$ in HF, as described in chapter 3.

When the use of cryoscopy in interpreting reactions in solution is dem-onstrated later in this chapter, it will be appreciated that these observations are of greatest value when integral or near-integral values of $v$ are obtained. Of course, this will not be the case for all solutes. For nitrobenzene as a solute in $H_2SO_4$, the experimentally determined value for $v$ lies between 1 and 2, indicating partial protonation of the solute. Although a value of unity is reported for a solution of $XeF_2$ in HF, precise conductivity measurements show a very slight enhancement of the conductance of HF when $XeF_2$ is added, indicating a very slight ionization to $XeF^+$ and $F^-$. However, within the experimental limits of accuracy in cryoscopic measurements, it is reasonable to regard as unity a $v$ value for $XeF_2$ which in reality must be very slightly greater than 1.

## 2.2 Selective Conductance Measurements: $\gamma$ Values

As stated earlier, it is difficult to obtain quantitative information on solvolytic reactions in any protonic solvent relying only on direct conductance measurements. The difficulties are exacerbated with a solvent such as $H_2SO_4$ which has an extensive and complex self-ionization, easily disturbed by minor amounts of impurities or reaction products.

In an extensive study of conductances of $H_2SO_4$ solutions, Gillespie and co-workers found a way of using these measurements selectively to overcome many of the interpretative problems. They observed values of limiting ionic mobilities for cations and anions in solution in $H_2SO_4$ as given in Table 2.1.[4] There are two major reasons for the very great differences in ionic mobilities between the values for the two ions of the fundamental self-ionization (namely, $H_3SO_4^+$ and $HSO_4^-$) and those for other ions in solution in $H_2SO_4$.

When a potential is applied to a solution containing $H_3SO_4^+$ (or $HSO_4^-$) ions, the entire solvated ionic entity does not migrate to the cathode (or anode). A "proton jump" mechanism operates, in which protons jump progressively from $H_3SO_4^+$ ions to neighboring $H_2SO_4$ molecules. $HSO_4^-$ anions do not migrate as such. Protons jump from neutral $H_2SO_4$ molecules to $HSO_4^-$ ions—the apparent migration of $HSO_4^-$ anions to the anode being in reality movement of protons to the cathode. This is a common mechanism for conduction in protonic solvents. For water, ionic mobilities for $H_3O^+$ and $OH^-$ are 349.8 and 198.6 $\Omega^{-1}$ $cm^2$ $mol^{-1}$, respectively, while for other singly charged cations and anions the values usually lie in the range 50–70. For

**Table 2.1.** Limiting Ionic Mobilities in 100% Sulfuric Acid at Infinite Dilution (25°C)

| Ion | Ionic mobility ($\Omega^{-1}$ $cm^2$ $mol^{-1}$) |
|---|---|
| $H_3SO_4^+$ | 242 |
| $HSO_4^-$ | 171 |
| $H_3O^+$ | ~5 |
| $Na^+$ | ~3 |
| $K^+$ | ~5 |

$HSO_3F$, values of mobilities for $H_2SO_3F^+$ and $SO_3F^-$ are 185 and 138 compared with typical values of 17 and 21 for $K^+$ and $NH_4^+$. In the less-viscous anhydrous HF, $H_2F^+$ and $F^-$ values of 350 and 273 can be compared with typical values for other singly charged cations or anions that are in the range about 150.[8]

For $H_2SO_4$ there is an additional factor causing a larger difference between the mobilities of the cation and anion of the solvent and those of other ions than occurs for the other solvents. Whereas the viscosities of $H_2O$, $HSO_3F$, and HF are all reasonably similar, that for $H_2SO_4$ is some 25 times higher. Consequently, cations and anions solvated by bulky $H_2SO_4$ molecules, and moving as such, have their movement under an applied potential reduced, by comparison with that in the other solvents, because of the much greater viscous drag of the medium.

It is worth noting from Table 2.1 that the water molecule protonated by the solvent $H_2SO_4$ is a cationic entity $H_3O^+$ in its own right. Solvated by $H_2SO_4$ molecules, its mobility is similar to that of $K^+$ or $Na^+$. It is not participating in any proton jump mechanism as it would in water. $H_3O^+$ in $H_2SO_4$ is not fundamentally different in its discrete ionic character from the $NH_4^+$ ion in water, which migrates under an applied potential as the aquated cation.

Reference to Table 2.1 shows that in acidic or basic $H_2SO_4$ solutions, most of the current would be due to the "movement" of $H_3SO_4^+$ or $HSO_4^-$ ions and that the contribution to the current from the mobility of other ions can, to a good approximation, be ignored. Experiment showed that, for a series of alkali metal monohydrogensulfates dissolved in $H_2SO_4$, the observed conductance was virtually independent of the nature of the counter cation of the solute in each case. This means that an acid-base or solvolysis reaction which produces 1 mol of $HSO_4^-$ ions per mol of solute would have at any particular concentration of solute virtually the same conductance as would be shown by the same concentration of the alkali metal monohydrogensulfate. A reaction in $H_2SO_4$ which produces 2 (or $n$) mols of $HSO_4^-$ per mol of solute would show the same conductance as $KHSO_4$, for example, at twice (or $n$ times) the concentration of the solute under investigation.

This led Gillespie to define a quantity $\gamma$ as the number of mols of $HSO_4^-$ (or, in acidic solutions, of $H_3SO_4^+$) produced when 1 mol of any solute interacts with $H_2SO_4$.[4] By measuring values of $\gamma$ after adding solutes to $H_2SO_4$ and considering these values in conjunction with values of $v$ obtained by cryoscopy for the same reactions, Gillespie was able to increase markedly the ability to deduce from simple physicochemical measurements the overall pattern of acid-base and solvolysis reactions in $H_2SO_4$, and subsequently in $HSO_3F$. This approach will be demonstrated in the next section.

Figure 2.3 illustrates an experimental approach to determining values of $\gamma$. The conductance of a solute like $KHSO_4$ which produces 1 mol of $HSO_4^-$ per mol of solute could be measured over a range of concentrations, c. Similar measurements could be made for $KHSO_4$, over a range of concen-

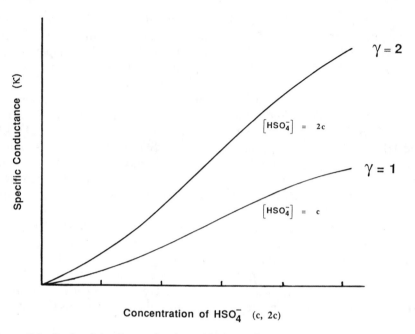

Concentration of $HSO_4^-$  (c, 2c)

**Figure 2.3.** Basis of the Determination of Values of $\gamma$ From Conductance Measurements.

trations, 2c, each of which was twice that in the first series. Matching the conductance of a solute under investigation at a particular concentration c with that for $NaHSO_4$ at the same concentration would indicate whether the acid-base or solvolysis reaction produces 1 or 2 (or more) mol of $HSO_4^-$ per mol of solute.

## 2.3 Determination of Number and Nature of Species in Solution

Table 2.2, drawn largely from reviews published by Gillespie[4,7] is presented to show the way in which he and his co-workers have used cryoscopy, which gives $v$ values, and conductance measurements, which lead to observed $\gamma$ values, to obtain preliminary but usually very reliable information on the course of a wide range of types of reaction in the superacids $H_2SO_4$ and $HSO_3F$. Table 2.2 is not meant to be a comprehensive account of that work. It uses some representative examples to demonstrate the experimental methodology adopted. It will be appreciated that $v$ and $\gamma$ values give the easiest and most reliabale information on products of reaction when integral or near-integral values are obtained. For this reason no $v$ and $\gamma$ measurements have been reported for acidic solutes in $H_2SO_4$ and $HSO_3F$, with one exception, $B(HSO_4)_3$. Bronsted acids are too weak to show $v$ values very different from unity, and Lewis acids usually interact in a very complex fashion, as will be

**Table 2.2.** Interpretation of Reactions in Superacids Based on $\nu$ and $\gamma$ Values

| Solute-solvent Interaction | Solvent | Solute(s) | $\nu^a$ | $\gamma^a$ | Equation |
|---|---|---|---|---|---|
| Lewis basicity | $H_2SO_4$ | $NaHSO_4$ | 2 | 1 | $NaHSO_4 \rightarrow Na^+ + HSO_4^-$ |
| | $HSO_3F$ | $KSO_3F$ | 2 | 1 | $KSO_3F \rightarrow K^+ + SO_3F^-$ |
| Bronsted basicity | $H_2SO_4$ | $C_6H_5COOH$ | 2 | 1 | $C_6H_5COOH + H_2SO_4 \rightarrow C_6H_5C(OH)_2^+ + HSO_4^-$ |
| | $HSO_3F$ | $C_6H_4(NH_2)_2$ | 3 | 2 | $C_6H_4(NH_2)_2 + 2HSO_3F \rightarrow C_6H_4(NH_3)_2^{2+} + 2SO_3F^-$ |
| | $H_2SO_4$ | $H_3PO_4$ | 2 | 1 | $H_3PO_4 + H_2SO_4 \rightarrow H_4PO_4^+ + HSO_4^-$ |
| | $H_2SO_4$ | $K_2SO_4$ | 4 | 2 | $K_2SO_4 + 2H_2SO_4 \rightarrow 2K^+ + 2HSO_4^-$ |
| Solvolysis | $H_2SO_4$ | $HNO_3$ | 4 | 2 | $HNO_3 + 2H_2SO_4 \rightarrow NO_2^+ + H_3O^+ + 2HSO_4^-$ |
| | $H_2SO_4$ | $KNO_3$ | 6 | 3 | $KNO_3 + 3H_2SO_4 \rightarrow K^+ + NO_2^+ + H_3O^+ + 3HSO_4^-$ |
| | $H_2SO_4$ | $C_2H_5OH$ | 3 | 1 | $C_2H_5OH + 2H_2SO_4 \rightarrow C_2H_5.HSO_4 + H_3O^+ + HSO_4^-$ |
| | $H_2SO_4$ | $(C_6H_5)_3COH$ | 4 | 2 | $(C_6H_5)_3COH + 2H_2SO_4 \rightarrow (C_6H_5)_3C^+ + H_3O^+ + 2HSO_4^-$ |
| | $H_2SO_4$ | $B(OH)_3$ | 6 | 2 | $B(OH)_3 + 6H_2SO_4 \rightarrow B(HSO_4)_4^- + 3H_3O^+ + 2HSO_4^-$ |
| Synthesis | $HSO_3F$ | $S_8, S_2O_6F_2$ (1:1) | 3 | 2 | $S_8 + S_2O_6F_2 \rightarrow S_8^{2+} + 2SO_3F^-$ |
| | $HSO_3F$ | $I_2, S_2O_6F_2$ (3:1) | 1.33 | 0.67 | $3I_2 + S_2O_6F_2 \rightarrow 2I_3^+ + 2SO_3F^-$ |

[a] Tabulated values have been rounded slightly from reported experimental values in some cases.

shown in chapter 3. However, the nature of $B(HSO_4)_3$ as a strong Lewis acid was indicated through the observation of $v$ and $\gamma$ values, as is seen from the interpretation of the solvolysis reaction of $B(OH)_3$ given below.

Most of the examples listed in Table 2.2 are drawn from $H_2SO_4$ chemistry. There has been a far wider experimental coverage of acid-base and solvolysis reactions in this solvent than in $HSO_3F$. It can be reliably assumed that $HSO_3F$, somewhat more acidic than $H_2SO_4$, will exhibit similar protonation reactions. $HSO_3F$ has proved a valuable medium for using $v$ and $\gamma$ values to follow synthetic reactions for reasons outlined below.

$HSO_3F$ is formed from interaction of HF and $SO_3$. Each of these compounds, when dissolved in $HSO_3F$, was found to be a nonelectrolyte ($\gamma = 0$). Cryoscopy gave $v = 1$ in each case; that is, they behave as molecular solutes in $HSO_3F$. Furthermore, the two freezing-point curves for $SO_3$ in $HSO_3F$ and for HF in $HSO_3F$ meet in a very sharp maximum indicating that self-dissociation of $HSO_3F$ into $SO_3$ and HF is very small.

The first two solutes listed in Table 2.2 are, simply, completely ionized *sources* of the Lewis bases $HSO_4^-$ and $SO_3F^-$. In the general description of the superacids $H_2SO_4$ and $HSO_3F$ given in chapter 1, it was stated that these solvents are so acidic that they can protonate a vast range of solutes containing O, S, N, or P atoms which are coordinately unsaturated, that is, atoms that present one or more electron pairs available for covalent bond formation. The third and fourth solutes exemplify this type of Bronsted basicity. It is not very surprising that $H_3PO_4$ or $OP(OH)_3$ is easily protonated to $P(OH)_4^+$. It was stated, without justification, in chapter 1 that $SO_4^{2-}$ is too strong a Bronsted base to exist as such in $H_2SO_4$ and that protonation occurs to produce $HSO_4^-$, the strongest base of the $H_2SO_4$ solvent system. Values of 4 and 2 for $v$ and $\gamma$ for $K_2SO_4$ support the following reaction:

$$SO_4^{2-} + H_2SO_4 \rightarrow 2HSO_4^- \qquad\qquad 2.1$$

Dissociation to two $K^+$ ions and one $SO_4^{2-}$ would give a value of 3 for $v$, whereas $2K^+$ and $2HSO_4^-$ ions are shown to be formed.

In early stages of the studies of $H_2SO_4$ solvent systems, it was believed that $HNO_3$ acted as a simple Bronsted base, being protonated to $H_2NO_3^+$ in similar fashion to the protonation of $H_3PO_4$. This may be a transient species in the interaction of $HNO_3$ and $H_2SO_4$, but for this reaction the values of $v$ and $\gamma$ would be 2 and 1, as for $H_3PO_4$. Experimental values of 4 and 2 showed that the overall reaction involved dehydration of the $HNO_3$ or $H_2NO_3^+$ and protonation of the resulting water, leading to the formation of the nitronium cation[4] as in the equation in Table 2.2. The value of 6 for $v$ for $KNO_3$ given in the previous section on experimental cryoscopy in $H_2SO_4$, when coupled with a value of 3 for $\gamma$, indicates a conceptual reaction of $H_2SO_4$ with $KNO_3$ to form $HNO_3$, $K^+$ and $HSO_4^-$, followed by the previously indicated reaction for $HNO_3$.

It might be expected that alcohols and carbinols would undergo simple protonation by $H_2SO_4$ as aldehydes, ketones, and carboxylic acids do. However, dehydration occurs with $C_2H_5OH$ to form ethyl hydrogen sulfate, protonated water, and the bisulfate ion. In the case of the carbinol, the carbonium ion $(C_6H_5)_3C^+$ is formed. An account of formation of carbonium ions in more acidic media, such as $HSO_3F$—$SbF_5$ mixtures, is given by Olah and co-authors.[9]

Dissolution of boric acid $H_3BO_3$ in $H_2SO_4$ provides a good example of the power of the use of cryoscopy and selective conductance measurements in deducing the progress of reactions in superacids.[4] On acid-base grounds, it might be expected that the $OH^-$ ligands of $B(OH)_3$ would be protonated first to $H_2O$ and then to $H_3O^+$, being replaced by $HSO_4^-$ ligands. This would give the following equation:

$$B(OH)_3 + 6HSO_4^- \rightarrow B(HSO_4)_3 + 3H_3O^+ + 3HSO_4^- \qquad 2.2$$

For this reaction, the value of $v$ would be 7 and that for $\gamma$ would be 3. Experimentally, values of 6 and 2 were observed. It was realized that $B(HSO_4)_3$ was acting as a very strong Lewis acid, bonding to one $HSO_4^-$ to form $B(HSO)_4)_4^-$ as in the equation in Table 2.2.

The last two examples in Table 2.2 demonstrate a very elegant extension of the use of measurement of $v$ and $\gamma$ values from following the direct interaction between a solvent and a single solute to monitoring the nature and extent of the interaction of two reacting solutes in a superacid. The examples involve synthesis of species which could not exist in media that are less acidic, that is, more basic, than the superacid, in this case $HSO_3F$.[7]

In chapter 4 it will be reported that for a considerable period of time, cations and other compounds of the halogens and of the chalcogens had been known or suspected to have been formed under oxidizing conditions in highly acidic media. Gillespie devised the procedure for using $S_2O_6F_2$, a very strong oxidant and a nonelectrolyte in $HSO_3F$, to oxidize solutes such as $S_8$ and $I_2$ to cationic species. The power of the approach lay in the fact that $S_2O_6F_2$, on reduction, gave $SO_3F^-$, the base of the $HSO_3F$ solvent system and, more importantly, the very mobile anion which allowed observation of $\gamma$ values in $HSO_3F$.

Table 2.2 gives the evidence obtained in $HSO_3F$ solution for the formulation of $S_8^{2+}$ from the reaction of equimolar amounts of the two nonelectrolytes $S_8$ and $S_2O_6F_2$. Cryoscopy showed that 3 mol of product per mol of $S_8$ were formed and selective conductance showed that 2 of these were $SO_3F^-$ ions. The remaining single entity, resulting from a two-electron redox reaction, could only have been $S_8^{2+}$. For the iodine system, 4 mol of particles were formed from 3 mol of iodine reacting with $S_2O_6F_2$. Of these 4, 2 were $SO_3F^-$ ions. Therefore there had to be two iodine-containing cations. The reaction stoichiometry indicated that these would be $I_3^+$. It should be noted that $v$ and $\gamma$ values are always quoted relative to 1 mol of reactant. Although in apparent contradiction to what was stated earlier, values of $v$ and $\gamma$ quoted in Table 2.2 for the iodine reaction are nonintegral when related to 1 mol of iodine, but they are integral numbers when considered for the 3 mol of reactant as written in the equation.

A vital experimental condition for the generation of the stable cations $S_8^{2+}$ and $I_3^+$ was that the medium $HSO_3F$ was very acidic; that is, its basicity was very low. As the basicity of the medium is increased, for example, if large amounts of $SO_3F^-$ were to be added to the solutions, the polyatomic cations would interact with the base and disproportionate into elemental $S_8$ and $I_2$ and compounds of sulfur and of iodine in higher oxidation states than those exhibited by the parent cations. Obviously they would disproportionate in the very basic solvent water. The conditions for generation and stability of a wide range of polyatomic cations of halogens and of chalcogens will be discussed in detail in chapter 4.

After using measurement of $v$ and $\gamma$ values to deduce the nature of species in solution after study of acid-base, solvolysis, and synthetic reactions in the superacids $H_2SO_4$ and $HSO_3F$, other techniques were used to confirm the initial deductions. Comparison of the Raman spectrum of the solid compound $NO_2^+ClO_4^-$ with that of a solution of $HNO_3$ in $H_2SO_4$ confirmed the formation of $NO_2^+$ in that solution.[4] X-ray crystallography proved the structure of $S_8^{2+}$ in the compound $S_8^{2+}(AsF_6^-)_2$ crystallized from liquid $SO_2$.[7] UV-visible spectroscopy, measurement of magnetic moments and, ultimately, crystallography were used to finally identify the iodine cations.

## 2.4  Acidity Measurements in Protonic Solvents

On purely physicochemical grounds, potentiometric measurement of acidity of solutions would probably be regarded as the most fundamental approach, particularly in aqueous solutions. The hydrogen electrode is the ultimate reference electrode. The glass electrode is the most convenient indicating electrode in aqueous systems and, in older work, the quinhydrone electrode was used for pH measurements. Each of these approaches has been adopted on a limited scale in certain areas of nonaqueous solution chemistry. As shown later in this chapter, two groups have investigated the hydrogen electrode in HF chemistry and the French groups at Saclay have used the glass electrode in anhydrous acetic acid and the chloranil electrode—a minor modification of the quinhydrone electrode—in HF, $HSO_3F$, and $CF_3SO_3H$.

None of these potentiometric approaches has been applied to all four of the superacids of particular interest in this book, namely, $H_2SO_4$, $CF_3SO_3H$, $HSO_3F$, and HF. However, determination of the Hammett Acidity Function $H_0$ has been carried out for all four solvents and their solutions. This method is based on spectrophotometric determination of acid and base forms of indicators in the appropriate solvents. In principle, this is the same approach as that for spectrophotometric pH determination in aqueous solutions. The same methodology has been used for all the superacids, the indicators being closely chemically related. Therefore, in this book $H_0$ values will be used throughout in interpretation and comparison of reactions in superacids, and the spectrophotometric approach to acidity measurement will be described first and in most detail. Potentiometric and other measurements will be discussed more briefly and, where possible, comparisons will be made between the acidity values obtained by the different methods.

## 2.4.1 Spectrophotometry

The working equation, given shortly, for the Hammett Acidity Function $H_0$, which was proposed by Hammett and Deyrup in 1932,[10] is very similar to that used in the spectrophotometric determination of the pH of an aqueous solution. Indeed, the $H_0$ scale becomes equal to the pH scale in dilute aqueous solution.

$$H_0 = pK_{HIn^+} - \log I \qquad\qquad 2.3$$

where $pK_{HIn^+}$ is the acidity constant of the conjugate acid $HIn^+$ of a weakly basic indicator In and I is the ionization ratio $[HIn^+]/[In]$, measured spectrophotometrically. In experimental terms I is equal to $(\varepsilon - \varepsilon_{In})/(\varepsilon_{HIn^+} - \varepsilon)$ where $\varepsilon_{In}$, $\varepsilon_{HIn^+}$, and $\varepsilon$ are, respectively, the extinction coefficients of the indicator fully in its base form, fully protonated, and in acid-base equilibrium in a medium of unknown acidity. The extinction coefficients are determined at the wavelength providing the maximum difference between $\varepsilon_{In}$ and $\varepsilon_{HIn^+}$, usually the wavelength for maximum absorption for $HIn^+$.

Hammett and Deyrup[10] commenced a study of the $H_2O$—$H_2SO_4$ system with a conventional acid-base indicator of the aqueous system, for which the value of $pK_{HIn^+}$ was known, and measured the ionization ratio for that indicator in dilute $H_2SO_4$. On solutions of increasing $H_2SO_4$ concentration, they repeated the ratio measurements with a succession of progressively weaker bases as indicators for which ionization ratios overlapped; that is, each pair of successive indicators had measurable concentrations in protonated and nonprotonated forms at the solution acidity being measured. This allowed them to calculate $pK_{HIn^+}$ values for successive indicators and then to determine $H_0$ values over the whole composition range 100% $H_2O$ to 100% $H_2SO_4$.

The reliability of this procedure rests on the assumption that the activity coefficient ratio of the indicators used, in acid and base form, is independent of the chemical nature of the indicators in any particular medium. Gillespie and co-workers[11,14] have shown this assumption to be valid within the limits of accuracy of the method for bases of similar chemical structure, such as nitroanilines and nitroaromatic compounds. They also report that some of the polynitroaromatic indicators can undergo a second protonation at very high acidities, allowing $H_0$ determinations to be made at acidity levels for which no sufficiently weak neutral bases are available.

In the early 1960s there was some refinement of the earlier study by Hammett and Deyrup of the $H_2O$—$H_2SO_4$ system because it was recognized that they had not used a consistent set of indicators which overlapped with each other in a reliable manner. In 1971 Gillespie et al.[11] extended $H_0$ determinations into acidic $H_2SO_4$ systems, those containing the Bronsted acids $HSO_3F$ and $HSO_3Cl$ and the Lewis acids $SO_3$ and $B(HSO_4)_3$. They prepared the latter in situ by the reaction.

$$B(OH)_3 + 3SO_3 \xrightarrow{\ H_2SO_4\ } B(HSO_4)_3 \qquad\qquad 2.4$$

$B(HSO_4)_3$ would then react with $HSO_4^-$ in solution as described in chapter 1 and as shown in Table 2.2 to form $H_3SO_4^+$ and $B(HSO_4)_4^-$. They also measured $H_0$ values in basic $H_2SO_4$. They did not adopt the strategy that they used later with the $HSO_3F$ and HF systems of adding the formal base of the solvent system. Instead they added $H_2O$ incrementally on the premise that it would be fully protonated and would yield the stoichiometric amount of the base $HSO_4^-$ in the solution.

Plotting of $H_0$ values in acidic solutions, resulting from progressive additions of $SO_3$, and in basic solutions gave a sigmoidal curve as shown in the insert in Figure 2.4. The point of inflection located at $-11.9$ was taken as the value for $H_0$ for the pure solvent. It will be shown later in this section that this method of arriving at a value of $H_0$ for a 100% pure superacid is far

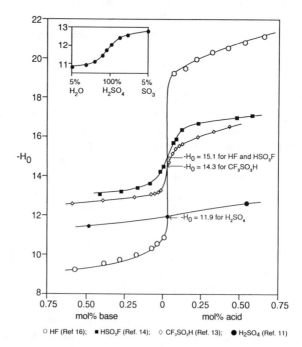

O HF (Ref 16);  ■ HSO₃F (Ref. 14);  ◇ CF₃SO₃H (Ref. 13);  ● H₂SO₄ (Ref. 11)

**Figure 2.4.** Dependence of $H_0$ Values on Acidity and Basicity for $H_2SO_4$, $CF_3SO_3H$, $HSO_3F$, and HF.

more reliable than carrying out a spectrophotometric determination on the supposedly pure superacid. Anhydrous HF provides a spectacular example. Most current textbooks quote $H_0$ for HF as being about $-11$. The "point of inflection value" is $-15.1$. The reasons for this enormous difference between the value obtained with supposedly pure HF and that obtained using Gillespie's experimental approach will be discussed in detail later in this section.

Of the four superacids presented here, trifluoromethylsulfuric acid ("triflic acid," $CF_3SO_3H$) has received the least attention from inorganic chemists, although a large body of organic chemistry has been conducted in this medium. Boron "triflate," $B(OSO_2CF_3)_3$, is a strong Lewis acid, accepting one $CF_3SO_3^-$ ion to form $B(OSO_2CF_3)_4^-$. In 1970 Engelbrecht and Tschager[12] measured $H_0$ values on $CF_3SO_3H$ solutions made progressively stronger in $B(OSO_2CF_3)_3$ until the solution contained about 20 mol% of the Lewis acid. Later work by Adrien[13] substantially reproduced the Engelbrecht and Tschager values to a $B(OSO_2CF_3)_3$ concentration of 0.4 mol%. More importantly, Adrien added $NaOSO_2CF_3$, the source of the base of the system, progressively to 2 mol%. His experimental values provide the data for $H_0$ values for $CF_3SO_3H$ in Figure 2.4. Taking the point of inflection as the value of $H_0$ for pure $CF_3SO_3H$, a value of $-14.33$ is obtained. The slope for the plot of $H_0$ vs. $[B(OSO_2CF_3)_3]$ near the neutral point is much less steep than the slope for $H_0$ vs. $[CF_3SO_3^-]$, indicating that $[B(OSO_2CF_3)_3]$ is not as strong as an acid as $(CF_3SO_3^-)$ is as a base. This suggests that if Lewis acids which are stronger than $B(OSO_2CF_3)_3$ are found, $H_0$ for pure $CF_3SO_3H$ will prove to be more negative than $-14.33$.

Gillespie and Peel[14] measured $H_0$ values in basic $HSO_3F$ using the solute $KSO_3F$ over a range of concentrations to 0.4 mol%. They studied $SbF_5$ and $SbF_5 \cdot 3SO_3$, that is, $SbF_2(SO_3F)_3$, amongst other Lewis acids (Figure 2.4) and quote a value of $-15.07$ for $H_0$ for 100% $HSO_3F$. Recently Aubke and Cicha[15] measured marginally higher $H_0$ values for $Ta(SO_3F)_5$ than Gillespie and Peel recorded for $SbF_2(SO_3F)_3$ for solute concentrations beyond 1.5 mol%. At this concentration $Ta(SO_3F)_5$ is about 1 $H_0$ unit more acidic than $SbF_5$ in $HSO_3F$. However, in more dilute solutions, up to 0.3 mol%, Aubke and Cicha's measurements show $Ta(SO_3F)_5$ to be somewhat less acidic than $SbF_5$ which in turn is considerably less acidic than $SbF_2(SO_3F)_3$.

Using trinitrated aniline, toluene, and benzene as indicators in basic and weakly acidic HF and the already-protonated forms of trinitrotoluene, trinitrobenzene, and dichlorotrinitrobenzene in strongly acidic HF, Gillespie and Liang[16] recorded $H_0$ values for the solutes $H_2O$, KF, $HSO_3F$, $PF_5$, $NbF_5$, $TaF_5$, $AsF_5$, and $SbF_5$ in anhydrous HF. Their values for the KF—HF and $SbF_5$—HF systems are used in Figure 2.4. They give a "point of inflection" value of $-15.1$ for $H_0$ for pure HF. Their value was based on the averages of several experimental $H_0$ values from different low concentrations of acids and base.

Figure 2.4 has been constructed not only to indicate the $H_0$ values for the four "pure" superacids and the way in which those values were arrived at but also to allow a direct comparison of the extents of the self-ionization reactions of the acids and a consideration of the implications of the differing self-ionizations for Hammett Function measurements.

As stated in chapter 1, the self-ionization of $H_2SO_4$ is complex and extensive. Consequently $H_2SO_4$ is substantially buffered against significant change in acidity or basicity resulting from impurities such as traces of water in the solvent or adventitious additions of other impurities in subsequent solution preparations. As can be seen from the figure, additions of up to 0.5 mol% of base or acid have little effect on measured $H_0$ values. In the insert in Figure 2.4, $H_0$ values are plotted against acid and base concentrations up to 5 mol% in order to show a reasonable sigmoidal curve.

Of the four superacids under consideration, HF is the most different from $H_2SO_4$ in its self-ionization. Its ionic product[16,17] is believed to be between $10^{-12}$ and $10^{-13}$, that is, not very different from that of water, which is $10^{-14}$ at 25°C. Consequently the plot of $H_0$ vs. acid-base concentrations is very similar to that for the variation of the pH of water with addition of acid or base.

A major implication is that if traces of water, the most likely impurity, remain in distilled HF, $H_2O$ will be protonated, thereby disturbing the self-ionization equilibrium of HF to provide an equivalent amount of $F^-$, rendering the system slightly basic. Additionally, in most experimental work, distilled HF is stored in KelF containers. O'Donnell and Cockman have shown[18] that, depending on its acidity level, HF can leach unsaturated chlorofluorocarbons out of KelF. (See section 6.1.2.1.) These are protonated, causing an enhancement of $F^-$ concentration, that is, an increase in basicity. The outcome is that HF, no matter how well it has been distilled, will become somewhat basic on storage. This explains why "single-value" $H_0$ determinations on supposedly pure HF gave values in the region $-10$ to $-11$, very different from the "point of inflection" value of $-15.1$. Even the use of an indicator in spectrophotometric determination of a "single-value" $H_0$ would increase the basicity of the system quite markedly.

By measuring $H_0$ values for $H_2SO_4$, $HSO_3F$, and HF solutions containing significant concentrations of bases and of acids, Gillespie and co-workers were able to "swamp" the effects of impurities and indicators in the superacids under study and to obtain reliable experimental values of $H_0$ for basic and acidic solutions which could then be extrapolated to give reliable values for the "pure" superacid.

Reference to Figure 2.4 shows that $HSO_3F$ with an ionic product[7] of about $10^{-8}$ has a much less steep curve around its inflection point than HF. $CF_3SO_3H$, as well as having a less negative value of $H_0$ than those for HF and $HSO_3F$, has an ionic product somewhat greater than for $HSO_3F$.

Chapter 10 contains a discussion of experimental strategies recommended for adoption in the synthesis of inorganic species in superacids. It deals, in part, with the possibility of the loss of necessary control of acidity or basicity during some syntheses and with procedures for buffering acidic media in order to optimize synthesis conditions.

## 2.4.2 Potentiometry

Although possible applicability of the hydrogen electrode has been explored experimentally quite well as an acidity indicator in superacids, it has been shown to have a very small role to play, and that role is restricted to a limited number of HF systems. It provides acidity and basicity measurements in basic HF and in HF solutions of medium acidity, but not those with strong Lewis acids such as $SbF_5$ and $AsF_5$ which oxidize $H_2$.[19] Carre and Devynck[20] say that S(VI), produced as $SO_3$ from the internal decomposition of $HSO_3F$, $CF_3SO_3H$, and $H_2SO_4$, also oxidizes hydrogen, making it impossible to use the hydrogen electrode in these superacids. Devynck and co-workers have used the chloranil and fluoranil electrodes in HF, $HSO_3F$, and $CF_3SO_3H$ to obtain an acidity scale $R(H)$ which, for similar superacid systems, is numerically different from the $H_0$ scale, but relative values for different superacid systems within the $R(H)$ and $H_0$ scales show interesting comparability, as demonstrated in a later section of this chapter.

### 2.4.2.1 The Hydrogen Electrode

Gut and Gautschi[21] constructed apparatus from PTFE, KelF, and FEP to conduct potentiometric titrations of solutions of Lewis acids of the HF system against NaF using a Pd-wire hydrogen electrode, the potential of which was measured against the reference electrode $Ag/AgBF_4$ (sat)/$KBF_4$ (sat) in HF. They reported that $SbF_5$—HF interacted chemically with the hydrogen electrode, but they reported titration curves for the acids $AsF_5$, $TaF_5$, $BF_3$, $NbF_5$, and $PF_5$, which were very like curves for pH change in titration of aqueous strong bases against aqueous acids of differing strength.

Subsequently Devynck and co-workers[19] examined the behavior of the hydrogen electrode potentiometrically and voltammetrically in HF solutions of KF and of Lewis acids. They demonstrated electrochemical reversibility in basic HF and in weakly acidic $BF_3$ and $TaF_5$ solutions, although for the latter, the equilibrium potential reached initially was not stable with time. The Pt electrode required frequent platinizing for reliable measurements. $AsF_5$ and $SbF_5$ both interacted with the hydrogen electrode.

In a later publication[17], Gut agreed that the acidity of the $AsF_5$—HF system could not be studied using the hydrogen electrode, but his earlier observations that acid strengths increase in the order $PF_5 < NbF_5 < BF_3 < TaF_5$ agree with those of workers using other measurements of acid strengths. His values for pF in solutions of the acids in HF are incorporated into the comparison of different acidity scales made later in section 2.4.5.

Gut and Gautschi[21] also reported that $HSO_3F$ and $CF_3SO_3H$ acted as non-electrolytes in HF; their addition to fluoride-containing HF solutions did not change the pF values.

### 2.4.2.2 Chloranil and Fluoranil Electrodes

Prior to the easy availability of the glass electrode—which, of course, cannot be used in HF solutions—the quinhydrone electrode had been fairly widely used for pH measurement in aqueous solutions, particularly in analytical chemistry. It was a system in which p-benzoquinone exists in a pH-dependent redox equilibrium with p-dihydroxybenzene:

$$O=\!\!\left\langle\bigcirc\right\rangle\!\!=O + 2H^+ + 2\varepsilon \rightleftharpoons HO-\!\!\left\langle\bigcirc\right\rangle\!\!-OH \qquad 2.5$$

A Pt wire, the potential of which relative to a calomel or Ag-AgCl electrode indicated the position of the redox equilibrium, acted also as a pH indicator because of the pH-dependent nature of the redox reaction.

In 1975 Masson et al.[22] carried out an investigation of 11 quinones or hydroquinones as potential pF indicating electrode systems in HF and established the weakly basic chloranil (tetrachloro-p-benzoquinone), represented here as Q, as the most reliable for HF studies. Using as reference electrode a silver wire in HF saturated with AgF and 1 M in KF, they established electrochemical reversibility in basic solution (KF—HF) for the two-electron step:

$$Q + 2H^+ + 2\varepsilon \rightleftharpoons QH_2 \qquad 2.6a$$

In strongly acidic HF (containing $SbF_5$) two one-electron steps were observed, depending on the level of acidity:

$$QH^+ + H^+ + 1\varepsilon \rightleftharpoons QH_2^{\cdot+} \qquad 2.6b$$

and $\quad QH_2^{\cdot+} + H^+ + 1\varepsilon \rightleftharpoons QH_3^+ \qquad 2.6c$

Later, members of that same Paris group used this indicator system to place HF solutions of the base $F^-$ and of several Lewis acids in the form of their buffer conjugate couples on an acidity scale, defined as $R(H)$, where $R(H) = 0$ corresponds with pH $= 0$ in aqueous solutions.[23] Their results are given in Table 2.3.

**Table 2.3.** $R(H)$ Values for Strong Base, Strong Acid, and Equimolar Buffers $(MF_n/MF_{n+1}^-)$ in HF

| Solute | $R(H)$ | Solute | $R(H)$ |
|---|---|---|---|
| 1 M KF | $-14.2$ | $TaF_5/TaF_6^-$ | $-23.3$ |
| $PF_5/PF_6^-$ | $-18.2$ | $AsF_5/AsF_6^-$ | $-25.8$ |
| $NbF_5/NbF_6^-$ | $-20.9$ | 1 M $SbF_5$ | $-27.9$ |
| $BF_3/BF_4^-$ | $-21.3$ | | |

More recently, Carre and Devynck have used tetrachloroquinone and the analogous tetrafluoroquinone as potentiometric indicators to measure acidities of solutions of the Lewis acid $SbF_5$ in $HSO_3F$ and in $CF_3SO_3H$ and of NaF and $H_2O$ as supposedly strong bases in $HSO_3F$ and $CF_3SO_3H$, respectively. They report $R(H)$ values of $-23.4$ and $-22.1$ for acidified $HSO_3F$ and $CF_3SO_3H$ and $-17.2$ and $-16$ for the two basic systems.[20]

## 2.4.3 Nuclear Magnetic Resonance Spectroscopy

If proton exchange between an indicator B and its conjugate acid is "slow," that is, the mean lifetime $\tau$ is $\geq 10^2$ s, the species $BH^+$ can be observed in the NMR spectrum as if no exchange were taking place and an ionization ratio $BH^+/B$ cannot be measured.

If, however, the exchange is fast, that is, $\tau$ for $BH^+$ is less than $10^{-4}$ s, the observed NMR band is the weighted average of those of the species in equilibrium. If the chemical shift $\delta_B$ of the indicator B and of its conjugate acid $BH^+, \delta_{BH^+}$, are known, the ionization ratio $BH^+/B$ can be calculated from values of $\delta_B$, $\delta_{BH^+}$, and $\delta_{obs}$, the average band observed in the superacidic environment under investigation. The value of $pK_{BH^+}$ can be determined using the curve obtained by plotting chemical shift variation against acidity, $H_0$. Then $H_0$ values can be determined from ionization ratios in different media.

The method is rapid and is not confined to measurements of shifts for $^1H$ spectra. Observed shifts of several nuclei, for example, $^{13}C$, $^{19}F$, and $^{17}O$, can be used for acidity measurements. Unlike spectrophotometry, the NMR method can be used over a very wide acidity range, that is, beyond the range where current spectrophotometric methods can be used because sufficiently weak bases are not available to act as indicators or where change in the nature of the medium would have marked effects on the absorption spectra and, in particular, on the position of $\lambda_{max}$ for the protonated indicator. Such changes in the nature of the medium would be expected as the concentration of $SbF_5$ in HF is increased, where, as will be shown in the next chapter, the monomeric anion $SbF_6^-$, which exists in reasonably dilute solutions, is progressively replaced by polymeric species: $Sb_2F_{11}^-$, $Sb_3F_{16}^-$, . . . $Sb_nF_{5n+1}^-$ on increase in $SbF_5$ concentration.

In systems where proton exchange rates are measurable, that is, intermediate in magnitude between the two cases discussed above, changes in NMR spectral line shapes can be related to the proton exchange or may be due to a separate exchange process related to the proton exchange, allowing a procedure known as dynamic nuclear magnetic resonance to be adopted. Sommer and co-workers at Strasburg,[24] as well as Gold in London,[25] have used this procedure for measuring acidities of $HSO_3F$ and of HF containing very high concentrations of $SbF_5$, their acidities being well beyond those for which Gillespie could find sufficiently weak bases to act as Hammett indicators.

Sommer[24] used protonated p-methoxybenzaldehyde as an indicator and two complementary NMR techniques to measure ionization ratios and obtain $H_0$ values a little lower than Gillespie's for $SbF_5$ solutions to concentrations of about 10 mol% in $HSO_3F$ and to 0.6 mol% in HF and then extended them in what was virtually an extrapolation of Gillespie's values to 25 mol% in $HSO_3F$ and to 4 mol% in HF, that is, to regions where reliable spectrophotometric measurements could not be made. Gold[25] also used dynamic NMR techniques "to indicate that the acidity of $HSO_3F$—$SbF_5$ increases monotonically at least up to 90 mol% $SbF_5$." He derived a value of about $-26.5$ for $H_0$ for 90 mol% $SbF_5$ in $HSO_3F$.

## 2.4.4 Chemical Kinetics

For a large number of acid-catalyzed organic chemical reactions, kinetic studies show that plots of measured rate constants against $H_0$ values for the acidic media under investigation have unit slopes. From the acid-dependent rates of conversion of the 5-ethyl-2-methyltetrahydrofuranyl cation to the 2,5,5-trimethyltetrahydrofuranyl cation, a reaction in superacidic media proceeding through disproportionated intermediates, Brouwer and Van Doorn[26] arrived at the following order of acidities for the systems given:

(1$SbF_5$/1HF):(1$SbF_5$/9HF):(1$SbF_5$/$HSO_3F$):(1$SbF_5$/5$HSO_3F$) as
$> 500:1:10^{-1}:10^{-5}$

This and other recent studies show that $SbF_5$/HF systems are much more acidic than $SbF_5$/$HSO_3F$ systems for similar Lewis acid concentrations.

## 2.4.5 Comparison of Methods of Acidity Measurement

Only for the HF system have there been studies using spectrophotometric and potentiometric procedures of a wide range of acidities and basicities. The strong acid $SbF_5$ and correspondingly strong bases have been investigated potentiometrically in $HSO_3F$ and $CF_3SO_3H$ and the reported $R(H)$ values can be compared with $H_0$ values determined spectrophotometrically for similar strongly acidic and basic solutes in these two superacids. However, there have been no potentiometric studies of intermediate acidities and basicities in $HSO_3F$ and $CF_3SO_3H$. In the main, NMR spectroscopic and chemical kinetic studies have been applied to exceptionally acidic $HSO_3F$ and HF, that is, those solutions containing up to 90 mol% of $SbF_5$. They have very great value, particularly in the investigation of kinetics and mechanisms of organic chemical reactions, but they have little relevance to the inorganic systems to be presented in this book.

Accordingly, Figure 2.5, based on a range of Lewis acidity–Lewis basicity measurements in HF, is offered as the main vehicle for comparison of the experimental methods used for acidity measurements in protonic superacids. The $H_0$ values for HF of Gillespie and Liang,[16] selected for 0.5 mol%

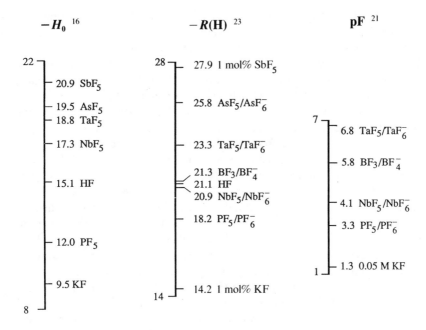

(i)   All values of $H_0$ for [acid] = [base] = 0.5 mol%. The value for NbF$_5$ was
      obtained by extrapolation.
(ii)  pF and $-R(H)$ scales constructed relative to $-H_0$ scale by placing
      basicity limits for each set of measurements at $-H_0$ values for
      appropriate concentrations of [F$^-$] used.
(iii) $-R(H)$ value of 27.9 set at $-H_0$ value for 1 mol% SbF$_5$.
(iv)  pF value of 6.8 set arbitrarily between $-H_0$ and $-R(H)$ values for TaF$_5$.
(v)   Superscript numbers in column heads are reference numbers.

**Figure 2.5.** Comparison of $H_0$, $R(H)$, and pF Values in HF.

of acid or base in each case, are taken as the reference set, because they will
be used in dicussion of generation and stability of inorganic species in this
book. The potentiometric measurements based on the use of the chloranil
electrode by Devynck and his co-workers[23] to give an $R(H)$ scale in HF are
aligned with the $H_0$ values by setting $R(H)$ for 1 M KF at the level of the
extrapolated $H_0$ value for 1 M KF, and by using a corresponding alignment
of $R(H)$ for 1 M SbF$_5$ with the extrapolated $H_0$ value for 1 M SbF$_5$. Gut and
Gautschi's H$_2$ electrode pF values[21] are included, with the value of pF for
0.05 M F$^-$ set at the $H_0$ value for the same base concentration. Because the
H$_2$ electrode gave meaningless pF values when HF was acidified with SbF$_5$
or AsF$_5$, the HF—TaF$_5$ system is the most acidic HF system for which a
reasonably reliable value of pF based on the H$_2$ electrode is available. It is
not possible to measure a pF value for SbF$_5$—HF to be aligned with a value
of $H_0$ for SbF$_5$—HF as a reference point. Therefore, quite arbitrarily, pF for
TaF$_5$—HF has been set between the $H_0$ and $R(H)$ values for this system.

It is immediately apparent that the order of Lewis acidity and basicity is the same for all three sets of measurements. Concerning the two sets that are directly comparable over the whole studied range of acidity and basicity, it is seen that, numerically, $-R(H)$ values are greater than $-H_0$ values by about 6 units. This is observed also in the more restricted range of measurements on $HSO_3F$ and $CF_3SO_3H$. This is recognized by both research groups that have played the major role in this area,[16,20] but at this stage there seems to be no basis for rationalizing the almost constant difference. It is apparent that the pF values arising from the hydrogen electrode work must not be taken as absolute values. Assuming a value of about 13 for $pK_i$ for HF, the $BF_3/BF_4^-$ system would be very weakly basic and $TaF_5/TaF_6^-$ would be very weakly acidic if the pF values as published were to be accepted.

There are some interesting anomalies to be observed when comparing systems on the three scales. On the basis of $H_0$ values, $NbF_5$ appears as a much stronger Lewis acid in HF than $R(H)$ and pF values would suggest. In fact, on the $-R(H)$ scale the $NbF_5/NbF_6^-$ system falls below the midpoint of the scale, which might be taken as $-R(H)$ for pure HF. Gillespie and Liang added the Lewis acids, as such, to HF. Except for $SbF_5$, Devynck and colleagues measured $R(H)$ values for equimolar buffers $AF_n/AF_{n+1}^-$. As will be shown in section 3.4.5, recent Melbourne work has shown that, even at low concentrations and in the absence of $MF_6^-$, metal pentafluorides other than $SbF_5$ form dimeric anions $M_2F_{11}^-$ to a large extent in HF. This dimerization would be favored in the presence of $MF_6^-$. It is possible that in the $R(H)$ studies the initial concentration of $NbF_6^-$ was slightly in excess of that for $NbF_5$ and that there was virtually no free Lewis acid $NbF_5$ in solution. In the hydrogen electrode work $AF_{n+1}^-$ concentrations were vastly in excess of the Lewis acids $AF_n$. In the paper on that work[21] Gut and Gautschi report that titration curves of $F^-$ vs. $AF_n/AF_{n+1}^-$ were shifted to lower pF values than those for $F^-$ vs. $AF_n$. This effect of formation of dimeric fluoroanions on potentiometric measurement of Lewis acid strength in HF solutions is discussed in detail in section 3.4.5.

$PF_5$ presents an interesting case. In many studies such as those on thermal stability of adducts, $PF_5$ is classed as a Lewis acid weaker than the other pentafluorides and $BF_3$. On both the $-H_0$ and $-R(H)$ scales, the $PF_5$ system is located about three units lower than the value, obtained by interpolation, for 100% HF. As stated toward the end of section 2.4.1, well-distilled HF, stored and used for solution preparation in KelF containers, will be markedly basic as a result of protonation of impurities remaining after distillation or leached from KelF. This small amount of adventitious $F^-$ present in the original solvent for the solutions used for acidity measurements with Lewis acids from the very strong $SbF_5$ to the reasonably weak $NbF_5$ will be neutralized by the Lewis acid and will have a negligible effect on the acidity measurement. Additionally, in Hammett Function determinations, protonation of indicators would lead to a marked increase in basicity of the supposedly pure solvent, as could any residues that remained from the acetone that was used in introducing the indicators into the KelF cells and was subsequently distilled away. It appears that $PF_5$ is vanishingly weak as a Lewis acid in HF. At no stage in the addition of $PF_5$ to HF is $PF_5$ able to neutralize all the base present in the working solvent, as a result of the presence of adventitious impurities or, in the case of measurement of $H_0$ values, as a result of protonation of the indicator.

It is probably of significance in this connection that Gillespie and Liang[16] found that $HSO_3F$ was somewhat less efficient than $PF_5$ in reducing the basicity as it was added to a working solution of HF containing indicators. For $HSO_3F$, $H_0$ values changed from $-10.87$ to $-11.19$ as the concentration was changed from 0.1 to 0.5 mol%, by comparison with $PF_5$, where the changes in $H_0$ were from $-11.58$ to $-12.02$ over the same concentration range. $HSO_3F$ is generally regarded as a nonelectrolyte in HF. Gut and Gautschi[21] described it as such in reporting that its addition to a fluoride-containing HF solution did not change the observed pF value. The observed $H_0$ values of $-11$ to $-12$ for solutions of $HSO_3F$ in a "neat" working solution of HF— for which values in this region are to be expected—is consistent with the fact that observed "point of inflection" values of $H_0$ are the same for both HF and $HSO_3F$, namely, $-15.1$. $HSO_3F$ is not likely to protonate HF or vice versa.

Some limited comparisons between values of $H_0$ and $R(H)$ can be made for the $HSO_3F$ and $CF_3SO_3H$ systems, but the studies in these superacids are much less comprehensive than in HF. Strongly acidic and strongly basic solutions have been investigated in these two solvents but there is very little reported study of acids and bases of intermediate strength. The data presented in Table 2.4 for $H_0$ values in $HSO_3F$ and $CF_3SO_3H$ are very fragmentary by comparison with those for $H_0$, $R(H)$ and pF in HF as presented in Figure 2.5. However, certain common features emerge. For example, there is a difference of about 6 units between the values of $H_0$ and $R(H)$ for 1 mol% $SbF_5$ in both $HSO_3F$ and $CF_3SO_3H$, as was seen for HF in Figure 2.5. Also, as indicated in Figure 2.4, the difference in $H_0$ values between solutions of

**Table 2.4.** Comparison of $H_0$ and $R(H)$ Values for $HSO_3F$ and $CF_3SO_3H$

| Superacid | Lewis acid | Base | Concentration (mol%) | $H_0$ | $R(H)$ |
|---|---|---|---|---|---|
| $HSO_3F$ | $Ta(SO_3F)_5$ | | 1.0 | $-18.1$[15] | |
| $HSO_3F$ | $SbF_2(SO_3F)_3$ | | 1.0 | $-18.2$[14] | |
| $HSO_3F$ | $SbF_5$ | | 1.0 | $-17.3$[14] | $-23.4$[20] |
| $HSO_3F$ | $TaF_5$ | | 1.0 | $-16.1$[33] | |
| $HSO_3F$ | $NbF_5$ | | 1.0 | $-15.5$[33] | |
| $HSO_3F$ | | $KSO_3F$ | 0.4 | $-13.2$[14] | |
| $HSO_3F$ | | $NaF$ | 1.0 | | $-17.2$[20] |
| $CF_3SO_3H$ | $B(SO_3CF_3)_3$ | | 0.4 | $-16.5$[13] | |
| $CF_3SO_3H$ | $SbF_5$ | | 1.0 | $-16.4$[33] | $-22$[20] |
| $CF_3SO_3H$ | | $H_2O$ | 1.0 | | $-16$[20] |
| $CF_3SO_3H$ | | $NaSO_3CF_3$ | 1.0 | $-12.5$[13] | |

the true base of the solvent system and the strongest acids measured in the system is smaller for $CF_3SO_3H$ than for $HSO_3F$; that is, self-ionization is somewhat greater in $CF_3SO_3H$ than in $HSO_3F$, but much greater for both than for HF.

One important aspect that is demonstrated in comparing acidity-basicity measurements from different research groups is the necessity to use acids and bases derived directly from the solvent when making measurements of extremes of acidity and basicity. In this respect measurements in HF are the most reliable to date. The strongest-measured Lewis acid is $SbF_5$, which interacts directly with $F^-$, the base of the self-ionization equilibrium, and the strongest base is $F^-$ itself. Gillespie and Liang showed that a 1 mol% solution of $H_2O$ in HF was less basic than KF at the same concentration by about 0.8 of an $H_0$ unit.[16] So Carre and Devynck's[20] basicity limits for $CF_3SO_3H$ would have been more reliable had they used $KOSO_2CF_3$ as their source of base instead of $H_2O$. Similarly, $KSO_3F$ would have been a better choice of base in $HSO_3F$ than NaF.[20] Apparently it was assumed that $F^-$ would be fully protonated in $HSO_3F$. If this was the basis for solute selection, that choice is flawed by the fact that $F^-$ is in competition with $SO_3F^-$ for protonation. $B(OSO_2CF_3)_3$ may not prove to be the strongest Lewis acid in $CF_3SO_3H$, but it is measurably stronger than $SbF_5$.[13] Similarly, Gillespie[14] demonstrated that $SbF_5 \cdot 3SO_3$, that is, $SbF_2(SO_3F)_3$, is a stronger Lewis acid in $HSO_3F$ than $SbF_5$ and, more recently, Aubke and Cicha[15] have shown that $Ta(SO_3F)_5$ is stronger than the binary fluoride $SbF_5$. It is unfortunate that polymerization in solid compounds like $Sb(OSO_2CF_3)_5$ and $Sb(OSO_2F)_5$ causes them to be so insoluble in their parent acids that they cannot be used for $H_0$ measurements because in all superacid systems (HX) in which comparative observations have been made, $SbX_5$ compounds have proved to be stronger Lewis acids than $TaX_5$.

In general NMR spectroscopic and chemical kinetic methods have been of value in estimating high acidities in superacid systems containing 50 or more mol% of $SbF_5$. What emerges clearly is that at these high concentrations $HF-SbF_5$ systems are very much more acidic than their $HSO_3F-SbF_5$ counterparts. Sommer did demonstrate a fairly close parallel between his NMR measurements of $H_0$ values for $SbF_5-HSO_3F$ solutions up to about 10 mol% with spectrophotometric values.[24] These methods have not yet been applied to as wide a range of Lewis acids as spectrophotometric and potentiometric methods as set out in Figure 2.5.

It will be demonstrated throughout this book that consideration of the effect of a range of Lewis acids at controlled concentrations in different protonic superacids can be used to rationalize the existence and stability of many known inorganic chemical species and can be used as the basis of synthetic strategies to generate these known species under optimal conditions or to set acidity-basicity levels that will favor the formation of new species. Spectrophotometric determination of $H_0$ values will form the basis of this discussion because the same general methodology has been applied to all four superacids of interest here, for example, a related series of acid-base indicators has been used throughout. Historically the sulfuric acid system is important in generating unusual cationic species, but there has been no application of potentiometric or NMR studies to the measurement of acidity levels in this solvent.

## 2.5  Acidity-Basicity in Chloroaluminate Systems

A major objective in writing this book is to show the dependence of the generation and stability of cationic and, to a less extent, anionic species on levels of acidity or basicity of the media in which they exist. For the four protonic superacids discussed, this can be done quantitatively or, at worst, semiquantitatively using Hammett Acidity Function values to define levels of acidity or basicity. These measurements rely ultimately on proton activities in the media under investigation, as do all the other acidity-basicity measurements outlined earlier, and therefore cannot be used to measure acidity levels in chloroaluminates of differing composition to relate chloroaluminate acidities to those in protonic superacids.

It will be shown in chapters 4–7 that the particular cationic or anionic species whose nature can be related to $H_0$ values for superacids can, in many instances, be shown to have been generated in chloroaluminates under acidic or basic conditions. As stated earlier, the acidity measurements described in this chapter cannot be applied to these nonprotonic media. However, the phenomenological evidence for preparation and stabilization of particular cations or anions in protonic solvents, for which $H_0$ values are available, and in chloroaluminates points in a qualitative way to a marked dependence of speciation on acidity or basicity levels of chloroaluminate solvent systems.

In chapter 1 it was stated that in chloroaluminates generally, the fundamental Lewis acid–Lewis base interaction is for the acid of the system $AlCl_3$ to accept the base $Cl^-$ to form, in the simplest situation, $AlCl_4^-$:

$$AlCl_3 + Cl^- \rightleftharpoons AlCl_4^- \qquad\qquad 2.7$$

For "high-temperature" melt mixtures, that is, those containing varying ratios of $AlCl_3$ and an alkali metal chloride, usually NaCl or KCl, there have been many potentiometric and Raman spectroscopic studies[27,28] which have led to identification of other anionic species than those given in equation 2.7, and these are represented in general form in Figure 2.6.

If one considers change in melt composition, starting from a 1:1 $AlCl_3$:$Cl^-$ mixture, that is, neutral $AlCl_4^-$, the addition of $Cl^-$ results in melts containing only differing proportions of $AlCl_4^-$ and $Cl^-$, the most basic system being pure NaCl or KCl in the case of high-temperature melts or, for room-temperature melts, a salt of $Cl^-$ and an organic cation such as N-butylpyridinium or 1-methyl-3-ethylimidazolium.

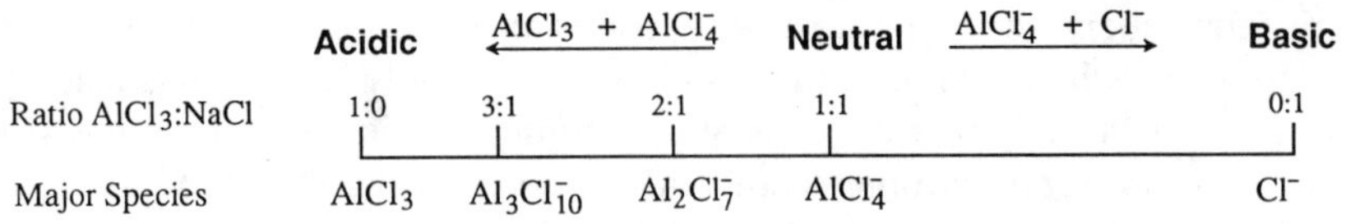

**Figure 2.6.** Acid-Base Dependence of Speciation in Chloroaluminates.

As the percentage of $AlCl_3$ increases beyond neutrality in high-temperature melts, $AlCl_4^-$ is gradually replaced by $Al_2Cl_7^-$ and $Al_2Cl_6$. These three species are reported[27] to be in comparable concentrations in the melt $AlCl_3$:NaCl, 2:1 at about 300°C. However in the room-temperature melts $AlCl_3$-1-methyl-3-ethylimidazolium chloride which are liquid for higher $AlCl_3$ concentrations than for the N-butylpyridinium chloride system, infrared spectroscopy[29] indicates the presence of $AlCl_4^-$, $Al_2Cl_7^-$, and $Al_3Cl_{10}^-$ as the predominant species as the $AlCl_3$:$Cl^-$ ratio increases from 1:1 through 2:1 and 3:1. These equilibria shifts imply a reduction in chloride concentration, that is, in basicity, as these polymeric anions are formed. Hussey has written an extensive review[30] of melt structure and haloaluminate equilibria in room-temperature molten salt systems. In Figure 3 of that review he presents a distribution diagram showing the fractions of $Al_2Cl_7^-$, $AlCl_4^-$, and $Cl^-$ as a function of composition in the $AlCl_3$—BuPyCl melt system. $Al_2Cl_7^-$ and $AlCl_4^-$ are shown as being at comparable concentrations at about 60% $AlCl_3$. In the immediate range of neutrality pCl is shown as changing by 8–10 units.

While potentiometric and spectroscopic measurement of shifts in equilibria for the species $Al_2Cl_6$, $Al_3Cl_{10}^-$, $Al_2Cl_7^-$, $AlCl_4^-$, and $Cl^-$ indicate a reduction in chloride concentration, that is, in basicity, as the polymeric anions are formed, there has been no *direct* experimental measurement of acidity or basicity of melts that are markedly different from the 1:1 system. Trémillon and Letisse[31] used on Al indicating electrode relative to an $Al/AlCl_3/Cl^-$ reference electrode to measure changes in pCl near the neutrality point and Kerridge[32] used their data to plot pCl vs. $AlCl_3$:NaCl ratio. As shown in Figure 2 of the Kerridge review, pCl changes by about 2.2 units between 49% $AlCl_3$ and 51 mol% $AlCl_3$ in these high-temperature melts. In chapters 6 and 7 much interesting inorganic chemistry will be related to this marked charge in acidity and basicity in passing through the neutral point for high-temperature chloroaluminates and to the greater pCl change at neutrality for room-temperature melts.

# References

1. A.I. Popov in *The Chemistry of Nonaqueous Solvents*, Vol. III. J.J. Lawgowski, ed. Academic Press, New York and London, 1970, pp. 241–337.

2. B. Trémillon, *Chemistry in Non-aqueous Solvents*. Reidel, Dordrecht and Boston, 1974, pp. 107–119.

3. R.J. Gillespie, E.D. Hughes, C.K. Ingold. *J. Chem. Soc.* 2473 (1950).

4. R.J. Gillespie in *Inorganic Sulfur Chemistry*, G. Nickless, ed., Elsevier, 1968 p. 566–579.

5. R.J. Gillespie, J.B. Milne, R.C. Thompson, *Inorg. Chem.*, *5*, 468 (1966).

6. R.J. Gillespie, D.A. Humphreys, *J. Chem. Soc (A)*, 2311 (1970).

7. R.J. Gillespie, *Accts. Chem. Res.*, *1*, 202 (1968).

8. M.F.A. Dove, A.F. Clifford in *Chemistry in Nonaqueous Ionizing Solvents*, Vol. II, Part 1. G. Jander, H. Spandau, C.C. Addison, eds. Pergamon Press, 1971, pp. 145–148.

9. G.A. Olah, G.K. Surya Prakash, J. Sommer, *Superacids*, Wiley-Interscience, 1985.

10. L.P. Hammett, A.J. Deyrup, *J. Am. Chem. Soc.*, *54*, 2721 (1932).

11. R.J. Gillespie, T.E. Peel, E.A. Robinson, *J. Am. Chem. Soc.*, *93*, 5083 (1971).

12. A. Engelbrecht, E. Tshager, *Z. Anorg. Allg. Chem.*, *433*, 19 (1977).

13. R. Adrien, Ph.D. Thesis, University of Melbourne (1992).

14. R.J. Gillespie, T.E. Peel, *J. Am. Chem. Soc.*, *95*, 5173 (1973).

15. W. Cicha, Ph.D. Thesis, University of British Columbia (1989).

16. R.J. Gillespie, J. Liang, *J. Am. Chem. Soc.*, *110*, 6053 (1988).

17. R. Gut, *J. Fluor. Chem.*, *15*, 163 (1980).

18. T.A. O'Donnell, *J. Fluor. Chem.*, *25*, 75 (1984).

19. J. Devynck, A.B. Hadid, P.-L. Fabre, *J. Inorg. Nucl. Chem.*, *41*, 1159 (1979).

20. B. Carre, J. Devynck, *Anal. Chem. Acta*, *159*, 149 (1984).

21. R. Gut, K. Gautschi, *J. Inorg. Nucl. Chem.*, Supplement, 95 (1976).

22. J.P. Masson, J. Devynck, B. Trémillon, *J. Electroanal. Chem.*, *64*, 175 (1975).

23. J. Devynck, A.B. Hadid, P.L. Fabre, B. Trémillon, *Anal. Chem. Acta*, *100*, 343 (1978).

24. J. Sommer, P. Canivet, S. Schwartz, P. Rummelin, *Nouv. J. de Chimie*, *5*, 45 (1981).

25. V. Gold, K. Laali, K.P. Morris, L.Z. Zdunek, *Chem. Commun.*, 769 (1981).

26. D.M. Brouwer, J.A. Van Doorn, *Rec. Trav. Chim.*, *91*, 895 (1972).

27. L.G. Boxall, H.L. Jones, R.A. Osteryoung, *J. Electrochem. Soc.*, *120*, 223 (1973).

28. E. Rytter, H.A. Øye, S.J. Cyvin, B.N. Cyvin, P. Klaeboe, *J. Inorg. Nucl. Chem.*, *35*, 1185 (1973).

29. C.J. Dynek, Jr., J.S. Wilkes, M.-A. Einarsrud, H.A. Øye, *Polyhedron*, *7*, 1139 (1988).

30. C.L. Hussey in *Advances in Molten Salt Chemistry*, Vol. 5. G. Mamantov and C.B. Mamantov, eds. Elsevier, Amsterdam, 1983, pp. 185–230.

31. B. Trémillon, G. Letisse, *J. Electroanal. Chem.*, *17*, 371 (1968).

32. D.H. Kerridge in *The Chemistry of Nonaqueous Solvents*, Vol. VB. J.J. Lagowski, ed. Academic Press, New York, 1978, p. 287.

33. J. Liang, Ph.D. Thesis, McMaster University (1976).

# CHAPTER 3

# Lewis Acidity in Protonic Superacids

In chapter 2, where methods of measurement and results of the determination of relative acidities of the pure protonic superacids $H_2SO_4$, $CF_3SO_3H$, $HSO_3F$, and HF were being presented, some very strong Lewis acids and bases of the superacid systems were introduced in order to illustrate the methodology for determining the value of the acidity of a particular superacid at 100% purity by interpolation between values obtained experimentally in both very strongly acidic and very strongly basic solutions of the superacid. Some Lewis acids of intermediate strength were discussed as reference points for comparison of the various methods of acidity measurement. However, the presentation of Lewis acids of the superacid systems was far from comprehensive and, in many cases, the acidity of the Lewis acid was discussed for a single value of concentration. Little effort was made to describe solute-solvent interaction for these Lewis acids with the appropriate protonic superacids. In this chapter, that information will be given, where available, and experimental evidence to support the proposed solution speciation will be alluded to.

There will be no formal presentation of the Lewis bases of the systems. As has been seen in the first two chapters, these are simply the anions occurring in the self-ionization equilibria of the superacids. They are most conveniently added in soluble form as the alkali metal or ammonium salts of the anion in question and are usually assumed to be completely ionized in these media of high dielectric constant. Sometimes "weaker" sources of the base are encountered in acid-base studies in protonic superacids, for example, $SF_4$, $BrF_3$, and ONF ionize in HF to produce differing amounts of solvated $F^-$ ion, but it is the $F^-$ ion which is the base of the solvent system.

In the case of Lewis acidity and basicity there is at present no easy or general way to establish absolute values of acid and base strengths of the acids and bases themselves. In aqueous solution, strengths of Bronsted acids and bases, that is, their ability to transfer protons to the solvent water or accept them from it, are easily measured, and it is relatively easy to transpose Bronsted acidity-basicity scales from water to many other protonic solvents. Various attempts have been made to determine quantitatively absolute strengths for Lewis acids and bases by measuring their ability to bond or coordinate to bases or acids, preferably in a series involving a common conjugate species. The task of establishing acidity levels is somewhat easier within the framework of this book in that Lewis acid strengths will almost invariably be being considered relative to a particular superacid solvent, that is, the extent to which, by interacting with the base of the solvent system and displacing the self-ionization equilibrium, they enhance the acidity of the system. It is emerging that the order of Lewis acidity strengths in different superacids, such as $CF_3SO_3H$, $HSO_3F$, and HF, appears to be the same, or very similar.

## 3.1 Sulfuric Acid

Of the four protonic superacids discussed in this chapter, sulfuric acid has been the one most widely studied. It has been an important chemical, industrially and economically, for a long time. Reasonably rigorous study of its physical and chemical properties dates back to early in this century. Gillespie and his research colleagues contributed immensely to the rigorous studies throughout a period of over a decade starting around 1950 by investigating self-ionization equilibria and quantitative aspects of solute-solvent interaction. Much of their work and that of others in the field has been reviewed by Gillespie and Robinson.[1]

### 3.1.1 Sulfur Trioxide

Most mechanistic and synthetic work that has been done in $H_2SO_4$ of enhanced acidity has been based on oleum systems, that is, solutions of $SO_3$ in $H_2SO_4$. Chemically, these are very complicated systems, as demonstrated in a Raman spectroscopic study reported by Gillespie and Robinson in 1962.[2] For additions of $SO_3$ to $H_2SO_4$ up to 11.6%, they report spectroscopic evidence for the gradual formation of $H_2S_2O_7$, with molecular $SO_3$ also present. This could also be interpreted as limited interaction of the Lewis acid $SO_3$ with $HSO_4^-$, the base of the self-ionization equilibrium of $H_2SO_4$, to form $HS_2O_7^-$. Gillespie and Robinson report that $H_2S_2O_7$ is a weak Bronsted acid of the $H_2SO_4$ system. So $HS_2O_7^-$ would also be in equilibrium with $H_3SO_4^+$ and $HSO_4^-$. Presumably the concentrations of these ions are small by comparison with the molecular species, $H_2S_2O_7$ and $SO_3$. With increasing $SO_3$ concentration in $H_2SO_4$, this study pointed to the formation of the trimer $(SO_3)_3$ and of the more extensively oxygen-bridged acids $H_2S_3O_{10}$ and $H_2S_4O_{13}$, possibly through interactions such as for $SO_3$—$HS_2O_7^-$. Molecular $SO_3$ or its trimer persisted throughout. It could be argued that $SO_3$ concentration is the principal determinant of speciation in oleum systems, as suggested in Figure 3.1. On this basis it seems reasonable to treat $SO_3$ as a Lewis acid of the $H_2SO_4$ system, but not a particularly strong one. Conductance measurements,[1] shown in Figure 3.2, support this proposition. Emphasis on $SO_3$ as a Lewis acid, which interacts directly with the Lewis base $HSO_4^-$, reduces the need to postulate an intermolecular reaction between $SO_3$ and $H_2SO_4$ to form $H_2S_2O_7$ which is then described as a weak Bronsted acid.

$$H_2SO_4 \;+\; H_2SO_4 \;\rightleftharpoons\; H_3SO_4^+ \;+\; HSO_4^-$$

$$+$$

$$SO_3$$

$$\updownarrow$$

$$H_2S_2O_7 \;+\; H_2SO_4 \;\rightleftharpoons\; H_3SO_4^+ \;+\; HS_2O_7^- \;+\; SO_3$$

$$+$$

$$SO_3 \qquad \rightleftharpoons \qquad (SO_3)_3$$

$$\updownarrow$$

$$H_2S_3O_{10} \;+\; H_2SO_4 \;\rightleftharpoons\; H_3SO_4^+ \;+\; HS_3O_{10}^- \;+\; SO_3 \;+\; (SO_3)_3$$

**Figure 3.1.** SO$_3$ as a Lewis Acid in Oleum Systems.

Formation of $H_2S_2O_7$, within the Lewis acid–Lewis base framework, becomes rather incidental, resulting from protonation of the strong Lewis base $HS_2O_7^-$. Similarly, $HS_3O_{10}^-$ can be regarded as being an equilibrium product of direct interaction between $SO_3$ and $HS_2O_7^-$, with some $H_2S_3O_{10}$ resulting from partial protonation of $HS_3O_{10}^-$.

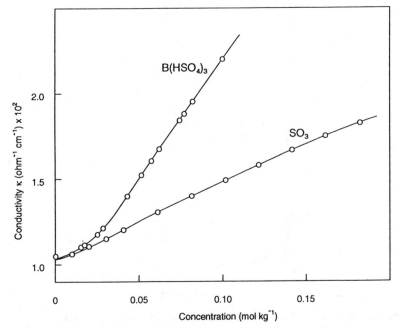

**Figure 3.2.** Conductivities of the Lewis Acids SO$_3$ and B(HSO$_4$)$_3$ in H$_2$SO$_4$.

## 3.1.2 Binary Hydrogen Sulfates

It was shown in chapters 1 and 2 that when $B(OH)_3$ is dissolved in $H_2SO_4$, the $B(HSO_4)_3$ formed initially reacts with $HSO_4^-$ in solution to form $B(HSO_4)_4^-$; that is, $B(HSO_4)_3$ acts as a strong Lewis acid, as:

$$B(OH)_3 + 6H_2SO_4 \rightarrow B(HSO_4)_3 + 3H_3O^+ + 3HSO_4^- \quad 3.1$$
$$\text{and} \quad B(HSO_4)_3 + HSO_4^- \rightarrow B(HSO_4)_4^- \quad 3.2$$

This gives an overall equation:

$$B(OH)_3 + 6H_2SO_4 \rightarrow B(HSO_4)_4^- + 3H_3O^+ + 2HSO_4^- \quad 3.3$$

It should be noted that the reaction mixture represented by equation 3.3 is basic—it contains 2 mol of $HSO_4^-$ per mol of $B(OH)_3$. The Lewis acid $B(HSO_4)_3$ can be generated in situ in $H_2SO_4$ without formation of $HSO_4^-$ according to the equation

$$B(OH)_3 + 3SO_3 \text{ (in } H_2SO_4) \rightarrow B(HSO_4)_3 \quad 3.4$$

The Lewis acid will then react with the solvent to form $H_3SO_4^+$ and $B(HSO_4)_4^-$, giving a much more highly conducting solution than for $SO_3$ in $H_2SO_4$, as indicated in Figure 3.2.

Measurements of Hammett Acidity Functions[3] confirm this difference in the acidity of the formal Lewis acids $SO_3$ and $B(HSO_4)_3$ at low concentrations but show differing behavior in more concentrated solutions as shown in Figure 3.3. The plot of $H_0$ vs. concentration for $B(HSO_4)_3$ begins to flatten

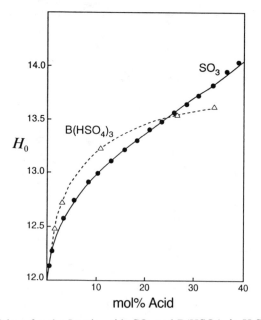

**Figure 3.3.** $H_0$ Values for the Lewis acids $SO_3$ and $B(HSO_4)_3$ in $H_2SO_4$.

markedly at 10–20 mol% and cannot be extended beyond about 30 mol% because "a complex polymeric boron sulfate precipitates from solution."[3] For $SO_3$—$H_2SO_4$ systems $H_0$ values continue to become significantly more negative up to and through the mixture 50 mol% $SO_3$.

Gillespie and Robinson[1] report a few hydrogensulfato Lewis acids in addition to $B(HSO_4)_3$. Dissolution of tin and lead tetraacetates in excess $H_2SO_4$ leads to the formation in situ of $Sn(HSO_4)_4$ and $Pb(HSO_4)_4$, each of which can accept two $HSO_4^-$ anions to form $[Sn(HSO_4)_6]^{2-}$ and $[Pb(HSO_4)_6]^{2-}$, leading to enhanced $H_3SO_4^+$ concentrations in the solvent $H_2SO_4$. $Sn(HSO_4)_4$ and $Pb(HSO_4)_4$ appear to be considerably weaker Lewis acids than $B(HSO_4)_3$, possibly comparable with the formal $H_2S_2O_7$. $As_2O_3$ dissolves in $H_2SO_4$ to form $As(HSO_4)_3$, which is a weak $HSO_4^-$ acceptor.

## 3.2 Trifluoromethylsulfuric ("Triflic") Acid

Very much less inorganic chemistry has been conducted in triflic acid, $CF_3SO_3H$, than in $H_2SO_4$, $HSO_3F$, and HF. Virtually nothing has been reported so far in the open literature on synthesis and reactions of inorganic compounds of the system in this acidic medium, and few Lewis acids have been studied quantitatively.

### 3.2.1 Triflates

The strongest Lewis acid studied to date is $B(OSO_2CF_3)_3$. Engelbrecht and Tschager[4] reported Hammett Function measurements for the $B(OSO_2CF_3)_3$—$CF_3SO_3H$ system for concentrations to about 20 mol% of the Lewis acid. Their values of $H_0$ were very similar to those determined subsequently by Adrien,[5] whose values for the Lewis acid and for $CF_3SO_3Na$ in the acid were presented in Figure 2.4 together with those for $H_2SO_4$, $HSO_3F$, and HF and demonstrate the overall acid-base domain of the $CF_3SO_3H$ solvent system. Work with the boron compound is difficult experimentally because of its marginal thermal stability when dissolved in the acid, particularly in moderately concentrated solutions of the Lewis acid at room temperature.

### 3.2.2 Fluorides

Carre and Devynck[6] determined R(H) acidity values for $SbF_5$ as a Lewis acid in $CF_3SO_3H$, but these are difficult to compare with the data of Engelbrecht and Tschager and of Adrien because of the numerical differences between values of acidities measured on the $H_0$ and R(H) scales. However, Liang[7] compared $H_0$ values using spectrophotometric and NMR chemical shift methods and demonstrated that $SbF_5$ was quite a strong Lewis acid in $CF_3SO_3H$. Based on comparison of $^{19}F$ NMR spectra of $SbF_5$ in $CF_3SO_3H$ and in $HSO_3F$ (see section 3.3.1), Liang postulated that two of the species

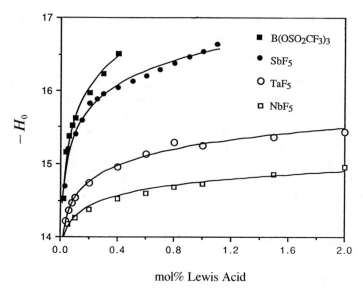

**Figure 3.4.** $H_0$ Values for Lewis Acids in $CF_3SO_3H$.

resulting from Lewis acid interaction of $SbF_5$ with the base of $CF_3SO_3H$ are the monomeric anion $[F_5Sb.OSO_2CF_3]^-$ and the triflate bridged species $[(F_5Sb)_2.SO_3CF_3]^-$ which would be expected to have a structure similar to that for the corresponding fluorosulfato entity shown in Figure 1.1.

Very recently Adrien[5] has recorded $H_0$ values for solutions of $SbF_5$, $TaF_5$, and $NbF_5$ in triflic acid and has compared them with his values for $B(OSO_2CF_3)_3$, shown earlier in Figure 2.4. Of these four Lewis acids, $B(OSO_2CF_3)_3$ is the strongest. The pentafluorides exhibit the order of acid strength shown in other superacids, namely, $SbF_5 > TaF_5 > NbF_5$, as illustrated in Figure 3.4.

## 3.3 Fluorosulfuric Acid

Initially, binary fluorides and, later, mixed fluoride-fluorosulfates derived from the binary fluorides were investigated for Lewis acidity in $HSO_3F$ by Gillespie and colleagues. These will be described first to provide a quantitative background to more recent work by Aubke and his co-workers, who have been preparing binary fluorosulfates and studying their behavior as Lewis acids in $HSO_3F$.

### 3.3.1 Fluorides and Derived Fluorosulfates

In 1969 Gillespie et al.[8] published a survey of the concentration-dependent conductances of many pentafluorides and of $TiF_4$ in $HSO_3F$ up to concentrations of about 0.1 M. This work is summarized in Figure 3.5. Within this concentration range, conductances of solutions indicate an order of Lewis acid strength for the binary fluorides as:

$$SbF_5 > BiF_5 > AsF_5 \sim TiF_4 > NbF_5 \sim PF_5$$

They found that addition of $SO_3$ considerably enhanced the conductances of the more strongly acidic systems, namely, $AsF_5$—$HSO_3F$ and $SbF_5$—$HSO_3F$. Thus, as shown in Figure 3.5, the conductance for the $AsF_5$ system is trebled for the solute $AsF_5:SO_3$, 1:1, and increased more than seven times for the ratio 1:3. Ratios for $SO_3:AsF_5$ larger than 3 had little effect on conductances. It was assumed that in the 1:1 mixture of $AsF_5$ and $SO_3$ the active Lewis acid was $AsF_4(SO_3F)$ and that, for the 3:1 mixtures, the acids were $AsF_2(SO_3F)_3$ and $SbF_2(SO_3F)_3$. As Figure 3.5 shows, $AsF_4(SO_3F)$ is somewhat weaker as a Lewis acid in $HSO_3F$ than $SbF_5$, but $AsF_2(SO_3F)_3$ and $SbF_2(SO_3F)_3$ are both much stronger than $SbF_5$.

Noting that replacement of fluoride by fluorosulfate appeared to cause an increase in acceptor strength, they apparently sought maximum acidity through the pentafluorosulfates but reported that these compounds "do not appear to be stable" and that "at the present time therefore it appears that $SbF_2(SO_3F)_3$ is the strongest known acceptor for fluorosulfate ion." In a summary of the conductivity work the authors noted that concentrated so-

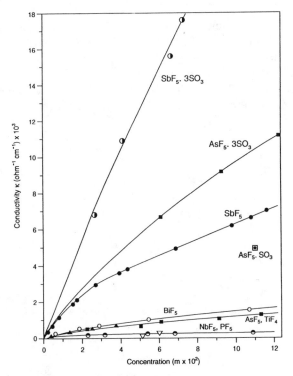

**Figure 3.5.** Conductivities of Lewis Acids in $HSO_3F$.

lutions of $SbF_5$ and $SO_3$ in $HSO_3F$ were of technological importance and pointed out that "in such concentrated solutions conductometric measurements cannot be interpreted with any certainty and other methods must be used to determine relative acidities (e.g., the measurement of the Hammett acidity function").

In 1973 Gillespie and Peel[9] published such a study and their results for Lewis acidities in $HSO_3F$ are shown in Figure 3.6. $SO_3$ is a very weak acid. It causes enhancement of the acidity of $HSO_3F$ by less than 0.5 unit, the Hammett function for 100% $HSO_3F$ being given as $-15.07$. $SO_3$ can be viewed as interacting to a small extent with $SO_3F^-$ to form $S_2O_6F^-$, the base of the formal acid $HS_2O_6F$ that was postulated by Gillespie and Robinson[10] as a weak Bronsted acid in $HSO_3F$. $AsF_5$ is seen as being reasonably weak while $SbF_5$ is quite strong, increasing the acidity of the system by about 2 to 3.5 $H_0$ units over the concentration range 1–10 mol%. [19]F NMR experiments on $SbF_5$—$HSO_3F$ systems have been interpreted as showing that $SbF_5$ acts at low concentration in $HSO_3F$ by accepting $SO_3F^-$ to form the octahedral complex $(F_5SbOSO_2F)^-$. With increasing $SbF_5$ concentration, a dimeric anion forms in which one $SO_3F^-$ is *cis*-bridged to two $SbF_5$ units, as shown in Figure 1.1. This dimerization would lead to a reduction in the possible effective Lewis acidity of $SbF_5$ with increase in $SbF_5$ concentration; that is, less $H_2SO_3F^+$ ions would be formed per $SbF_5$ than in the absence of dimerization.

The 1969 conductometric work by the McMaster group[8] had indicated qualitatively that the acidity of both the $AsF_5$—$HSO_3F$ and $SbF_5$—$HSO_3F$ systems was markedly increased by the addition of $SO_3$ up to the point where it was in threefold excess over the pentafluoride. The Hammett Function measurements[9] bore this out, the $SbF_5.3SO_3$—$HSO_3F$ system giving $H_0$ values more negative than $-19$. Conductometric titrations, supported by [19]F NMR studies, indicate that as 1, 2, and 3 mol of $SO_3$ per mol of $SbF_5$ are

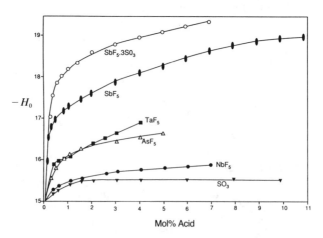

**Figure 3.6.** $H_0$ Values for Lewis Acids in $HSO_3F$.

added in turn, the species $SbF_4(SO_3F)$, $SbF_3(SO_3F)_2$, and $SbF_2(SO_3F)_3$ are produced, each capable of forming one additional $SO_3F^-$ ligand and being a progressively stronger Lewis acid in $HSO_3F$.[11]

Until recent work with $Ta(SO_3F)_5$, reported in section 3.3.2, $SbF_2(SO_3F)_3$ was considered the strongest acid of the $HSO_3F$ system. Liang, another member of the McMaster group, subsequently measured $H_0$ values for $TaF_5$ and $NbF_5$ in $HSO_3F$[7] and found $NbF_5$ to be a very weak acid, only slightly stronger than $SO_3$, and $TaF_5$ to be somewhat stronger than $AsF_5$, as shown in Figure 3.6.

## 3.3.2 Binary Fluorosulfates

Aubke and his group at the University of British Columbia have synthesized several noble-metal fluorosulfates by reacting the appropriate metals with bis(fluorosulfuryl)peroxide $S_2O_6F_2$ in $HSO_3F$. This is a beautifully "clean" oxidant in $HSO_3F$, being reduced to $SO_3F^-$, the base of the solvent. A minimum at the 1:1 stoichiometry for a conductometric titration of $Au(SO_3F)_3$[12] against $KSO_3F$ in $HSO_3F$ indicates the formation of the anion $Au(SO_3F)_4^-$ by interaction of the Lewis acid with the base of the system. The concentration-dependent specific conductivities of solutions of $Au(SO_3F)_3$ in $HSO_3F$ were found to be somewhat less than those of $SbF_2(SO_3F)_3$ of comparable strengths in $HSO_3F$ but considerably greater than those of $SbF_5$ (Figure 3.7). An indication of the level of strength of $Au(SO_3F)_3$ as an acid in $HSO_3F$ lies in the report by Lee and Aubke[13] that the cation $Br_3^+$ is stable in $Au(SO_3F)_3$—$HSO_3F$ whereas $Br_2^+$ is not. This is consistent with data from Table 4.2 of chapter 4 that show that $Br_3^+$ is stable in $SbF_5 \cdot 3SO_3$—$HSO_3F$, but that $Br_2^+$ is only marginally stable at this level of acidity.

Following the gold work, Lee and Aubke oxidized metallic platinum to $Pt(SO_3F)_4$, showed by conductometric titration against $KSO_3F$ that it could accept two $SO_3F^-$ ions, and measured a conductance in $HSO_3F$ (Figure 3.7) very similar to that of $Au(SO_3F)_3$.[14] Their preparation of the compound $(Br_3)_2[Pt(SO_3F)_6]$, even though it is described by them as being "thermally marginally stable," is further indication of the very low basicity of a system containing $Pt(SO_3F)_4$. Lee and Aubke also produced ternary and binary fluorosulfato compounds of Pd(II) and Pd(IV)[15] and of Ir(III) and Ir(IV).[16] Although the iridium fluorosulfates are soluble in $HSO_3F$ there was no study of their possible Lewis acidity by conductance measurements because of limited thermal stability of the compounds.

Through comparison with other superacidic solvent systems, it would be expected that $Sb(OSO_2F)_5$ might be the strongest Lewis acid of the $HSO_3F$ system. There have been no reports of preparation of solutions of $Sb(SO_3F)_5$ in $HSO_3F$. It is probable that, if the solute has been prepared, it is polymeric and therefore very insoluble in $HSO_3F$. Cicha and Aubke prepared $Nb(SO_3F)_5$ and $Ta(SO_3F)_5$ by oxidation with $S_2O_6F_2$ of the metals in $HSO_3F$.[17] Measurement of specific conductances (Figure 3.8) showed that, at very low

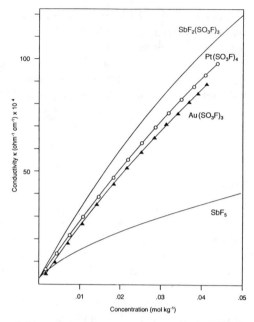

**Figure 3.7.** Conductivities of Au(SO₃F)₃ and Pt(SO₃F)₄ in HSO₃F.

concentrations, Ta(SO₃F)₅ appeared to be marginally more acidic than SbF₅ in HSO₃F but considerably less so than Au(SO₃F)₃ and therefore less than SbF₂(SO₃F)₃ as shown from Figure 3.6. These conductances, interpreted as reflecting the proton-jump enhanced conductance of the $H_2SO_3F^+$ cation, indicated that Nb(SO₃F)₅ was weaker than Ta(SO₃F)₅ but stronger than

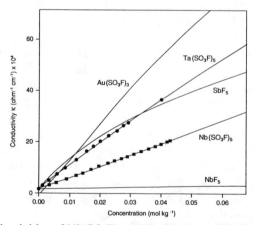

**Figure 3.8.** Conductivities of Nb(SO₃F)₅ and Ta(SO₃F)₅ in HSO₃F.

$NbF_5$. Subsequently Cicha and Aubke measured Hammett functions for $Ta(SO_3F)_5$ in $HSO_3F$[18] and compared the values with those for $SbF_5$ and $SbF_2(SO_3F)_3$ in $HSO_3F$. Their results showed that, for concentrations up to about 0.4 mol%, $Ta(SO_3F)_5$ is a weaker acid than $SbF_2(SO_3F)_3$ or $SbF_5$. At 0.5 mol%, $Ta(SO_3F)_5$ is stronger than $SbF_5$ but weaker by about 0.5 $H_0$ unit than $SbF_2(SO_3F)_3$. At concentrations a little beyond 1 mol%, $Ta(SO_3F)_5$ appears to be marginally more acidic than $SbF_2(SO_3F)_3$.

## 3.4 Hydrogen Fluoride

Despite the difficulties associated with experimentation in anhydrous HF that were alluded to briefly in chapter 1, there has been quite comprehensive study of Lewis acid strengths and of the speciation of the Lewis acids in solution in HF, more so than in the other superacids. This is particularly true for the two Lewis acids $SbF_5$ and $AsF_5$.

## 3.4.1 Antimony Pentafluoride

Historically, the first acid to be studied in detail has proved to be the strongest Lewis acid of the HF system, namely $SbF_5$. In 1961 Hyman et al.[19] reported conductances and Raman spectra over the whole composition range for the HF—$SbF_5$ system. Addition of $SbF_5$ to HF with an initial specific conductance of about $10^{-6}\ \Omega^{-1}\ cm^{-1}$ at 0°C caused a very rapid increase in conductivity, providing a maximum value of $0.35\ \Omega^{-1}\ cm^{-1}$ at about 8 mol% $SbF_5$. Progressive addition of $SbF_5$ led to a surprising reduction in conductivity to about one-half of the maximum value by the stage at which about 20 mol% of $SbF_5$ had been added. In dilute solutions of $SbF_5$, they observed Raman signals consistent with the formation of $SbF_6^-$, but these Raman intensities decreased as $SbF_5$ concentration was increased. On the basis of their conductance measurements, Hyman and co-workers postulated that $SbF_5$ was a relatively weak acid in HF, giving a value of $4.4 \times 10^{-4}$ at 20 mol% $SbF_5$ for the equilibrium constant for the reaction

$$2HF + SbF_5 \rightleftharpoons H_2F^+ + SbF_6^- \qquad\qquad 3.5$$

To explain the very great reduction in conductance with increase in $SbF_5$ concentration, they further postulated an ion-pairing equilibrium with a constant of $1.7 \times 10^2$, the ion-pairing becoming more significant as the concentration of $SbF_5$ increased:

$$H_2F^+ + SbF_6^- \rightleftharpoons H_2F^+.SbF_6^- \qquad\qquad 3.6$$

Any significant amount of ion-pairing seems an unlikely possibility in a solvent for which the dielectric constant of 83.6 at 0°C is greater than that for water.

In 1966 Gillespie and Moss[20] repeated conductance measurements for the whole composition range of the HF—$SbF_5$ system and obtained values essentially the same as those obtained by Hyman and co-workers. They put forward a model to explain the change in conductance of the system with increasing $SbF_5$ concentration by proposing that in reasonably dilute solutions (up to about 10 mol% $SbF_5$) the principal anionic species was $SbF_6^-$, which, because of hydrogen bonding, might be expected to be hexa-solvated as below:

$$8HF + SbF_5 \rightarrow H_2F^+ + [SbF_6(HF)_6]^-  \qquad\qquad 3.7$$

The overall system 11 mol% $SbF_5$—89 mol% HF would then have the composition, 11 mol $H_2F^+$:11 mol $[SbF_6(HF)_6]^-$:1 mol HF. Obviously no further ionization according to equation 3.7 can occur and further addition of $SbF_5$ will lead to the formation of solvated $Sb_2F_{11}^-$ and $Sb_3F_{16}^-$ and to anions of greater-complexity $Sb_nF_{5n+1}^-$. In their paper they argue the case for a reasonable constancy of solvation of the polymeric anions and point out that, up to about 11 mol% $SbF_5$, the number of ions in the systems is increasing. They suggest that the number would be approximately constant until the mole fraction of HF reaches 22%, at which point there would not be sufficient HF to form $H_2F^+$ and solvated anions. From that point the number of ions in the system would decrease and their complexity would increase.

Gillespie and Moss made one other important point by constructing a predicted curve for conductivity vs. $SbF_5$ concentration. Up to about 11 mol% $SbF_5$, there is sufficient free HF in the system to allow conductance to be dominated by the high mobility of $H_2F^+$ resulting from a "proton-jump" mechanism; however, with decrease in free HF beyond that point, conductance will be due to the normal electrolytic ionic migration mechanism, and even that will be reduced by the increasing viscosity of the system as it approaches the $SbF_5$-rich region. By combining these factors diagrammatically, Gillespie and Moss were able to construct a predicted conductivity curve "that has a very similar shape to that actually observed."[20]

Gillespie and Moss supported their model of progressive polymerization of $Sb_nF_{5n+1}^-$ anions by demonstrating the existence of $Sb_2F_{11}^-$, at least, by [19]F NMR spectroscopy. The [19]F spectrum of a dilute HF solution, 2 mol% in $SbF_5$, observed at room temperature and at lower temperatures gave only a broad peak assigned to $SbF_6^-$ in addition to the line for the solvent. At 20 mol% the room-temperature spectrum gave the broad peak, which at lower temperatures split into three peaks, A, B, and C, each with some fine structure, in addition to a solvent peak and a peak D attributed to $SbF_6^-$, as shown in Figure 3.9. The relative areas of the three peaks A, B, and C occurred in the ratio 1:2:8. In broad outline this spectrum is consistent with that to be expected for the dimeric fluorine-bridged anion also in Figure 3.9 which shows three different sets of fluorine atoms—namely, the single bridging atom ($F_a$), eight equivalent atoms ($F_b$) that are *cis* to $F_a$, and two equivalent atoms ($F_c$) *trans* to $F_a$. The fine structure of the spectrum can be interpreted as further demonstration of the formation in 20 mol% $SbF_5$—80 mol% HF of the dimeric anion $Sb_2F_{11}^-$. The signal of largest area due to $F_b$ is a doublet-doublet indicating coupling of $F_b$ to both $F_a$ and $F_c$. The $F_c$ signal is a quintet,

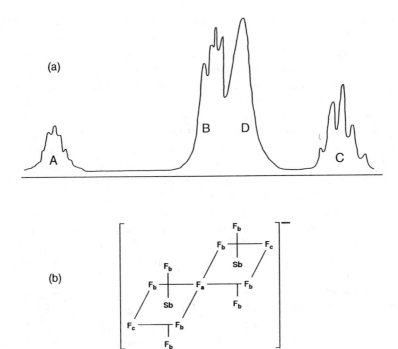

**Figure 3.9.** (a) $^{19}F$ NMR Spectrum of 20% $SbF_5$—HF.
(b) Idealized Structure of $Sb_2F_{11}^-$ Anion.

showing coupling to the four nearest $F_b$ atoms, with coupling to $F_a$ being near zero. Nine lines would be expected for the small $F_a$ peak, reflecting coupling to the eight equivalent $F_b$ atoms. Full resolution of this peak was not obtained although the five central lines can be seen.

Subsequently Bacon et al.[21,22] isolated a solid adduct, $CsF:SbF_5$, 1:2, from $SO_2$ and 1:3 and 1:4 adducts from stoichiometric amounts of $(CH_3)_3CF$ and $SbF_5$ in $SO_2FCl$, which is less basic than $SO_2$. Anions more complex than the dimer $Sb_2F_{11}^-$ could not be isolated from $SO_2$ because that solvent is sufficiently basic to form a competitive adduct $F_5SbOSO$. The NMR spectra of these adducts, in $SO_2$ and $SO_2ClF$ and in the solid state, confirm the proposal by Gillespie and Moss that $Sb_2F_{11}^-$, $Sb_3F_{16}^-$, and so on form in HF. The anions $Sb_2F_{11}^-$ and $Sb_3F_{16}^-$ have been identified by X-ray diffraction crystallography in several compounds, two of which, to be presented in chapter 4, are $I_2^+Sb_2F_{11}^-$[23] and $Br_2^+Sb_3F_{16}^-$[24]

Gillespie and Moss had proposed that $SbF_5$ is a very strong acid in HF yielding $SbF_6^-$ almost exclusively in relatively dilute solution, whereas Hyman and co-workers had proposed that it was rather weak. Two pieces of experimental evidence point to the great strength of $SbF_5$. The first of these was cryoscopy. Gillespie and Humphreys[25] developed a very reliable cryoscope constructed, where necessary, from the HF-resistant materials Teflon, KelF, and platinum. They and co-workers[26] determined a curve with a slope corresponding with $\nu = 2$ for freezing points of $SbF_5$—HF solutions up to about 0.4 molal. The solutes KF and $Et_4NSbF_6$, both ionic in HF, gave virtually the same curves. It should be noted that this evidence for total ionization of $SbF_5$ to $H_2F^+$ and $SbF_6^-$ is demonstrated at temperatures below $-83°C$, a point that will be discussed later in reporting cryoscopic measurements for the $AsF_5$—HF system. It will be shown there that low temperatures favor the formation of the fluoroanions at the expense of the pentafluoride; but it must be said that, at the present time, there is no evidence for the presence of any significant amount of free $SbF_5$ in dilute solutions of $SbF_5$ in HF. In fact, synthetic work presented in chapter 10 suggests that the $SbF_5$ is effectively totally ionized; that is, it is a very strong Lewis acid, except in quite concentrated solutions. It will be shown that $Ti^{2+}$ is stable in HF which is 2–3 molal in $SbF_5$, where the $SbF_5$ is present as fluoroantimonate anion, but is oxidized in more concentrated $SbF_5$ solutions containing free $SbF_5$.

The great strength of $SbF_5$ as a Lewis acid relative to other pentafluorides is clearly demonstrated in the recent Hammett Acidity Function measurements reported by Gillespie and Liang[27] and summarized in Figure 3.10. In that work it was shown that even in very dilute solutions (0.05 mol%) the value of $H_0$ was $-19.53$, greater than the values of $-18.94$ and $-19.35$ for 10 mol% $SbF_5$—$HSO_3F$ and 7 mol% $SbF_5$.$3SO_3$—$HSO_3F$, respectively. There is, as seen from Figure 3.10, a steady increase in acidity of $SbF_5$—HF to $-21.13$ at 0.6 mol% $SbF_5$, the limit of measurement dictated by the availability at that time of reliable indicators for the spectrophotometric method for $H_0$ measurements.

## 3.4.2 Arsenic Pentafluoride

The first measurement of conductances in the $AsF_5$—HF system[28] showed that at moderate concentrations (e.g., about 0.1 molal or 0.2 mol%) the specific conductance at 0°C was 30–40% of that of the $SbF_5$—HF system. The plots of molar conductance vs. (concentration)$^{1/2}$ showed slight curvature for $SbF_5$, indicating strong acid behavior, and the marked curvature of a relatively weak acid system for $AsF_5$. Both systems approached values of limiting conductances slightly in excess of 600 $\Omega^{-1}$ $cm^2$ $mol^{-1}$ at very high dilution.*

Cryoscopic observations on the $AsF_5$—HF system by the McMaster group[26] gave a reliable value of $\nu = 1$ for $AsF_5$ for concentrations from 0.08 to about 0.5 molal while $SbF_5$ gave $\nu = 2$. A simplistic interpretation of the

---

*The conductance values given here are not available from the reference cited, which is an abstract. They can be obtained from Ref. 29.

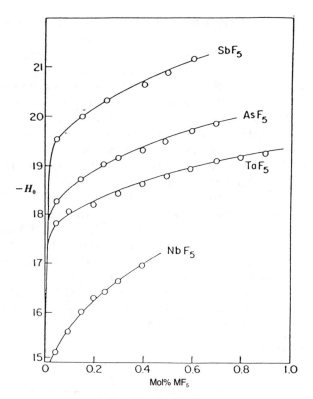

**Figure 3.10.** $H_0$ Values for Pentafluorides in HF.[27]

cryoscopy would be that $AsF_5$ is dissolved as the molecular, nonionized species. However, this runs counter to the knowledge that $AsF_5$ provides conducting solutions in HF. Cryoscopy was followed by conductivity measurements in HF at $-83.6°C$.[26] For similar concentrations of $SbF_5$ and $AsF_5$ up to about 1 molal, conductivity values for $AsF_5$ were almost exactly one-half of those for $SbF_5$, leading to the conclusion that whereas, in this concentration range, $SbF_5$ yielded $H_2F^+$ and $SbF_6^-$ ($v = 2$), $AsF_5$ gave $H_2F^+$ and $As_2F_{11}^-$, that is, 2 mol of conducting particles for 2 mol of $AsF_5$ ($v = 1$). In this same paper Gillespie and co-workers carried out a cryoscopic titration of $Et_4NAsF_6$ against $AsF_5$, below $-83°C$, of course, and found a maximum in the titration curve consistent with the formation of $Et_4NAs_2F_{11}$ at the 1:1 point. A conductimetric titration *at* $-63.5°C$ gave a minimum at the 1:1 addition, again indicating the formation of $As_2F_{11}^-$ under these conditions.

Isolation of solids containing the $As_2F_{11}^-$ anion proved difficult, but Gillespie and colleagues prepared $Et_4NAs_2F_{11}$ and $Bu_4NAs_2F_{11}$ and showed that in the solvent $SO_2ClF$ at $-140°C$ the $^{19}F$ NMR spectrum was very similar in relative peak areas and coupling patterns to that observed for $Sb_2F_{11}^-$ in HF.[21] Infrared spectra of the two solids showed bands in the $650–770$ cm$^{-1}$ region, characteristic of terminal fluorine-to-arsenic bonds, with a single band at 481 cm$^{-1}$ for Et$_4$NAs$_2$F$_{11}$. Bands in the 480–520 cm$^{-1}$ region, usually weaker than those for stretches for terminal fluorines, have become diagnostic for dimeric and polymeric species where one fluorine atom bridges two atoms such as Sb, As, Nb, Ta, and so on in complex fluoroanions.

In a much later study at Melbourne, O'Donnell, and co-workers used conductance measurements and Raman spectroscopy to investigate the AsF$_5$—HF system at ambient or near-ambient temperatures.[29] Their Raman spectrum, obtained at room temperature by computer addition of 40 spectral scans, for a solution 0.1 molal in AsF$_5$, is shown in Figure 3.11. The peak at 732 cm$^{-1}$ corresponds with that at 733 cm$^{-1}$ reported as the strongest Raman peak for liquid AsF$_5$. The broader peak around 685 cm$^{-1}$ can be related to a peak at 685 cm$^{-1}$ reported for CsAsF$_6$. So the largest peak at 703 cm$^{-1}$ might then be expected to be associated with As$_2$F$_{11}^-$. This was demonstrated by recording in turn Raman spectra of NaAsF$_6$ in HF and of HF solutions 0.3 molal in NaAsF$_6$ to which 1, 3, and 7 mol proportions (relative to NaAsF$_6$) of AsF$_5$ were added. The peak at 686 cm$^{-1}$ in the first spectrum (AsF$_6^-$) was joined and ultimately overshadowed by a strong sharp peak at 703 cm$^{-1}$, indicating the formation of As$_2$F$_{11}^-$ at the expense of AsF$_6^-$. From the first addition of AsF$_5$, a peak grew in progressively at 736 cm$^{-1}$, indicating the presence of free AsF$_5$ in an equilibrium

$$AsF_6^- \ + \ AsF_5 \rightleftharpoons As_2F_{11}^- \qquad\qquad 3.8$$

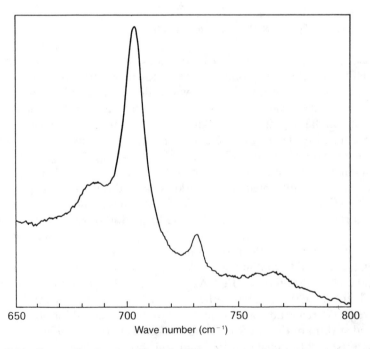

**Figure 3.11.** Raman Spectrum of AsF$_5$ in HF.

The next step in this HF—$AsF_5$ study was to record Raman spectra of quite concentrated solutions (2, 4, 6, and 8 molal) of $AsF_5$ in HF containing no $NaAsF_6$. There was little or no evidence for $AsF_6^-$ and the height of the $AsF_5$ peak was approximately 27, 35, 50, and 90% that of the $As_2F_{11}^-$ peak in each case as the $AsF_5$ molality was increased. While it is difficult to relate peak heights to solute concentrations using Raman spectroscopy, it is evident that the concentration of free $AsF_5$ relative to $As_2F_{11}^-$ increases markedly as the concentration of total $AsF_5$ is increased at room temperature. It was stressed above that Gillespie's very good experimental evidence for the predominance of $As_2F_{11}^-$ in HF—$AsF_5$ systems applied to solutions at low temperatures, for example, $-60$ to $-80°C$. The Melbourne group recorded Raman spectra for a solution of 0.18 molal in $AsF_5$ at 20, 0, and $-20°C$ and demonstrated a temperature-dependent shift in the equilibrium in equation 3.8. A sharp peak for free $AsF_5$ at 20°C was not detectable at $-20°C$ and there was diminution in the small $AsF_6^-$ peak with decrease in temperature. Gillespie is correct in stating that the predominant species in HF—$AsF_5$ systems at very low temperature is $As_2F_{11}^-$. However, much free $AsF_5$ and some $AsF_6^-$ exist in equilibrium with $As_2F_{11}^-$ at ambient temperatures. It should be noted that in synthetic work, $AsF_5$—HF solutions are more usually at ambient temperatures, and the implications of having free $AsF_5$ as an oxidant in such solutions will be discussed in chapter 10.

Figure 3.12 shows that values of specific conductances at 0°C for $AsF_5$

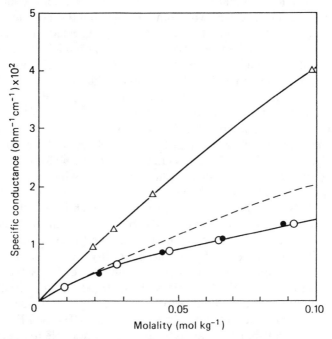

**Figure 3.12.** Conductivities of $SbF_5$ and $AsF_5$ in HF at 0°C.
Δ $SbF_5$ (Ref. 20), • $AsF_5$ (Ref. 26), ○ $AsF_5$ (Ref. 29).

solutions in HF reported in 1971 by the McMaster group[26] and in 1988 by the Melbourne group[29] are virtually identical; but, over the concentration range, they lie well below the dotted curve plotted for values of conductances that are exactly one-half those for $SbF_5$—HF. These would be the $AsF_5$ values if $SbF_5$ gave $H_2F^+$ and $SbF_6^-$ and $AsF_5$ gave $H_2F^+$ and $As_2F_{11}^-$ exclusively. To the extent that $AsF_6^-$ were to be formed, $AsF_5$—HF values should be *greater* than those given by the dotted curve. Conductance values from both groups show that there must be a large concentration of free, un-ionized $AsF_5$ in HF solution at or near ambient temperature.

The effect of equimolar amounts of $SbF_5$ and of $AsF_5$ on the $H_2F^+$ concentration of HF solutions is seen from $H_0$ values in Figure 3.10. $SbF_5$ provides $H_2F^+$ concentrations virtually equal to its own. To the extent that it ionizes, $AsF_5$ yields about half the $H_2F^+$ concentration, but there is a large concentration of $AsF_5$ in HF as the free, molecular entity providing no $H_2F^+$ ions. Between 0.05 mol% and 0.6 mol% values of $H_0$ for $SbF_5$ are more negative by between 1.2 and 1.4 units than those for $AsF_5$.[27] Potentiometric determination of $R(H)$ values also showed $AsF_5$ to be a much weaker Lewis acid than $SbF_5$, as shown in Figure 2.5.

## 3.4.3 Phosphorus Pentafluoride

Dean et al. included $PF_5$ in their cryoscopic and conductance studies of $SbF_5$ and $AsF_5$.[26] The low solubility of $PF_5$ in HF resulted in less reliable values of $v$ for $PF_5$ than for the other two pentafluorides, but the slope for the freezing-poing depression vs. molality curve strongly suggested a $v$ value of unity for $PF_5$. Even at $-83.6°C$, where maximum solubility might be expected for $PF_5$, the specific conductivity of a solution 0.73 molal in $PF_5$ was only $5.4 \times 10^{-4} \, \Omega^{-1} \, cm^{-1}$, little above the background conductivity of the solvent. So they concluded that $PF_5$ is essentially a nonelectrolyte in HF.

Measurement of Hammett Acidity Functions by Gillespie and Liang[27] gave values for $PF_5$ between $-11.34$ at 0.02 mol% and $-12.02$ at 0.5 mol%. These values are less negative than the extrapolated value of $H_0$ for pure HF, namely, $-15.1$. Figure 2.5 shows that the $R(H)$ value for the $PF_5$ system is less negative than the "median" value of $-21.1$, which can be taken as $R(H)$ for pure HF. In chapter 2 there was a discussion of the fact that the best-distilled HF will always contain impurities which lead to this HF being considerably more basic than "pure" HF by anything up to about four $H_0$ units because of the very steep slope of the curve for $H_0$ values of the HF system in the immediate region of true neutrality. Addition of indicators, whether spectroscopic or electrochemical, which are themselves partially protonated will enhance the basicity of the distilled HF, as will all experimental manipulations carried out before a measurement is made.

The best interpretation of the values of $H_0$ and $R(H)$ for $PF_5$ is that the pentafluoride is vanishingly weak as a Lewis acid, not able to neutralize the basicity of the working solvent even when present at about 0.5 mol%.

Stronger acids in minute amount neutralize the basic impurities and, even at low concentrations, exhibit characteristic acidities in HF solution. Whereas $SbF_5$, $AsF_5$, $NbF_5$, $TaF_5$, and others act as fluoride acceptors in a very wide range of solid Lewis acid–Lewis base adducts, $PF_5$ forms very few such adducts and when it does so is frequently a fluoride donor.

### 3.4.4  Boron Trifluoride

$BF_3$ has been widely used to provide acidic HF solutions in synthetic work. For example, the HF-solvated cations $Ni^{2+}$, $Co^{2+}$, $Pr^{3+}$, $Nd^{3+}$, $U^{3+}$, and so on, were generated in solution by applying a pressure of gaseous $BF_3$ to suspensions of the insoluble binary fluorides in HF, as will be shown in detail in chapters 6 and 7. Although it is a weak Lewis acid with low solubility in HF, it has special value in systems like $U^{3+}$—$BF_3$—HF in that, unlike some of the pentafluorides, it is nonoxidizing.

Despite all this, there has been little detailed or quantitative study of its role as a Lewis acid in HF. After very early investigation of its conductance in HF,[30] it was included in the potentiometric studies summarized in Figure 2.5. Measured values of $R(H)$ and pF indicate that it is a rather weak acid, stronger than $NbF_5$ but weaker than $TaF_5$.

### 3.4.5  Transition-Metal Pentafluorides and Oxidetetrafluorides

Several compounds within this classification have been studied as Lewis acids of the HF solvent system—some rather cursorily, others like $NbF_5$ and $TaF_5$ in somewhat greater depth. In 1976 Paine and Quarterman[31] published conductivities for several pentafluorides and oxidetetrafluorides, measured at a single value of concentration in each case, usually about 0.1 M. They reported that "the metal oxidetetrafluorides and $NbF_5$ and $MoF_5$ are essentially non-ionized in anhydrous HF while $TaF_5$, $ReF_5$ and $OsF_5$ are partially ionized." Combining their measured conductivities with those previously published for $SbF_5$, $AsF_5$, and $BiF_5$, they arrived at an order of Lewis acid strengths for pentafluorides:

$$SbF_5 > AsF_5 \geq OsF_5, ReF_5, TaF_5 > MoF_5 > NbF_5 \geq BiF_5$$

Several members of the Melbourne group have measured conductances of transition-metal pentafluorides and of oxidetetrafluorides in HF at 0°C and have recorded Raman spectra of these HF solutions and of the parent compounds and their fluoroanions.[32] Some of these conductance measurements are summarized in Figure 3.13 and are compared with previously measured conductances for $SbF_5$ and $AsF_5$. $TaF_5$, shown in other studies to be more highly conducting than $NbF_5$, is comparable in acid strength with $WOF_4$, and $ReF_5$ emerges as an acid of medium strength.

Figure 3.13 shows an initial sharp rise in conductance of $WOF_4$,[32b] fol-

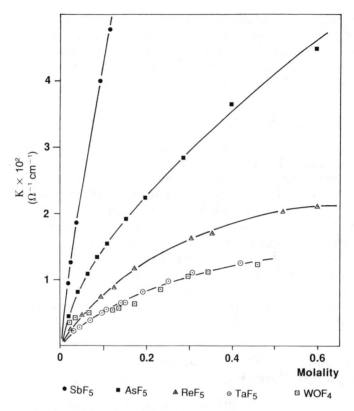

**Figure 3.13.** Conductivities of Pentafluorides and $WOF_4$ in HF.

lowed by a "plateau" in conductance values and then a steady rise, all of which probably reflects a real effect. In sufficiently dilute solutions, the ionic entities would be $H_2F^+$ and $WOF_5^-$, to the extent that $WOF_4$ acts as a fluoride acceptor. As the $WOF_4$ concentration is increased, the dominant ionic entities would be $H_2F^+$ and $W_2O_2F_9^-$, half the number of ions per $WOF_4$ solute particle as in dilute solution. Linden[32a] found a similar effect in measuring lowering of the vapor pressure of HF with progressive addition of $NbF_5$. O'Donnell and Peel,[33] realizing that cryoscopy gave information dealing only with solutions in the region of $-83$ to $-84°C$, had developed a reliable method of measuring lowering of the vapor pressure of HF at $0°C$, a colligative property able to be related to depression of freezing point, but having the advantage that the working temperature at which vapor pressure lowering is observed is closer to conditions for syntheses in HF than those of cryoscopy. Linden's interpretation of a similar "plateau" effect in the plot of lowering of vapor pressure vs. concentration was that $NbF_6^-$ is the dominant anionic species at low $NbF_5$ concentrations, with $Nb_2F_{11}^-$ becoming more significant at higher concentrations, for example, above about 0.25 molal. The conductance values for $WOF_4$[32b] quoted above suggest monomeric anion-dimeric anion equilibrium at concentrations of $WOF_4$ less than 0.1 molal. Canale's Raman spectra[32b] show evidence only for $W_2O_2F_9^-$. Any $WOF_5^-$ is probably formed at $WOF_4$ concentrations too low to be observed by Raman spectroscopy.

Hoskins et al.[34] identified by X-ray crystallography the dimeric anion $W_2O_2F_9^-$ in the compound $H_3O^+W_2O_2F_9^-$ isolated after the reaction of $WF_6$ with $H_2O$ in HF. This does not necessarily imply that $W_2O_2F_9^-$ is the major anionic species in such a solution. It could mean that, for solubility reasons, $H_3O^+W_2O_2F_9^-$ is the preferred species to be crystallized from solution. Linden's conductance measurements[32a] and Rigoni's Raman spectra[32c] are consistent with $Ta_2F_{11}^-$ being the predominant anionic species down to quite dilute solutions of $TaF_5$ in HF. Raman spectra suggest that the nonionized $TaF_5$ is present in solution in HF in the same form in which it exists in the solid state, namely, as $Ta_4F_{20}$.[32a]

There has been little reported study of Lewis acid strengths of the molybdenum compounds $MoF_5$ and $MoOF_4$. Conductance measurements[31,32a] show that $MoF_5$ is a very much weaker fluoride acceptor than $TaF_5$, for example. There does not appear to be any record of comparison of conductance measurements in HF of $MoF_5$ and $NbF_5$, although qualitative observations suggest that $MoF_5$ is the weaker acid of the two. $^{19}F$ NMR spectroscopic observations in HF solution show that $MoOF_4$ is a weaker Lewis acid than $WOF_4$.[35]

Only two of the pentafluorides and oxidetetrafluorides discussed in this section, $NbF_5$ and $TaF_5$, have been the subject of spectrophotometric and potentiometric measurements of acid strength. The $H_0$ values in Figure 3.10 show $NbF_5$ to be very much weaker than $TaF_5$, which in turn is significantly weaker than $AsF_5$. A solution containing 0.05 mol% of $NbF_5$ has a value of $H_0$ virtually identical with that of 100% pure HF—it appears to be just capable of neutralizing the $F^-$ resulting from protonation of the impurities and of the indicator in the solvent HF. It will be recalled that $PF_5$ does not appear to be sufficiently acidic to bring about this neutralization. A solution containing 0.4 mol% of $NbF_5$ is less than 2 $H_0$ units more acidic than 100% HF. The $H_0$ value for this solution is $-16.98$. Even the very dilute solution, 0.05 mol% $TaF_5$, with $H_0 = -17.83$, is nearly 3 $H_0$ units more negative than 100% HF, and a solution containing 0.9 mol% of $TaF_5$ has a value for $H_0$ of $-19.32$, quite close to the acidity of 0.05 mol% $SbF_5$ and equal in acidity to 0.4 mol% $AsF_5$.

Potentiometric measurements of Lewis acidity levels in HF based on the chloranil electrode[36] and on the hydrogen electrode[37] led to reported acidity values $R(H)$ and pF, which were summarized in Figure 2.5. Experimentally determined values of $H_0$, $R(H)$, and pF, where available, all give the same order of acid strengths for the transition-metal and main-group pentafluorides in HF:

$$SbF_5 > AsF_5 > TaF_5 > NbF_5 > PF_5$$

However, on the basis of the $R(H)$ and pF values in Figure 2.5, $NbF_5$ appears to be a much weaker Lewis acid relative to $TaF_5$, $AsF_5$, and $SbF_5$ than is the case when $H_0$ values are considered. Furthermore, the measured $R(H)$ for the $NbF_5$—$NbF_6^-$ buffer system which Devynck and colleagues studied lies below the average of $R(H)$ values for 1 M $SbF_5$ and 1 M KF, that is, below the expected $R(H)$ for pure HF, whereas the $H_0$ value for 0.5 mol% $NbF_5$ in HF which contained no added $NbF_6^-$, is about 2 units more negative than $H_0$ for pure HF. The probable clue to these inconsistencies lies in the ability of transition-metal pentafluorides to dimerize even when the pentafluoride is the only solute added. The presence of the hexafluorometallate(V) anion would greatly favor the formation of $M_2F_{11}^-$ and would mean that much less pentafluoride is available to interact with the solvent as a Lewis acid than if the pentafluoride alone were available. It could even be that there was slight excess of $NbF_6^-$ over $NbF_5$, leading to a basic solution. There is no evidence that $BF_3$ and $PF_5$ form dimeric anions in HF solution.

The decision by the French and the Swiss groups to measure acidities of buffer systems has probably led to greater lowering of acidities than occurs for Bronsted buffer systems in water because their is no chemical interaction of the weak Bronsted acid and its conjugate base in aqueous solutions.

## 3.5  Summary

There is no doubt that the amount of general chemistry, particularly organic chemistry, studied in $H_2SO_4$, and probably in $HSO_3F$ and $CF_3SO_3H$ as well, is greater than that in HF. It is then somewhat intriguing that, despite the fact that experimental manipulation of HF and its solutes is considered to be very much more difficult than that required for studies in the other three superacids, the quantitative physicochemical study of speciation, individual strengths, and comparative strengths of Lewis acids is far more extensive in HF than in the other three acids. Hence the HF work will be summarized first.

The order of Lewis acid strengths for pentafluorides in HF, as given toward the end of section 3.4.5, is very reliable, being based on $H_0$, $R(H)$, and pF values and on conductance measurements supported by Raman and NMR spectroscopy. The order can be widened a little, but with somewhat less reliability, by including Lewis acids which have been studied less comprehensively. For some of these, for example, $OsF_5$, $MoF_5$ and $BiF_5$, a conductance, measured at a single concentration, is all that is available and no attempt is made here to list their strengths relative to those for which acidity values at different concentrations have been measured spectrophotometrically and potentiometrically. For others, such as $ReF_5$ and $WOF_4$, conductances have been measured over a range of concentrations and there have been additional physicochemical measurements made on their solutions in HF. The order for $BF_3$ has been established by potentiometric measurement of $R(H)$ and pF values and it can be included fairly reliably as below:

$$SbF_5 > AsF_5 > ReF_5 > TaF_5 \approx WOF_4 > BF_3 > NbF_5 > PF_5$$

Reliable conductance measurements over a range of concentrations do not appear to be available for $NbF_5$ or $MoOF_4$, but the observations that have been made suggest that they are of comparable Lewis acid strength.

The thermal stability of solid Lewis acid–Lewis base adducts is frequently cited as providing orders of Lewis acid (or of Lewis base) strengths, where adducts formed by a range of Lewis acids with a particular fluoride donor (or vice versa) are considered. For example Fox and colleagues[38] examined the fluoride-donor capability of trifluoramine oxide $ONF_3$ toward some of the common Lewis acids of the HF system. They established an order $SbF_5 > AsF_5 > BF_3 > PF_5$ by showing that $ONF_2^+SbF_6^-$ is stable at room temperature, whereas a vapor pressure of 5 mm of $AsF_5$ is in equilibrium with $ONF_2^+AsF_6^-$ at room temperature. $ONF_2^+BF_4^-$ is stable only below $-50°C$, and it was not possible to isolate an adduct from the $PF_5$ reaction. Care must be exercised in this type of comparison, in that relative volatilities of the Lewis acids are also important factors in the adduct stabilities.

Bartlett and Gillespie independently have determined crystal structures of a large number of adducts formed by fluoride donation to a range of penta-fluorides by the noble gas fluorides $XeF_2$, $XeF_4$, and $XeF_6$. The distance $Xe \cdots F$ for $F_nXe^+ \cdots FMF_5^-$ gives an indication of the acid strength of $MF_5$. When $Xe \cdots F$ is long compared with the distance of Xe from its terminal F atoms and the formally bridging F atom is closely comparable in distance from M with the M—F terminal bonds, the solid has a large degree of ionic character and $MF_5$ is acting as a strong Lewis acid; the $MF_6^-$ anion is approximately octahedral and can be viewed as a discrete entity, as can the $XeF_n^+$ cation. When the bridging F atom is closer to Xe and removed somewhat from M, the adduct is less ionic in character and $MF_5$ is a weaker acid. On these criteria, $PtF_5$[39] and $RuF_5$[40] appear as acids of comparable strength but somewhat weaker than $AsF_5$[41] when the crystal structures of the adducts $XeF_6.MF_5$ are examined. Similar analysis of the $XeF_4.BiF_5$[42] and $XeF_4.SbF_5$[43] adducts shows $BiF_5$ to be considerably weaker than. The noble metal pentafluorides have low solubilities in HF, making solution studies difficult. These comparisons based on crystallography allow some qualitative comparisons of acid strengths to be made. However, there is not available a sufficiently large number of crystal structures of adducts from a single fluoride donor with a range of Lewis acids to allow useful semiquantitative comparisons. The situation is further complicated by the fact that Lewis acids like $SbF_5$ and $NbF_5$ form adducts containing either the $MF_6^-$ or the $M_2F_{11}^-$ anions, depending on the conditions of adduct formation. There can be further modification of the bond lengths to bridging F atoms depending on whether the anion is monomeric or dimeric. $M_2F_{11}^-$ adducts appear to be more ionic than their $MF_6^-$ counterparts. There are no crystallographic data available to allow similar semiquantitative estimations of the relative strengths of the Lewis acids of the $H_2SO_4$, $CF_3SO_3H$, and $HSO_3F$ systems.

When Gillespie and Moss showed in 1966[20] that, as $SbF_5$ was added to HF, the $SbF_6^-$ anion formed during the addition of the first 10 mol% or so and was replaced in turn by $Sb_2F_{11}^-$, $Sb_3F_{16}^-$, and so on as the solute was added progressively, $SbF_5$ was regarded as an extraordinary Lewis acid. Those working with Lewis acid—HF systems at that time were probably in fairly general agreement that $AsF_5$, $BF_3$, and so on would all be expected to give monomeric anions exclusively. Gillespie's group showed in 1971[26] that at lower temperatures, that is, those near the freezing point of HF, dimeric $As_2F_{11}^-$ anions predominated in the HF—$AsF_5$ system even at low concentration. In 1988 O'Donnell and colleagues[29] were to modify that view, especially for ambient temperature HF—$AsF_5$ systems, and show that, when acidity measurements are made at ambient temperature, the observed weakness of $AsF_5$ relative to $SbF_5$ is as much a consequence of the presence of free, molecular $AsF_5$ in solution as it is of the formation of $As_2F_{11}^-$.

Subsequent work[32] indicates that, especially for transition-metal pentafluorides and oxidetetrafluorides, dimeric anion formation is probably the norm rather than the exception, except in very dilute solutions. The implication of this in interpretation of measurements of relative acidities has already been sketched out in section 3.4.5 and the further implication when selecting Lewis acids for synthetic work in HF will be discussed in chapter 10.

Conductances, some NMR spectroscopy, and spectrophotometric determination of Hammett Acidity Functions have been used to establish relative Lewis acidities in $H_2SO_4$. Gillespie et al.[3] showed that $HSO_3F$ and $HSO_3Cl$ are weak Bronsted acids; that is, they have very limited ability to donate protons to the highly acidic solvent. $SO_3$ is a moderately strong Lewis acid, values of Hammett Acidity Functions increasing from $-11.93$ in pure $H_2SO_4$ to about $-15$ on addition of 75 mol% $SO_3$. In moderate concentrations, up to about 10 mol%, $B(HSO_4)_3$ is a somewhat stronger Lewis acid than $SO_3$, but $H_0$ values taper off at about $-13.6$ as "complex polymeric boron sulfate precipitates from solution."[3] There is some qualitative evidence[1] for compounds such as $Sn(HSO_4)_4$ and $Pb(HSO_4)_4$ acting as Lewis acids in $H_2SO_4$.

Until the work of Cicha and Aubke on $Ta(SO_3)_5$,[18] there was no $H_0$ measurement in $HSO_3F$ of an acid which was truly a Lewis acid *of the solvent system*, that is, one in which each of ligands was an entity which is the base of the solvent system. By comparison, $BF_3$ and the pentafluorides can be regarded as true Lewis acids *of the HF system*. Lee and Aubke had previously used conductance measurements to show that $Au(SO_3F)_3$[13] and $Pt(SO_3F)_4$[14] were strong Lewis acids in $HSO_3F$.

Prior to Aubke's work, the Lewis acids investigated in $HSO_3F$ were a range of binary fluorides—$SbF_5$, $BiF_5$, $AsF_5$, $NbF_5$, $PF_5$, and $TiF_4$—studied conductometrically, or a selection of these, with or without insertion of $SO_3$ into the central atom–ligand bond, first studied conductometrically[8] and then spectrophotometrically to yield quantitative $H_0$ values.[9] Of these $SbF_2(SO_3F)_3$ was the strongest, being nearly one $H_0$ unit more negative than

$SbF_5$ at typical concentrations of 1–5 mol%, while $AsF_5$ was weaker than $SbF_5$ by about 1.5 $H_0$ units in this concentration range.

Boron triflate, $B(OSO_2CF_3)_3$, is the only fully triflated Lewis acid whose strength has been measured in $CF_3SO_3H$.[4,5] Attempts have been made to synthesize $Sb(OSO_2CF_3)_5$ and $Ta(OSO_2CF_3)_5$.[5] Like some of their fluorosulfato- counterparts they have very low solubilities and are probably polymeric. It is not surprising that pentafluorides with monodentate ligands accept fluorides from HF to form discrete soluble anionic species whereas $SO_3F^-$ and $CF_3SO_3^-$, both potentially bidentate, tend to form polymeric insoluble solids. The acidity of $SbF_5$ in $CF_3SO_3H$ has been measured potentiometrically to give a single value of $R(H)$,[6] and Liang[7] reported values for $H_0$ for a range of concentrations of $SbF_5$ in $CF_3SO_3H$. Subsequently Adrien[5] compared the strengths of $B(OSO_2CF_3)_3$, $SbF_5$, $TaF_5$ and $NbF_5$ by measuring values of $H_0$ for each acid at several concentrations, as shown in Figure 3.4.

# References

1. R.J. Gillespie, E.A. Robinson in *Advances in Inorganic Chemistry and Radiochemistry*, Vol. 1. H.J. Emeléus, A.G. Sharpe, eds. Academic Press, New York, 1959, pp. 385–423.

2. R.J. Gillespie, E.A. Robinson, *Can. J. Chem.*, *40*, 658 (1962).

3. R.J. Gillespie, T.E. Peel, E.A. Robinson, *J. Am. Chem. Soc.*, *93*, 5083 (1971).

4. A. Engelbrecht, E. Tshager, *Z. Anorg. Allg. Chem.*, *433*, 19 (1977).

5. R. Adrien, Ph.D. Thesis, University of Melbourne (1992).

6. B. Carre, J. Devynck, *Anal. Chim. Acta*, *159*, 149 (1984).

7. J. Liang, Ph.D. Thesis, McMaster University (1976).

8. R.J. Gillespie, K. Ouchi, G.P. Pez, *Inorg. Chem.*, *8*, 63 (1969).

9. R.J. Gillespie, T.E. Peel, *J. Am. Chem. Soc.*, *95*, 5173 (1973).

10. R.J. Gillespie, E.A. Robinson, *Can. J. Chem.*, *40*, 675 (1962).

11. R.C. Thompson, J. Barr, R.J. Gillespie, J.B. Milne, R.A. Rothenbury, *Inorg. Chem.*, *4*, 1641 (1965).

12. K.C. Lee, F. Aubke, *Inorg. Chem.*, *18*, 389 (1979).

13. K.C. Lee, F. Aubke, *Inorg. Chem.*, *19*, 119 (1980).

14. K.C. Lee, F. Aubke, *Inorg. Chem.*, *23*, 2124 (1984).

15. K.C. Lee, F. Aubke, *Can. J. Chem.*, *59*, 2835 (1981).

16. K.C. Lee, F. Aubke, *J. Fluorine Chem.*, *19*, 501 (1982).

17. W.V. Cicha, F. Aubke, *J. Am. Chem. Soc.*, *111*, 4328 (1989).

18. W.V. Cicha, Ph.D. Thesis, University of British Columbia (1989).

19. H.H. Hyman, L.A. Quarterman, M. Kilpatrick, J.J. Katz, *J. Phys. Chem.*, *65*, 123 (1961).

20. R.J. Gillespie, K.C. Moss, *J. Chem. Soc. (A)*, 1170 (1966).

21. J. Bacon, P.A.W. Dean, R.J. Gillespie. *Can. J Chem.*, *47*, 1655 (1969).

22. J. Bacon, P.A.W. Dean, R.J. Gillespie, *Can. J. Chem.*, *48*, 3413 (1970).

23. C.G. Davies, R.J. Gillespie, P.R. Ireland, J.M. Sowa, *Can. J. Chem.*, *52*, 2048 (1974).

24. A.J. Edwards, G.R. Jones, R.J.C. Sills, *Chem. Commun.*, 1527 (1968).

25. R.J. Gillespie, D.A. Humphreys, *J. Chem. Soc. (A)*, 2311 (1970).

26. P.A.W. Dean, R.J. Gillespie, R. Hulme, D.A. Humphreys, *J. Chem. Soc. (A)*, 341 (1971).

27. R.J. Gillespie, J. Liang, *J. Am. Chem. Soc.*, *110*, 6053 (1988).

28. H.H. Hyman, J. Lane, T.A. O'Donnell, *145th Meeting, Am. Chem. Soc., Abstracts*, p. 63J.

29. C.G. Barraclough, J. Besida, P.G. Davies, T.A. O'Donnell, *J. Fluorine Chem.*, *38*, 405 (1988).

30. M. Kilpatrick, F.E. Luborsky, *J. Am. Chem. Soc.*, *76*, 5863 (1954).

31. R.T. Paine, L.A. Quarterman, *J. Inorg. Nucl. Chem.*, Supplement 85 (1976).

32. B.Sc. (Honors) Reports, University of Melbourne: (a) A. Linden (1979), (b) G. Canale (1982), (c) P. Rigoni (1984).

33. T.A. O'Donnell, T.E. Peel, *J. Inorg. Nucl. Chem.*, *40*, 381 (1978).

34. B.F. Hoskins, A. Linden, T.A. O'Donnell, *Inorg. Chem.*, *26*, 2223 (1987).

35. R. Bougon, T. Bui Huy, P. Charpin, *Inorg. Chem.*, *14*, 1822 (1975).

36. J. Devynck, A.B. Hadid, P.L. Fabre, B. Trémillon, *Anal. Chem. Acta*, *100*, 343 (1978).

37. R. Gut, K. Gautschi, *J. Inorg. Nucl. Chem.*, Supplement 95 (1976).

38. W.B. Fox, C.A. Wamser, R. Eibeck, D.K. Huggins, J.S. McKenzie, R. Juurik, *Inorg. Chem.*, *8*, 1247 (1969).

39. N. Bartlett, F. Einstein, D.F. Stewart, J. Trotter, *J. Chem. Soc. (A)*, 1190 (1967).

40. N. Bartlett, M. Gennis, D.D. Gibler, B.K. Morrell, A. Zalkin, *Inorg. Chem.*, *12*, 1717 (1973).

41. N. Bartlett, B.G. De Boer, F.J. Hollander, F.O. Sladky, D.H. Templeton, A. Zalkin, *Inorg. Chem.*, *13*, 780 (1974).

42. R.J. Gillespie, D. Martin, G.J. Schrobilgen, D.R. Slim, *J. Chem. Soc., (Dalton Trans)*, 2234 (1977).

43. P. Boldrini, R.J. Gillespie, P.R. Ireland, G.J. Schrobilgen, *Inorg. Chem.*, *13*, 1690 (1974).

# CHAPTER 4

# Polyatomic Cations of Nonmetallic Elements

Nearly 200 years ago, it was observed that elemental sulfur dissolved in oleums to give highly colored solutions—yellow, blue, or red depending on the $SO_3$ content of the oleum. Subsequently selenium and tellurium were found to give highly colored solutions and, later, deep-blue solutions were noted when iodine was dissolved in strong oleums. There was conjecture, not based on experiment, that the colored species might be oxoanions or oxocations of the nonmetals concerned or might be polymeric in solution. In 1938 Masson postulated the existence of the electrophilic entities $I_3^+$ and $I_5^+$ in discussing the iodination of aromatic organic compounds. Subsequently it was postulated, incorrectly, that the intense deep-blue color resulting from dissolution of $I_2$ in strong oleums arose from the formation of $I^+$ in solution. This historical background has been presented by Gillespie and Passmore in reviews dealing with the homopolyatomic cations of the elements of groups VI and VII.[1-3]

In the second half of the 1960s Gillespie and co-workers at McMaster University showed by simple direct physicochemical measurements that the colored species obtained in oleums by oxidation of $I_2$ by the solvent or by controlled oxidation in $HSO_3F$ were the cations $I_2^+$, $I_4^{2+}$, $I_3^+$, and $I_5^+$, with no evidence for $I^+$. They demonstrated the existence in solution of $Br_2^+$ and $Br_3^+$, by oxidation of $Br_2$ in superacidic media. Subsequently, they and others isolated from media of very-low-basicity crystalline compounds, suitable for X-ray diffraction structural studies, containing several of these cations. Within this same general period the McMaster group showed good evidence for the generation of homo- and heteropolyatomic cations of S, Se, and Te by oxidation of the elements in superacidic media and later characterized the cations structurally in compounds isolated from media of low basicity. From the mid-1970s on Corbett, Mamantov, Bjerrum, and others were demonstrating the formation of polyatomic cations of I, S, Se, and Te in chloroaluminate melts.

An essential condition for the generation and stabilization of the cations of groups VI and VII, as for many other cationic species in unusual oxidation states, is that they should be prepared under experimental conditions of very low basicity. Frequently, particularly for purposes of structural analysis, compounds containing these cations have been isolated from volatile solvents of very low basicity, for example, $SO_2$, $SO_2ClF$, and $AsF_3$. The counteranions in these compounds must also be very weakly basic. In many instances the anions have been $AsF_6^-$ or $Sb_2F_{11}^-$. The cations are all strongly electrophilic and will interact with nucleophilic bases of sufficient strength and undergo disproportionation, as will be illustrated in detail in this chapter. A convenient experimental approach to investigation of these cations, particularly for solution studies, is to generate them in superacid media, where strengths and concentrations of bases are very low.

Reference to Tables 4.2 and 4.3 shows that for any group certain generalizations can be made about the ease of stabilization of the cations in media of increasing superacidity or conversely of decreasing basicity. Thus, for any cation, $X_n^+$ of a group VII element, the order of ease of stabilization is $I_n^+ > Br_n^+ > Cl_n^+$; that is, higher acidities (lower basicities) are required to stabilize the cations of elements of lower atomic number. For cations of the same complexity, electrophilicity is in the order $I_n^+ < Br_n^+ < Cl_n^+$. For any element within that group, stabilization becomes more difficult as the ratio of the charge to the number of atoms in the cation increases; that is, the order of ease of generation of the cations is $I_5^+ > I_3^+ > I_2^+$. Similar generalizations apply for cations of Br and of Cl and for those of the chalcogen elements.

Most of the discussion in this chapter will be restricted to homopolyatomic cations—the halogen cations $X_n^+$ or typical chalcogen cations $Ch_n^{2+}$. Some heteropolyatomic chalcogen cations, for example, $Se_8S_2^{2+}$, $Te_2Se_2^{2+}$, and so on, will be included because they are very close analogues of the homopolyatomic cations $Se_{10}^{2+}$ and $Se_4^{2+}$ and require similar acid-base experimental conditions for their generation and stabilization.

The term "heterocation" has been used in an earlier review[4] to describe polyhalide cations such as $BrF_2^+$, $BrCl_2^+$, $ClF_4^+$, and $IF_6^+$, and the concept would include halocations of other elements, for example, $SF_3^+$, $SeCl_3^+$, $PBr_4^+$, and $XeF_5^+$, as well as a very wide range of cations which do not contain halogen atoms as ligands, for example, $ClO_2^+$, $NO_2^+$, and $VO_2^+$. These can all be regarded, in a formal sense at least, as resulting from a Lewis acid–Lewis base transfer of a ligand from a parent molecule that has been formed by oxidation of the central element. To take only the first example, $Br_2$ is oxidized by $F_2$ to $BrF_3$, which can interact with the strong Lewis acid $SbF_5$ to form $BrF_2^+Sb_2F_{11}^-$. In all of these heterocations, the central element has a formal oxidation state greater than 1. In all of the homopolyatomic cations to be discussed in this chapter, the formal oxidation state of the element is less than 1—often considerably so. The same remark applies to the elements in the few heteropolyatomic cations to be included here. These cations in very low or fractional oxidation states can exist only in a medium of high superacidity—or, conversely, of very low basicity.

## 4.1 Homopolyatomic Cations of Group VII Elements

No cation of fluorine has been prepared in solution or in the solid state. Homopolyatomic cations of iodine, which can be prepared relatively easily in solution and in solids, were characterized before those of bromine and chlorine. For the latter element, only one cation $Cl_3^+$ has been prepared in the solid state, there being no evidence for chlorine cations in solution. The amount of solution and solid-state work carried out on iodine cations far exceeds that on cations of other nonmetals. For these reasons, cations of iodine will be presented initially.

## 4.1.1 Generation of Iodine Cations

Several polyatomic cations of iodine, $I_2^+$, $I_4^{2+}$, $I_3^+$, and $I_5^+$, have been generated in protonic superacidic media, in the solid state and in molten chloroaluminates. No reliable evidence has been put forward for the existence of the simple monatomic cation $I^+$ in solution or in a solid. However, iodine(I) cations are relatively easy to prepare when $I^+$ is coordinated to strong donor molecules such as pyridine and acetonitrile in the linear species $Ipy_2^+$ and $I(NCCH_3)_2^+$.

### 4.1.1.1 In Protonic Superacids

Gillespie and Morton[5] have reviewed the formation and stability of polyatomic cations of iodine in $HSO_3F$, $H_2SO_4$, and oleums. They report the 1966 work by Gillespie and Milne in which $I_2$ and the powerful oxidant $S_2O_6F_2$ were reacted in the ratio 2:1 in $HSO_3F$. Conductance measurements ($\gamma$ values, see section 2.2) indicating the number of moles of $SO_3F^-$ produced from the oxidant $S_2O_6F_2$ per mole of $I_2$ were coupled with cryscopic data, in a method to be fully interpreted below, to suggest that the *room-temperature* deep-blue solution contained $I_2^+$. This solution had the same spectrum as that reported much earlier for solutions of $I_2$ in oleums, where it had been believed that the absorbing species was $I^+$.

Investigation of the 1:1 reaction mixture for $I_2$:$S_2O_6F_2$ provided no evidence for the previously postulated species $I^+$,[5] even though this redox stoichiometry would have favored the formation of the monatomic cation.

Coulometric titration of iodine at controlled potential in $HSO_3F$ showed that $I_2$ is oxidized quantitatively to $I_2^+$ in a one-electron step,[6] producing the characteristic deep-blue solution with absorption bands at 640, 495, and 415 nm, essentially similar to those reported by Gillespie and Morton[5] for $I_2^+$ in $HSO_3F$. Coulometric generation of $I^+$ from $I_2$ would have shown a two-electron oxidation.

Cooling of the deep-blue solution of $I_2^+$ in $HSO_3F$ from room temperature to $-70$ or $-80°C$ caused a reversible color change to red. Gillespie and Milne had observed that, for the reaction $I_2$:$S_2O_6F_2$, 2:1, the observed value of $\gamma$ was unity (i.e., 1 mol of $S_2O_6F_2$ gave 2 mol of $SO_3F^-$ in oxidizing 2 mol of $I_2$). However, the observed value of $\nu$ from cryoscopy was 1.5, not 2, as would be the case for the reaction

$$2I_2 + S_2O_6F_2 \rightarrow 2I_2^+ + 2SO_3F^- \qquad\qquad 4.1$$

These experimental observations led to the proposal, suggested by the color change on cooling a room-temperature solution containing $I_2^+$, that oxidation of $I_2$ with $S_2O_6F_2$ led to the formation of $I_4^{2+}$, rather than $I_2^+$, at the freezing point of $HSO_3F$, that is, under the experimental conditions for cryoscopy, as in the equation

$$2I_2 + S_2O_6F_2 \rightarrow I_4^{2+} + 2SO_3F^- \qquad\qquad 4.2$$

Subsequently, the structure of the rectangular cation $I_4^{2+}$ has been determined.[7]

Gillespie and Peel[8] have reported a value for the Hammett Acidity Function $(H_0)$ of $-15.07$ for pure $HSO_3F$. When the oxidant $S_2O_6F_2$ is reduced to $SO_3F^-$ by $I_2$ in $HSO_3F$, the medium is made slightly more basic. Assuming $[SO_3F^-] \approx 0.05$ M, $H_0$ for the reaction medium would be about $-14.1$.[8] Gillespie and Morton[5] report that under these conditions $I_2^+$ is not completely stable, interacting with $SO_3F^-$ (the base of the solvent system) and disproportionating partially to $I_3^+$ and $I(SO_3F)_3$. They state that in 100% $H_2SO_4$, for which $H_0$ is about $-11.9$,[9] this disproportionation is virtually complete. They base this statement on Raman spectroscopic evidence. Identification of $I_2^+$ by conventional Raman spectroscopy was unsuccessful because of absorption of the exciting radiation by the highly colored solutions. However, they demonstrated that resonance Raman spectroscopy gives intense signals for $I_2^+$ in $HSO_3F$ and is capable of detecting $I_2^+$ down to about 1 part in $10^6$. They showed by this technique that there are only traces of $I_2^+$ in $H_2SO_4$. It is highly significant that when the $SO_3F^-$ concentration in the solvent $HSO_3F$ is substantially reduced by addition of $SbF_5$ to form species like $[F_5SbOSO_2F]^-$, $I_2^+$ is quite stable in this medium. Values of $H_0$ for such solutions would be likely to be $-16$ to $-18$, depending on the amount of $SbF_5$ added. $I_2^+$ is also reported to be stable[5] in 65% oleum for which $H_0$ is $-14.84$.[9]

By setting the stoichiometric ratio of the reactants $I_2:S_2O_6F$ at 3:1 and 5:1 in $HSO_3F$, Gillespie and Milne[5] obtained evidence for the formation of $I_3^+$ and $I_5^+$. These had previously been postulated as the cationic species in 100% $H_2SO_4$. $(H_0 = -11.93.)$

Gillespie's work has shown that $I^+$ cannot be generated in $HSO_3F$, whereas $I_2^+$ is marginally stable. In the less-acidic medium $H_2SO_4$, $I_3^+$ and $I_5^+$ are the iodine cations of highest formal charge to be stable, $I_2^+$ disproportionating to $I_3^+$ or $I_5^+$ on interaction with the "high" concentrations of the base $HSO_4^-$. Of course, in the more acidic $HSO_3F$ where the concentration of the $SO_3F^-$ is too small to cause disproportion, $I_3^+$ and $I_5^+$ can exist as stable entities, if the reactant stoichiometry is such as to generate these species.

The implicit strategy behind Gillespie's work in $HSO_3F$ was to use a medium of sufficient acidity such that the polyatomic cations sought would be stable in it and then to adjust the oxidant stoichiometry to generate the cations $I_5^+$, $I_3^+$, and $I_2^+$. This strategy is effective if a powerful oxidant, such as

$S_2O_6F_2$, is used. Later Melbourne studies in HF and in $CF_3SO_3H$ have shown that reactant stoichiometry dictates speciation, providing the medium is sufficiently acidic *and* providing the oxidant is very strong, for example, $F_2$ in HF. If, however, a mild oxidant is used, for example, one of the products of disproportionation of a cation $I_n^+$, it is the level of acidity or basicity of the medium alone which determines cation speciation.

Recent work by Besida and O'Donnell at Melbourne[10] has determined quite sharply the levels of acidity and basicity in anhydrous HF that determine the equilibria that exist between the iodine cations $I_2^+$, $I_3^+$, and $I_5^+$ and the final disproportionation products $I_2$ and $IF_5$. Two series of experiments were conducted. In the first, excess $I_2$ was reacted with controlled amounts of $F_2$ in a set of HF solutions to which the base NaF and the Lewis acids of increasing strength $NbF_5$, $TaF_5$, and $SbF_5$ were added at controlled concentrations to provide reaction media for which accurate values of the Hammett Acidity Function $H_0$ were known as a result of the work of Gillespie and Liang,[11] described in chapter 2. The reductant $I_2$ was maintained in excess in this series of experiments because the very strong oxidant $F_2$ would oxidize $I_2$ to a stoichiometrically determined cationic product virtually regardless of the level of acidity of the HF. In the second series of experiments, the mild oxidant $IF_5$, demonstrated by Raman spectroscopy to be a disproportionation product of the iodine cations, was maintained in tenfold excess over the reductant $I_2$ as the acidity-basicity levels were fixed.

The results in Table 4.1 drawn from the first series of experiments show that $I_2^+$ is the only species detected spectroscopically in very acidic solutions, that is, for those in which $H_0$ is more negative than $-16.20$. A very

**Table 4.1.** Dependence of Formation of Iodine Cations on Level of Acidity of Hydrogen Fluoride[a]

| HF acidity-basicity (solute, molality) | $-H_0^b$ | Species in solution |
|---|---|---|
| NaF, 0.1 | 9.95[c] | $I_2 + IF_5$ |
| NaF, 0.01 | 10.80[c] | $I_2 + IF_5$ |
| NaF, 0.0075 | 10.95[d] | $I_5^+ + I_3^+$ |
| NaF, 0.001 | 11.80[d] | $I_5^+ + I_3^+$ |
| $NbF_5$, 0.05 | 15.65[c] | $I_3^+ + $ trace $I_2^+$ |
| $NbF_5$, 0.075 | 15.90[c] | $I_3^+ + I_2^+$ |
| $NbF_5$, 0.091 | 16.05[e] | $I_3^+ + I_2^+$ |
| $NbF_5$, 0.11 | 16.20[e] | $I_2^+$ |
| $TaF_5$, 0.4 | 19.20[c] | $I_2^+$ |
| $SbF_5$, 0.2 | 20.65[c] | $I_2^+$ |

[a]Excess $I_2$ was reacted with controlled amounts of $F_2$ under the conditions listed earlier.
[b]Values rounded to 0.05 unit.
[c]Experimentally determined values from Ref. 11.
[d]Calculated from experimental values from Ref. 11, an autoprotolysis constant of $5 \times 10^{-13}$, and a value of $-15.10$ for $H_0$ for pure HF.
[e]Obtained by interpolation between experimental values from Ref. 11.

acidic HF solution which was 5 molal in $SbF_5$ ($H_0 \approx -22$) gave no spectral peaks other than those for $I_2^+$; that is, there was no spectroscopic evidence for $I^+$. Between $H_0$ values of $-16.05$ and $-15.65$ $I_3^+$, a disproportionation product of $I_2^+$ was observed spectroscopically to "grow in" as the acidity scale diminished, with $I_2^+$ present in trace amount at $-15.65$. $I_3^+$ and $I_5^+$ co-exist in equilibrium with each other and with the excess reagent $I_2$ for $H_0$ values between about $-15.6$ and $-10.95$. In solutions more basic than that for which $H_0 = -10.80$, there was no spectroscopic evidence for any iodine cation. Elemental $I_2$ was characterized by a UV-visible band at 508 nm and a Raman spectrum for a typical solution for this basicity range showed $IF_5$ as the major product with a small amount of $IF_3$. This means that any attempt to generate iodine cations by reacting $F_2$ and excess $I_2$ in HF solutions with $H_0$ values significantly more positive than $-10.9$ would result in formation of the formal disproportionation products $I_2$ and $IF_5$.

In the second series of experiments, the possible interaction of $I_2$ and $IF_5$, which had been shown to be the products of disproportionation of iodine cations as a result of addition of the base $F^-$ to HF solutions of the cations, was studied to demonstrate the effect of change in acidity as $SbF_5$ was added on mutual oxidation and reduction, that is, conproportionation of $I_2$ and $IF_5$. $IF_5$, at 0.011 molal, was in tenfold excess over $I_2$ and, initially, it was shown that $I_2$ was the only entity present in HF other than $IF_5$. This situation persisted until the solution had been made 0.007 molal in $SbF_5$, a neutralization reaction leading to the formation of $IF_4^+$ and $SbF_6^-$ in the HF solution, with no excess $SbF_5$. $IF_5$ is a weak base of the HF solvent system and was in excess of the added $SbF_5$ at this stage. When the solution was 0.009 molal in $SbF_5$, that is, the solution was weakly basic, spectroscopy showed the presence of $I_3^+$ and $I_5^+$, and on further lowering of basicity, by making the solution 0.010 molal in $SbF_5$, $I_3^+$ and $I_2^+$ were observed. When $SbF_5$ was in slight excess—0.013 molal compared with an initial 0.011 molal $IF_5$—the principal species was $I_2^+$ with only a trace of $I_3^+$. Finally, when $SbF_5$ was vastly in excess of $IF_5$—0.02 molal—and this excess of $SbF_5$ was leading to the formation of $H_2F^+$, at the expense of $F^-$, $I_2^+$ was the only cation of iodine observed. The results of the second series of experiments, where the oxidant $IF_5$ was in excess, are in broad agreement with those of the first series, where the reductant was in excess. This leads to the conclusion that as long as a very strong oxidant is not present, the level of acidity of the medium is the principal determinant of cation speciation, regardless of whether oxidant or reductant is in excess.

It should be stressed that, regardless of the nature of iodine cations present in HF solutions, the addition of excess of the base $F^-$ always led to the formation of the disproportionation products $I_2$ and $IF_5$—the UV-visible spectrum characteristic of the particular cation was replaced by that of $I_2$ and $IF_5$ was shown to be present by Raman spectroscopy. Gradual reduction of acidity leads to disproportionation of $I_2^+$ to $I_3^+$ and $I_5^+$, in turn, with $I_2$ and $IF_5$ as the final disproportionation products.

The generality of the potential importance of acidity-controlled conproportionation reactions for the synthesis of cations was demonstrated by taking $IO_3^-$ and $I_2$ in HF solutions of differing acidity. HF—$SbF_5$ gave $I_2^+$ whereas $GeF_4$, an extremely weak Lewis acid of the HF system, gave a mixture of $I_5^+$ and $I_3^+$. While this synthetic approach may provide convenience in choice of reactants, it has the disadvantage that each of the oxoligands of $IO_3^-$ would be triply protonated on release from the oxidant, generating much of the unnecessary base $F^-$. Besida and O'Donnell[10] reported in their paper on iodine cations in HF that sulfur cations could be generated by reacting $S_8$ with $SO_3^{2-}$ or $SO_4^{2-}$ in HF solutions of appropriate acidity.

Also working with the Melbourne group, Adrien[12] has generated iodine cations in triflic acid by conproportionation reactions between $ICl_3$ and $I_2$. Regardless of whether oxidant or reductant was in tenfold excess, acidity level proved to be the determinant of the nature of cation speciation. Spectroscopy showed the presence of $I_3^+$ as the dominant species, with a trace of $I_2^+$, in triflic acid containing the reasonably weak Lewis acid $TaF_5$. When the strong Lewis acids $SbF_5$ and $B(OSO_2CF_3)_3$ were used to increase the acidity of the triflic acid, $I_3^+$ with a trace of $I_2^+$ was observed spectroscopically for low concentrations of the Lewis acids (0.08 m $SbF_5$ and 0.33 m $B(OSO_2CF_3)_3$) and $I_2^+$ became the dominant species at higher acid concentrations (5.18 m $SbF_5$ and 3.30 m $B(OSO_2CF_3)_3$).

## 4.1.1.2 In Chloride Melts

By comparison with the range of experimental observations in protonic solvents at or near room temperature, studies of homopolyatomic cations of iodine in chloride melts at the necessarily elevated temperatures are few. However, the data reported for these media support the trends of polyatomic cation stability shown for superacids as developed above.

Corbett and co-workers[13] conducted studies of phase equilibria and used nuclear quadrupole resonance spectroscopy in systems that they described as "neat" (i.e., stoichiometric) mixtures of $I_2$, $ICl$, and $AlCl_3$. As shown in chapter 1, these are best regarded as neutral chloride melts. They contain neither excess $AlCl_3$ nor excess $Cl^-$. It is highly significant that the cations observed when the ratios of $I_2$:$ICl$:$AlCl_3$ were 1:1:1 and 2:1:1 were $I_3^+$ and $I_5^+$ in the isolated compounds $I_3^+AlCl_4^-$ and $I_5^+AlCl_4^-$ and that in *these neutral media* there was no evidence for formation of $I_2^+$; that is, acidic melts would be required to provide stable compounds of $I_2^+$.

Much of the investigation of iodine systems in room-temperature chloroaluminates, that is, mixtures of $AlCl_3$ and butylpyridinium chloride (BuPyCl), seems to have been restricted to interaction of iodine in oxidation states $+I$, 0, and $-I$ with the anions of the media for acidic, neutral, and basic chloroaluminates, as exemplfied by the spectrophotometric study of iodine complexes in $AlCl_3$—BuPyCl mixtures.[14] Electrochemical investigations in room-temperature chloroaluminates[15,16] have been interpreted in terms of generation of these three oxidation states rather than of the fractional oxidation states exhibited in the cations $I_n^+$.

Interpretation of Mamantov's early work on iodine systems at Tennessee in "high-temperature" melts,[17] that is, NaCl—$AlCl_3$ at about 175°C, was restricted essentially to consideration of iodine in oxidation states $+I$ and $-I$. In fact, the species $I_3^+$ and $I_5^+$ were specifically eliminated from the discussion. Later, the Tennessee group wrote, in reporting a resonance Raman and UV-visible spectroscopic study of iodine oxidation in chloroaluminate melts,[18] that in acidic melts, "where the oxidation of iodine becomes less reversible as the melt acidity increases, the oxidation of iodine to I(I) is either a slow process or is more complex than originally thought." In this work, iodine was shown by UV-visible spectroscopy to be soluble in an acidic eutectic (63 mol% $AlCl_3$—37 mol% NaCl) and in neutral melts (50% $AlCl_3$—50% NaCl). When $I_2$ in each of these melts was oxidized anodically or by elemental $Cl_2$ there was spectroscopic evidence, as discussed below, for formation of $I_2^+$ in the acidic, but not in the neutral melts.

Since both oxidative paths generate $I_2^+$, it is unnecessary to regard $Cl_2$ as a specific chemical oxidant in these reactions. The generation of $I_2^+$ can be represented simply as

$$I_2 - 1e \rightarrow I_2^+ \qquad\qquad 4.3$$

With decrease in acidity of the medium, that is, with increase in concentration of the base of the system $Cl^-$, $I_2^+$ would disproportionate according to the equation

$$2I_2^+ + 2Cl^- \rightarrow 2ICl + I_2 \qquad\qquad 4.4$$

although it must be recognized that there could be immediate species such as $I_3^+$ and $I_2Cl^+$.

In their chemical oxidation of $I_2$ by $Cl_2$, Mamantov and co-workers had a difficult experimental system. They report incomplete dissolution of $I_2$ in molten chloroaluminates but do not comment on the solubility of $Cl_2$, which might be expected to be less than that of $I_2$ given the relative volatilities of the two similar diatomic molecular species. When they reacted a supposedly equimolar mixture of $I_2$ and $Cl_2$ in the 63:67 melt, the UV-visible spectrum showed a broad absorption in the region of 370–550 nm with a weaker band at ~690 nm. With bands at 640 nm and 626 nm reported for $I_2^+$ in $HSO_3F$[5] and in HF,[10] the 690-nm band can be assumed to be characteristic of $I_2^+$. Shifts to somewhat longer wavelengths are to be expected for inorganic species in moving from room-temperature protonic superacids to high-temperature chloroaluminate melts, as shown in chapter 6. With a supposed 2:1 ratio for $I_2$:$Cl_2$, the band at 690 nm occurred as well as an absorption characteristic of $I_2$, whereas there was no 690-nm band when a large excess (6.5:1) of chlorine to iodine was used.

Assuming a smaller solubility in the melt of $Cl_2$ than of $I_2$, these observations are all consistent with the formation of $ICl$ with a large excess of $Cl_2$ present, the formation of a mixture of $ICl$ and $I_2^+$ when the supposedly equimolar reaction mixture was used, and the formation of $I_2^+$ with residual $I_2$ when the reacting mixture $I_2:Cl_2$ was believed to be 2:1.

Of much greater importance in the context of this discussion was that the 690 nm was absent when $Cl_2$ and $I_2$ were reacted in the neutral (50:50) melt. The disproportionation reaction in equation 4.4 would lead to ultimate formation of $ICl$ under these conditions.

In a spectroelectrochemical cell a constant potential was applied by means of a platinum screen electrode to the acidic $AlCl_3$—$NaCl$ (63:37) melt containing iodine[18] and, over a period of 30 minutes, the band for $I_2$ was replaced by that for $I_2^+$, assigned on the basis of the resonance Raman spectroscopic identification of $I_2^+$ in $HSO_3F$.

Mamantov reported that extended anodic oxidation of $I_2$ beyond the formation of $I_2^+$ involved the formation of $ICl$. This is consistent with the fact that $I_2^+$ would not be observed even though the acidity of the medium was favorable to its formation. It would have been further oxidized from iodine in a formal oxidation state of 0.5 to iodine (I) in $ICl$. Further strong oxidation could have produced $ICl_3$. The acidity of the medium was no longer the determinant of the nature of the iodine-containing species.

## 4.1.1.3 As "Naked" Cations in Solids

In all of the solution studies of polyatomic cations of iodine, it can be shown that they must be generated in the absence of significant concentrations of base. Otherwise, varying degrees of disproportionation will occur, depending on the availability and strength of basic species. A convenient experimental approach to synthesis of these cations in media of very weak basicity has been to produce and study them in highly acidic solutions. However, there is a significant body of work in which the iodine cations—and other polyatomic cations of nonmetallic and metallic elements—are produced as solids, either directly or from a noninteracting solvent of very low basicity. In either case the counteranion must be very weakly basic.

Deep-blue solids, formulated as $I_2^+Sb_2F_{11}^-$ and $I_2^+TaF_{11}^-$ on the basis of analytical figures and comparison of spectra with that of $I_2^+$ in $HSO_3F$, were reported in 1968 by Kemmitt et al.[19] as a result of reaction of $I_2$ and $IF_5$ with $SbF_5$ and with $TaF_5$. They reported that, at that stage, "it had not proved possible to obtain single crystals suitable for X-ray structural measurements."

Somewhat later Gillespie and co-workers obtained dark-blue crystals of $I_2^+Sb_2F_{11}^-$, suitable for structure determination, by treating $I_2$ in liquid $SO_2$ with an approximate threefold excess of $SbF_5$.[20] The pentafluoride acted as an oxidant *and* as a Lewis acid *and* produced the very weakly basic anion

$Sb_2F_{11}^-$. Neither the reduction product $SbF_3$ nor the solvent $SO_2$ is sufficiently basic to cause disproportionation of the $I_2^+$ cation. The diatomic cation has a bond length (2.56 A°), shorter than that in $I_2$ (2.66 A°) and the frequency of the stretching vibration for $I_2^+$ (238 cm$^{-1}$) is greater than that for $I_2$ (213 cm$^{-1}$), consistent with the formation of a stronger bond as a result of removal of an electron from an antibonding $\pi^*$ orbital of the $I_2$ molecule in forming the $I_2^+$ cation.

Subsequently, Passmore and co-workers isolated and structurally characterized compounds containing the $I_3^+$ and $I_5^+$ cations under "naked" (i.e, nonbasic) conditions. $I_2$ and $AsF_5$ were reacted in stoichiometric proportions in $SO_2$ solution to give crystalline $I_3^+ AsF_6^-$.[21] Neither the solvent $SO_2$ nor $AsF_3$, the product of reduction of the oxidant $AsF_5$, is sufficiently basic to cause disproportionation of the cation $I_3^+$. For similar reasons, they were able to isolate $I_5^+ SbF_6^-$ from a stoichiometric mixture of $I_2$ and $SbF_5$ in liquid $AsF_3$.[22]

A 1980 paper by Aubke and co-workers[23] on improved synthetic procedures for preparing $[I_2^+][Sb_2F_{11}^-]$ and $[Br_2^+][Sb_3F_{16}^-]$ provides an elegant example of the dependence of the nature of the polyatomic cations of iodine on base strength of the medium. They report equilibrating a 2:1 mixture of $I_2$ (4.0349 mmol) and $S_2O_6F_2$ (2.054 mmol). In principle, this stoichiometry could have produced $[I_2^+][SO_3F^-]$. Instead, it gave an equimolar mixture of $ISO_3F$ and $I_3^+ SO_3F^-$. They then added a vast excess of $SbF_5$ (31.256 mmol), heated the mixture to 50°C, and removed all volatile materials. The residue (4.0706 mmol) was a dark blue-black solid which analyzed according to the formula $[I_2^+][Sb_2F_{11}^-]$. They report this second reaction as resulting from solvolysis of $SO_3F^-$ to $Sb_2F_{11}^-$. While this reaction could occur, it does not appear to be the main driving force. It seems better to view the potential $I_2^+$ from the first reaction as disproportionating to $I_3^+$ and $ISO_3F$ in the presence of the base $SO_3F^-$, whereas the weaker base $Sb_2F_{11}^-$ allows the conproportionation reaction of $I_3^+$ and $ISO_3F$ to form $I_2^+$ in the second reaction. This can be represented as an overall reversible reaction:

$$2I_2^+ + X^- \underset{X^- = Sb_2F_{11}^-}{\overset{X^- = SO_3F^-}{\rightleftharpoons}} I_3^+ + IX \qquad 4.5$$

## 4.1.1.4 Disproportionation Equilibria for Iodine Cations

Reported observations on generation, stability, and disproportionation of polyatomic cations of iodine in a range of protonic solvents, in chloroaluminate melts, and in solids where the cations are generated as "naked" entities are all consistent with the pattern developed in the study that demonstrated the dependence of cation stability on the levels of acidity and basicity in anhydrous HF.[10] $I_2^+$ was shown to be the cation of highest charge-to-atom ratio in the system and to exist only in very highly acidic media. Base-induced disproportionation occurred in HF solutions made progressively less acidic and led to the formation, in turn, of $I_3^+$, $I_5^+$, and of the final disproportionation products $I_2$ and $IF_5$.

That these disproportionation reactions are acidity-dependent equilibria was shown by allowing the final disproportionation products $I_2$ and $IF_5$ to mutually oxidize and reduce, that is, to conproportionate, in HF media of controlled acidity. The stepwise disproportionations can be represented as

$$14I_2^+ + 5F^- \rightleftharpoons 9I_3^+ + IF_5 \qquad\qquad 4.6a$$
$$12I_3^+ + 5F^- \rightleftharpoons 7I_5^+ + IF_5 \qquad\qquad 4.6b$$
$$5I_5^+ + 5F^- \rightleftharpoons 12I_2 + IF_5 \qquad\qquad 4.6c$$

These lead to an overall reaction:

$$10I_2^+ + 10F^- \rightleftharpoons 9I_2 + 2IF_5 \qquad\qquad 4.6d$$

Small increases in the amount of base available cause disproportionation of $I_2^+$ to $I_3^+$ and to $I_5^+$. When the *available* $F^-$ is comparable with or somewhat greater than the concentration of $I_2^+$, total disproportionation will occur to elemental iodine and an essentially covalent compound of iodine in a higher oxidation state than in the parent cation.

The important generalization that emerges is that, providing an extremely strong oxidant, for example, $F_2$, is not used, it is the level of acidity or basicity of the medium which determines whether iodine cations can be formed and if so, what the nature of the cationic speciation will be. An oxidant which will fit within this framework is one which is comparable in oxidant strength with the "high oxidation state" disproportionation product. If, for experimental convenience, a strong oxidant is to be used, the only further constraint on acid-base predetermination of speciation is that the oxidant must not be in excess. If it is in excess, redox reaction stoichiometry can and probably will have a determining effect on speciation. Gillespie's work on generation of iodine cations in $HSO_3F$ using the strong oxidant $S_2O_6F_2$ demonstrates this. Whether he generated $I_5^+$, $I_3^+$, or the marginally stable $I_2^+$, the solvent was sufficiently acidic to prevent disproportionation—for $I_5^+$ and $I_3^+$ in particular—and the reaction ratio $I_2:S_2O_6F_2$ determined speciation. Use of a great excess of a very strong oxidant or application of high anodic potentials will oxidize $I_2$ through the range of oxidation states exhibited by the cations to compounds corresponding with $IF_3$, $IF_5$, and even $IF_7$, regardless of the acidity levels in the medium.

The magnitude of the oxidation states of the higher disproportionation products will affect in a minor way the nature of the equations written for disproportionation—conproportionation reactions for acidic media other than HF. There seems to be general acceptance that $I(HSO_4)_3$, $I(SO_3F)_3$, and $I(SO_3CF_3)_3$ are the common stable "higher"-oxidation-state compounds of iodine in the $H_2SO_4$, $HSO_3F$, and $CF_3SO_3H$ solvent systems. On this basis idealized equations could be written

$$8I_2^+ + 3X^- \rightleftharpoons 5I_3^+ + IX_3 \qquad\qquad 4.7a$$
$$7I_3^+ + 3X^- \rightleftharpoons 4I_5^+ + IX_3 \qquad\qquad 4.7b$$
$$3I_5^+ + 3X^- \rightleftharpoons 7I_2 + IX_3 \qquad\qquad 4.7c$$

where $X^- = HSO_4^-$, $SO_3F^-$ or $CF_3SO_3^-$

In Aubke's method for synthesizing $I_2^{+}$[23] the disproportionation equilibrium is written as in equation 4.5 because the oxidant stoichiometry dictated the formation of $ISO_3F$ rather than $I(SO_3F)_3$ in the first step. For disproportionation of $I_2^+$ in chloroaluminates, ICl is the expected, and observed, chloride of iodine found under conditions where there is not strong oxidation. Equation 4.4 then represents the disproportionation reaction for $I_2^+$ in the presence of a significant amount of the base $Cl^-$.

In this section, disproportionation of iodine cations has been presented in terms of availability of base, usually of the anion of the solvent system, although this is not a necessary criteria. Any base of suitable strength that interacts with an iodine cation will cause disproportionation. The ultimate case is that in which water, often adventitious moisture in a synthetic procedure, will cause the desired product to disproportionate. The obverse of this proposition is that as the acidity of a medium, such as a protonic superacid or an acidic melt, increases cations of progressively higher formal charge per iodine atom can be stabilized.

## 4.1.2 Generation of Bromine Cations

Very much less study has been carried out on generation and stability of polyatomic cations of bromine than is the case for the formally corresponding iodine systems, and the bulk of the work has been related to isolation of solid compounds rather than to study of bromine cations in solution.

### 4.1.2.1 In Solid Compounds

The first reliable report of the formation of $Br_3^+$ as such was as postulated by McRae on the basis of a tensiometric study of the reaction of $Br_2$ and $SbF_5$.[24] She observed a rise in vapor pressure of bromine when the $Br_2$ content of the mixture exceeded 30 mol%. The brown color of this system ($\lambda_{max} = 375$ nm, later shown to be characteristic of $Br_3^+$) migrated to the cathode on electrolysis. No single compound was isolated from the reaction mixture because the desired product was contaminated with $SbF_3$; $SbF_5$ is both the oxidant and the source of the very weakly basic anion according to the following equation:

$$3Br_2 + 7SbF_5 \rightarrow 2Br_3^+ Sb_3F_{16}^- + SbF_3 \qquad\qquad 4.8$$

Subsequently, and again significantly in the terms of the synthetic principles being put forward in this book, Edwards reacted $Br_2$ and $SbF_5$ in the *oxidiz-*

*ing* solvent $BrF_5$ and generated $Br_2^+ Sb_3F_{16}^-$ under oxidizing, weakly basic conditions.[25]

$$9Br_2 + 2BrF_5 + 30SbF_5 \rightarrow 10Br_2^+ Sb_3F_{16}^- \qquad 4.9$$

His crystal structure showed a bond length of 2.15 A° in $Br_2^+$ compared with 2.27 A° in $Br_2$, a similar difference to that observed for $I_2^+$ and $I_2$.

In the same paper[23] in which they reported improved synthetic procedures for the isolation of $[I_2^+][Sb_2F_{11}^-]$, Aubke and co-workers described the formation of $[Br_2^+][Sb_3F_{16}^-]$. They took the 2:1 reaction mixture of $Br_2$ (6.148 mmol) and $S_2O_6F_2$ (3.102 mmol) and observed $Br_2$ and $BrSO_3F$ as products, whereas simple stoichiometry considerations would predict the formation of $[Br_2^+][SO_3F^-]$. Heating the mixture of $Br_2$ and $BrSO_3F$ with excess $SbF_5$ (48.055 mmol) resulted in a conproportionation reaction in the presence of the very weak base $SbF_5$ and the derived anion $Sb_3F_{16}^-$, which is even more weakly basic than $Sb_2F_{11}^-$. By analogy with equation 4.5, these reactions can be summarized as

$$Br_2 + 2BrX \underset{X^- = SO_3F^-}{\overset{X^- = Sb_3F_{16}^-}{\rightleftharpoons}} 2Br_2^+ + 2X^- \qquad 4.10$$

In other work directed primarily toward the demonstration that $Au(SO_3F)_3$[26] and $Pt(SO_3F)_4$[27] are strong Lewis acids of the $HSO_3F$ solvent system, Aubke and Lee provided further information on the generation and stability of $Br_3^+$ and $Br_5^+$. In a solution of $Au(SO_3F)_3$ in $HSO_3F$ they conproportionated $BrSO_3F$ and $Br_2$ to give $Br_3^+$ in the compound $[Br_3^+][Au(SO_3F)_4^-]$ and then reacted this compound with equimolar $Br_2$ to produce $[Br_5^+][Au(SO_3F)_4^-]$. They found that when metallic Pt reacted with excess $BrSO_3F$ to produce $Pt(SO_3F)_4$ in situ and $Br_2$ was added to the reaction mixture, a compound $(Br_3^+)_2[Pt(SO_3F)_6^{2-}]$, of limited thermal stability, could be isolated—a result of the type of conproportionation reaction between $BrSO_3F$ and $Br_2$ observed in the $Au(SO_3F)_3$—$HSO_3F$ system.

### 4.1.2.2 In Superacids

Gillespie and Morton have done much of the experimental work and have reviewed the experimental data for the existence of $Br_3^+$ and $Br_2^+$ in superacids.[28] $Br_3^+$, unstable in 100% $H_2SO_4$, is not completely stable in basic $HSO_3F$; that is, $HSO_3F$ containing $SO_3F^-$ from reduction of the oxidant $S_2O_6F_2$. $Br_3^+$ can be represented as disproportionating according to the equation

$$Br_3^+ + SO_3F^- \rightarrow Br_2 + BrSO_3F \qquad 4.11$$

In $SbF_5.3SO_3$—$HSO_3F$ the appropriate amount of $S_2O_6F_2$ oxidizes $Br_2$ to brown $Br_3^+$, which is quite stable. Further oxidation with $S_2O_6F_2$ gives a cherry-red solution for which there is the very sensitive resonance Raman spectroscopic evidence for $Br_2^+$. There is also some spectroscopic evidence[29] for a complex disproportionation of $Br_2^+$ into $Br_3^+$, $BrSO_3F$, and $Br(SO_3F)_3$. This disproportionation was rationalized by showing that dissolution of $BrSO_3F$ in $SbF_5.3SO_3$—$HSO_3F$, itself an oxidizing solution, gives a solution providing UV-visible spectroscopic evidence for $Br_2^+$ and $Br_3^+$ and resonance Raman spectroscopic evidence for $Br_2^+$. This solution also contains $Br(SO_3F)_3$.

## 4.1.3 Generation of Chlorine Cations

The progression from iodine, which has a rich cationic chemistry in superacids, through bromine is complete at chlorine, for which there does not appear to be any reliable evidence for the existence of any homopolyatomic cation, as such, in solution. Gillespie and Morton[30] have summarized their own experimental observations and those of others on work directed toward showing the existence of polychlorine cations in solution. Conductivity and Raman spectra of solutions of $ClOSO_2F$ and $Cl_2$ in $SbF_5.3SO_3$—$HSO_3F$ gave no evidence for the existence of polychlorine cations in solution at 25°C, whereas Aubke has shown that the analogous reaction between $BrOSO_2F$ and $Br_2$ produces $Br_2^+$ under highly acidic conditions at 25°C,[23] and progressive oxidation of $Br_2$ by $S_2O_6F_2$ in this same medium, $SbF_5.3SO_3$—$HSO_3F$, gave $Br_3^+$ and $Br_2^+$ in turn.[29] Gillespie and Morton,[30] while discussing the formation of $Cl_3^+$ in unstable solids, report that there is no evidence for the existence of $Cl_2^+$ even in solids, whereas $Br_2^+$ and $I_2^+$ can be generated relatively easily as stable crystalline compounds.

Gillespie and Morton[30] showed that reaction of $ClF$ and $Cl_2$ in $SbF_5$—$HF$ gave a yellow solid at $-76°C$ identified from its Raman spectrum (see later) as containing $Cl_3^+$. This solid, on warming to room temperature, is reported to have given $Cl_2$ and salts containing $ClF_2^+$. In the presence of the strong Lewis acid $SbF_5$, which readily accepts $F^-$ from $ClF_3$, this represents an overall disproportionation:

$$3Cl_3^+ + 3F^- \rightarrow 4Cl_2 + ClF_3 \qquad\qquad 4.12$$

The $ClF_3$ then undergoes fluoride transfer to $SbF_5$ to form $ClF_2^+ SbF_6^-$. The stepwise reaction probably involves a preliminary dissociation of $Cl_3^+$ into $Cl_2$ and $Cl^+$ followed by disproportionation of $Cl^+$ which, in the presence of the base $F^-$ from the solvent, would be far too unstable to exist:

$$3Cl^+ + 3F^- \rightarrow Cl_2 + ClF_3 \qquad\qquad 4.13$$

The cation $Cl_3^+$ was characterized by the Raman spectrum of the solid isolated after direct reaction at $-76°C$ of a mixture of $Cl_2$, $ClF$, and $AsF_5$, no HF being used. In addition to the Raman-active bands for $AsF_6^-$ there were three relatively intense bands at 490, 255, and 508 cm$^{-1}$, assigned to $v_1$, $v_2$, and $v_3$ for the bent $Cl_3^+$ cation by comparison with the isoelectronic molecule

$SCl_2$, which has a bond angle of 103° and vibrational frequencies of 514, 208, and 525 $cm^{-1}$.

The solid $Cl_3^+ AsF_6^-$, stable at $-76°C$, decomposed to $Cl_2$, ClF, and $AsF_5$ at room temperature. ClF is a thermally stable entity in the gas phase and so the complex disproportionations proposed earlier for reaction in HF solution would not occur. When $Cl_2$, ClF, and the weaker Lewis acid $BF_3$ were mixed at temperatures down to $-130°C$ there was no evidence of adduct formation.

## 4.1.4 Stabilities of Group VII Cations

It is obvious from the preceding sections of this chapter that the ease of formation of compounds containing $I_2^+$ and $I_3^+$ and the stabilities of these cations in superacidic solutions are far greater than for the corresponding $Br_2^+$ and $Br_3^+$. For chlorine the only cation prepared to date is $Cl_3^+$, and that in a solid compound but not in solution.

For each element, there is a greater attenuation of the single positive charge over $X_3^+$ than over $X_2^+$. $X_3^+$ is less electrophilic than $X_2^+$; that is, $X_3^+$ can exist with a greater availability of base, either in solid compounds or in solution than can $X_2^+$. Within group VII, the species $X_n^+$, for a particular value of n, is more electrophilic for X = Cl than for X = Br and I, in turn. That is, $Cl_n^+$ reacts more readily than $Br_n^+$ and $I_n^+$ with bases such as $HSO_4^-$, $SO_3F^-$, and $F^-$ (or with the base of any other medium, such as that of a molten salt).

These trends are summarized in Table 4.2, from which it can be seen that stabilization of $I_3^+$ requires a medium for which $H_0$ can be $-10$ to $-12$. $Br_3^+$ requires much higher acidity, a solution with $H_0$ more negative than about $-14$, and $Cl_3^+$ can be prepared only at low temperature in the solid state, that is, with no basic species available from a solvent. The logical extension of this is that $I_2^+$, more electrophilic than $I_3^+$, requires a medium with $H_0 \approx -15$ for stability. $Br_2^+$ is only marginally stable at $H_0 \approx -19$, and $Cl_2^+$ has not been formed.

## 4.1.5 Factors Governing Synthesis and Stability of Polyatomic Halogen Cations

There is general agreement in the literature of the inorganic chemistry dealing with halogen cations and superacidity that the formation of a particular cation $X_n^+$ will depend on (1) the ratio of oxidant to halogen (or other compound oxidized) and (2) the degree of acidity of the medium—more significantly, on the extent of availability of basic species. In this book it is postulated that the level of acidity of the medium is far more important in determining cation stability than the oxidant-reductant stoichiometry, providing that the oxidant is not too strong. If the oxidant is the ultimate base-

**Table 4.2.** Homopolyatomic Cations of Halogens

| General Formula | | | $X_2^+$ | $X_3^+$ |
|---|---|---|---|---|
| Oxidation State | | | 0.5 | 0.33 |
| *Iodine* | Solids: | (i) | $I_2^+ Sb_2F_{11}^-$ from stoichiometric $I_2$ + $SbF_5$ in $SO_2$ | $I_3^+ AsF_6^-$ from stoichiometric $I_2$ + $AsF_5$ in $SO_2$[a] |
| | | (ii) | $I_2^+ Sb_2F_{11}^-$ from $2I_2$ + $S_2O_6F_2$ in excess of $SbF_5$ | |
| | Solutions: | (i) | Stable in 60% oleum ($H_o \approx -14.8$)[b] | Stable in $H_2SO_4/HSO_4^-$ ($H_o \approx -11.9$) |
| | | (ii) | Marginally stable in $HSO_3F/SO_3F^-$ ($H_o \approx -13.8$) | Stable in $HSO_3F/SO_3F^-$ ($H_o \approx -13.8$) |
| | | (iii) | Stable in acidic HF ($H_o \approx -16$ to $-22$) | Stable in $HF$—$F^-$ ($H_o \approx -11$ to $-16$)[c] |
| | | (iv) | Stable in acidic melt (63% $AlCl_3$—37% NaCl) | Stable in neutral $AlCl_3$ melt[d] (50% $AlCl_3$—50% NaCl) |
| *Bromine* | Solids: | (i) | $Br_2^+ Sb_3F_{16}^-$ from $Br_2$ + $BrF_5$ + $SbF_5$ | $Br_3^+ Sb_3F_{16}^-$ from $Br_2$ + $SbF_5$ |
| | | (ii) | $Br_3^+ Sb_3F_{16}^-$ from $2Br_2$ + $S_2O_6F_2$ in excess $SbF_5$ | $Br_3^+ Au(SO_3F)_4^-$ from $Br_2$ + $BrSO_3F$ + $Au(SO_3F)_3$ in $HSO_3F$[e] |
| | Solutions: | | Marginally stable in $SbF_5 \cdot 3SO_3$—$HSO_3F$ ($H_o \approx -19$) | Marginally stable in $HSO_3F$—$SO_3F^-$ ($H_o \approx -13.8$) |
| | | | | Stable in $SbF_5 \cdot 3SO_3/HSO_3F$ ($H_o \approx -19$) |
| *Chlorine* | Solids: | | Not isolated | $Cl_3^+ AsF_6^-$ from $Cl_2$ + $ClF$ + $AsF_5$ at $-78°C$ |
| | Solutions: | | No evidence | No evidence |

[a]Solid $I_2^+SbF_6^-$ was isolated from stoichiometric reacting proportions of $I_2$ and $SbF_5$ in liquid $AsF_3$.

[b]Numbers in parenthesis are values of Hammett Acidity Functions for the different media, adjusted to experimental conditions, that is, containing appropriate concentrations of Lewis acids or bases or deemed to be about 0.05 M in base, where a base is produced in the synthesis, for example, $F^-$ from $F_2$ in HF, $SO_3F^-$ from $S_2O_6F_2$ in $HSO_3F$ or $HSO_4^-$ in $H_2SO_4$. The values have been calculated from Refs. 8, 9, and 11.

[c]Spectroscopic evidence for some $I_3^+$ as well as $I_2^+$ in this basic HF medium at values of $H_0$ nearer to $-11$.

[d]$I_3AlCl_4$ was crystallized from an equimolar mixture of $I_2$, ICl, and $AlCl_3$, that is, a neutral melt. An appropriate stoichiometry yielded $I_5AlCl_4$ under neutral conditions. $I_2^+$ is not stable in neutral $AlCl_3$ melts.

[e]$Br_3^+ Au(SO_3F)_4^-$ was isolated from reaction of excess $Br_2$ with $Br_3^+ Au(SO_3F)_4^-$ at $70°C$.

induced highest-oxidation-state disproportionation product of the cation, or is comparable in oxidant strength with that disproportionation product, the level of acidity of the medium—that is, the level of availability of base—is the only determinant of the nature of the stable cation, or cations in an equilibrium mixture.

Neither the chemical nature nor the temperature domain of the system has a significant effect on this generalization, which can be shown to hold for protonic superacids over a range of temperatures around ambient and for high-temperature and room-temperature melts—chloroaluminates in the case of this presentation.

For different protonic superacids there will be differences in the level of acidity, as measured by Hammett Acidity Functions, at which a particular halogen (or other nonmetal) cation is stabilized; however, this simply reflects differences in bond strength between the halogen of the cation and the base of the solvent, that is, differences in relative electrophilicity and nucleophilicity for that cation and the bases of the different solvents. This point will be demonstrated in detail in the discussion associated with Table 4.3 in section 4.2.4.

If a polyatomic cation of a particular element is stable above a certain level of acidity and the basicity of the system is then increased progressively by adding the anion which is the base of the solvent system, or by adding any compound which will increase the concentration of that base, the cation will disproportionate into a cation of lower formal charge per atom of the element and a compound, essentially covalent, formed between one or more of the basic anionic entities and an atom of the element concerned, that atom being in a higher formal positive oxidation state than in the original cation. Ultimately there will be disproportionation to the element itself and an essentially covalent compound formed between the element and the base, depending on the nature of the element itself and the degree of acidity of the medium, that is, depending on the availability and strength of the basic species in the medium. In sufficiently basic media, for example, in the hydrolysis of iodine cations, the higher-oxidation-state compound formed in addition to $I_2$ may be anionic. The obverse is to state that as the concentrations or the strengths of bases available to a halogen under mildly oxidizing conditions are progressively decreased, polyatomic halogen cations of increasing ratio of charge to halogen atom will be formed.

These factors determining halogen cation formation and stabilization have been presented in detail in section 4.1.1.4 for iodine cations for which most experimental information is available. Some examples are drawn from the cationic chemistry of all the halogens in this section.

Aubke[23] demonstrated very nicely that when 2 mmol of $I_2$ were oxidized by 1 mmol of $S_2O_6F_2$, the species generated *in the presence of $SO_3F^-$*, the reduction product of the oxidant, were an equimolar mixture of $ISO_3F$ and $I_3^+ SO_3F^-$. In the corresponding reaction with $Br_2$, which is less easy to oxidize than $I_2$, 2 mol of $BrSO_3F$ were formed and 1 mol of $Br_2$ remained unreacted. When each of the reaction mixtures was treated with a large excess of $SbF_5$ and all volatile products were removed, the residual products were $[I_2^+][Sb_2F_{11}^-]$ and $[Br_2^+][Sb_3F_{16}^-]$. These overall reactions have been represented above by equations 4.5 and 4.10. It appears that $I_3^+$ is the cation of highest charge that can be is formed in the presence of the base $SO_3F^-$ and is formed with the essentially covalent compound $ISO_3F$, but in the weakly basic medium $SbF_5$, these two species conproportionate to form $I_2^+$ which is isolated with the counteranion $Sb_2F_{11}^-$, which is less basic than $SO_3F^-$. A similar rationalization of the formation of $Br_2^+$ (equation 4.10) has the added strength that the anion required to isolate $Br_2^+$, which is more difficult to stabilize than $I_2^+$, is $Sb_3F_{16}^-$, which is less basic than $Sb_2F_{11}^-$.

A similar example involves the properties of compounds containing the cation $Cl_3^+$ as characterized by Gillespie and Morton.[30] The solid $Cl_3^+AsF_6^-$ isolated at $-76°C$ dissociates to $Cl_2$, $ClF$, and $AsF_5$ as it warms to room temperature. However, $Cl_3^+SbF_6^-$ in $SbF_5$—HF gives $Cl_2$ and $ClF_2^+$ adducts as products,[30] implying formation of $ClF_3$. Earlier, in section 4.1.3.1, it has been proposed that $Cl_3^+$ *in HF solution* would dissociate initially into $Cl_2$ and $Cl^+$ and that the highly unstable $Cl^+$ would then interact with the minute amount of $F^-$ in $SbF_5$—HF solution to disproportionate to $Cl_2$ and $ClF_3$. The $ClF_3$ would then undergo fluoride exchange with the Lewis acid $SbF_5$ to form $ClF_2^+$.

In section 4.1.1.2, it was stated that Mamantov and co-workers[18] reported UV-visible spectroscopic evidence for molecular iodine when iodine was added to $AlCl_3$—NaCl melts. When these solutions were oxidized anodically or by elemental chlorine in $AlCl_3$-rich melts, there was definitive resonance Raman spectroscopic evidence for $I_2^+$, but in chloride-rich melts the Raman signal for $I_2^+$ was not observed. This suggests that regardless of whether the oxidation is anodic or chemical, $I_2^+$ is formed in $AlCl_3$-rich melts, where the chloride formed in chemical oxidation-reduction would be converted predominantly into chloroaluminate anions. However, as the ratio of $Cl^-$:$AlCl_3$ in the melt is increased to and beyond the point of neutrality, disproportionation occurs by reaction with the base chloride which is now available, as in equation 4.4.

The basis of the electrochemical experiment appears to be perfectly straightforward. In an acidic melt, $I_2$ would be oxidized to $I_2^+$, according to equation 4.3, when controlled potential oxidation is used as in the Mamantov experiment in the spectroelectrochemical cell. More drastic electrochemical oxidation would generate $ICl$, as reported, even with the small equilibrium concentration of $Cl^-$ available in the medium.

$$I_2^+ + 2Cl^- - 1\varepsilon \rightarrow 2ICl \qquad\qquad\qquad 4.14$$

The most interesting of Mamantov's chemical oxidation experiments, from an acid-base point of view, is that in which $I_2$ and $Cl_2$ were added in the proportions 2:1 to the acidic melt contained in a spectral cell. This reactant stoichiometry should have produced $I_2^+$ and $Cl^-$ in equimolar amounts:

$$2I_2 + Cl_2 \rightarrow 2I_2^+ + 2Cl^- \text{ (as } 2AlCl_4^-) \qquad\qquad 4.15$$

However, in addition to the expected peak in the UV-visible spectrum characteristic of $I_2^+$, there was a peak for $I_2$, suggesting that there was a deficiency of $Cl_2$ available as oxidant in the melt, as described in section 4.1.1.2. With limited $Cl_2$ available, some $I_2$ remained unreacted. Only $I_2^+$, the species dependent on the acidity of the medium, was formed. It did not react with residual $I_2$ to form $I_3^+$ or $I_5^+$. In the neutral melt, no $I_2^+$ was detected spectroscopically, supporting the proposition summarized by equation 4.4 that $I_2^+$, if formed, would disproportionate in the presence of a significant concentration of chloride.

The schemes proposed above for redox reactions of $I_2$ and $Cl_2$ in chloroaluminates appear to be more consistent with the generalized scheme proposed for the interaction of cationic species and bases in protonic solvents than the suggestion by the authors that the entity undergoing disproportionation is ICl, which would require the unlikely formation of $ICl_3$ in order to generate $I_2^+$. Their other proposal that $Cl_2$ oxidizes $I_2$ to form ICl, which then reacts with more $I_2$ to form $I_2^+$ and $Cl^-$, is really a conproportionation reaction which is not favored in very low $Cl^-$ concentrations, and the proposal is not consistent with the electrolytic production of $I_2^+$ in $AlCl_3$-rich (i.e., chloride-deficient) melts.

Some apparent anomalies in the generalized schemes for stabilization of unusual cations presented in this section are easily explained. For example a cation of low formal oxidation state can be generated in a medium of an acidity which is capable of maintaining a higher formal oxidation state for the element concerned, providing a limited amount of a very strong oxidant is used. Thus Gillespie[5] generated stable $I_5^+$ and $I_3^+$ in $HSO_3F$ in which $I_2^+$ is also marginally stable by limiting the amount of the oxidant $S_2O_6F_2$.

Conversely, many of the investigations in molten salts involving nonmetal cations have involved electrolytic oxidation, and obviously such a technique can "override" oxidation states, which are marginally stable in the medium, under the forcing condition of electrolysis—particularly under voltammetric conditions where the applied potential is being changed continuously and where the absolute amount of product being produced by oxidation (or reduction) and sensed in the region of the working electrode is very small. Under these conditions, a cation could be produced which has a higher formal charge than that expected for the bulk acidity of the medium.

An example of the way in which a very strong chemical oxidant can "override" a particular oxidation state being sought occurs in Edwards' original synthesis and structural characterization of $Br_2^+ Sb_3F_{16}^-$.[25] It was found that the ratio of 9:2 for $Br_2$:$BrF_5$ required for the overall stoichiometry had to be adhered to very closely. A slight deficiency of $BrF_5$ led to some formation of $Br_3^+$; but, more importantly in this context, when excess of $BrF_5$ was used, $BrF_2^+$ was observed in the reaction products; that is, oxidation through to Br(III) occurred, with $BrF_3$ than acting as a fluoride donor to $SbF_5$. It is interesting to note that excess $IF_5$, a weaker oxidant than $BrF_5$, did not cause further oxidation of $I_2^+$ in the system $HF$—$I_2$—excess $IF_5$.[10]

It must be stressed that the conditions put forward to relate charge to acidity or basicity apply to *cationic* species. Thus a bromine(III) compound $Br(SO_3F)_3$, which is essentially covalent, is stable under superacidic conditions where $Br_3^+$ (formal oxidation state 0.33) and $Br_2^+$ (0.5) are in equilibrium.[29] Indeed, the whole postulation about disproportionation of cationic species involves the formation of nonionic species containing the element concerned in a higher formal oxidation state than that of the cation from which it was derived.

A point developed in chapter 9 of this book is that in many systems the necessary *concomitant* and *complementary* roles of oxidation and of acidity have been overlooked in discussion of the formation of unusual cationic species in a wide range of solvents and reaction media. It is true that the oxidant $S_2O_6F_2$ has been used deliberately and imaginatively in the solvent $HSO_3F$, but frequently there has been too little recognition of the strongly oxidizing properties of $SO_3$ in oleums or of $SbF_5$ in HF or in $HSO_3F$ when these have been added as Lewis acids and they have then been recognized only as increasing the acidity of the medium. Their roles not only in acting as oxidants but, frequently also, in providing counteranions of very low basicity, such as $Sb_2F_{11}^-$ and $Sb_3F_{16}^-$, have not been spelled out explicitly in many accounts of synthesis of potentially unstable cations.

## 4.2 Homopolyatomic Cations of Group VI Elements

Historically iodine, the heaviest member of group VII, was the first to be subjected to systematic investigation for formation of cationic species. In the case of group VI, there has been more extensive study of sulfur cations than of those of selenium and tellurium and so the sulfur systems will be presented in detail before those of the congeners, even though a reasonably large body of research has shown that it is easier to prepare and stabilize cations of Te and Se than of S, particularly in melts. In group VII stability of cations was shown to decrease in passing from iodine to bromine to chlorine. Not surprisingly, the same trend holds in group VI—the attainability of a higher ratio of charge-to-element in polyatomic cations becomes greater with increase in atomic number of the element.

The chemistry of group VI elements differs from that of group VII in that the first member, oxygen, can form a cationic species—namely, $O_2^+$. It was characterized definitively in a solid compound by the oxidation of $O_2$ by $PtF_6$, the resultant cation being stabilized in the compound $O_2^+ PtF_6^-$ in association with the weakly basic $PtF_6^-$.[31] Subsequently it has been isolated in solids with many weakly basic anions in compounds such as $O_2^+ Sb_2F_{11}^-$, $O_2^+ SbF_6^-$, and $O_2^+ AsF_6^-$.[1,2] As this is the only oxygen cation characterized to date and is always isolated *as a solid* in association with a weakly basic anion, and it has not been shown to exist in acidic solutions, it will not be discussed further. It should be noted, however, that even the small amount of information that is available fits the general pattern for relation of stability of cations to the atomic number of an element in a group of the periodic classification. Thus in group VII, the first member forms no cations and the second member has produced the single cation $Cl_3^+$ only in a solid compound. It can be seen by comparing Tables 4.2 and 4.3 that cations with "high" charge-to-atom ratios can be formed more easily—at lower acidities—for group VI elements than for corresponding members of group VII.

## 4.2.1 Sulfur Cations

As stated in the introduction to this chapter, it was observed early in the 19th century that dissolution of elemental sulfur in oleums gave highly colored solutions, the color depending on the strength of the oleum. Similar observations were recorded later as selenium and tellurium were dissolved in oleums.[1-3] The nature of the colored sulfur species began to be resolved when Gillespie and co-workers[32] reacted $S_8$ with either $AsF_5$ or $SbF_5$ in the reacting proportions 2:3 in anhydrous HF and reported the products $S_{16}(AsF_6)_2$ and $AsF_3$ or $S_{16}(SbF_6)_2$ and $SbF_3$. When the reacting proportions in HF were 1:3, the products were $S_8(AsF_6)_2$ and $S_8(SbF_6)_2$. $S_8$ and $SbF_5$ in the reacting proportions 1:5 in HF produced $S_8(Sb_2F_{11})_2$ and $SbF_3$. They found that by reacting $S_8$ with excess $SbF_5$ either *directly* or in $SO_2$ they produced $S_4^{2+}$ and ascribed this to the greater oxidant strength of $SbF_5$ than $AsF_5$. It seems more likely that with a suitable excess of oxidant present, the lower basicity of the medium containing excess $SbF_5$ allowed stabilization of the cation with higher charge per sulfur atom, when associated with the anion of low basicity $Sb_2F_{11}^-$. Preliminary identification of these solid products and characterization of the $S_8^{2+}$ cation by X-ray structural analysis provided the background for a detailed study within Gillespie's group of the preparation and study of the stability of cations of sulfur in superacidic media.

### 4.2.1.1 In Protonic Superacids

In Table 2.2 and the supporting text of chapter 2, it was shown how Gillespie's group had used $S_2O_6F_2$ as an oxidant in $HSO_3F$ and reported reaction patterns for the generation of sulfur cations based on cryoscopic observations and on monitoring by conductance methods the number of moles of $SO_3F^-$ produced in a reaction and thence the extent of oxidation of $S_8$, $SO_3F^-$ being both the reduction product of the oxidant $S_2O_6F_2$ and the highly conducting base of the solvent $HSO_3F$.

A 1:1 reaction mixture of $S_8$ and $S_2O_6F_2$ in $HSO_3F$ yielded values of 3 for $\nu$ from cryoscopy and 2 for $\gamma$ from conductance measurements, indicating the formation of 3 mol of product particles per mol of $S_8$, 2 of which were $SO_3F^-$ ions. This shows that the third particle had been produced by a two-electron oxidation of $S_8$ to $S_8^{2+}$ reported then as being blue in solution.[32a] They observed that $S_8^{2+}$ was marginally stable in $HSO_3F$ (made somewhat basic because of the $SO_3F^-$ from the oxidant) and that sulfur slowly precipitated as a result of disproportionation of $S_8^{2+}$ to $S_8$ and $SO_2$. Excess $S_2O_6F_2$ with $S_8$ in liquid $SO_2$ at low temperature gave solid $S_4^{2+}(SO_3F^-)_2$; that is, $S_4^{2+}$ was formed in the absence of excess base. However, in $HSO_3F$, with the base $SO_3F^-$ available, $S_4^{2+}$ gradually changed to $S_8^{2+}$. It is significant that $S_4^{2+}$ is stable in the less basic protonic medium $SbF_5$—$HSO_3F$. It does not disproportionate under these more acidic conditions.

In an early stage of their program on the synthesis of polyatomic cations of sulfur, Gillespie and co-workers reported that a 2:1 mixture of $S_8$ and $S_2O_6F_2$ in $HSO_3F$ produced the cation $S_{16}^{2+}$.[33] They based this on observed values of $v$ and $\gamma$, on analysis of a compound which they believed to be $S_{16}(AsF_6)_2$, and on the fact that no unreacted $S_8$, initially suspended in $HSO_3F$, was observed on gradual addition of $S_2O_6F_2$ when the ratio of the reactants $S_8$ and $S_2O_6F$ reached 2:1. They subsequently showed by structural analysis that the large cation in solution was probably not $S_{16}^{2+}$ but the unexpected $S_{19}^{2+}$. They followed the procedure they had adopted earlier when they had determined the structure of $S_8^{2+}(AsF_6^-)_2$ in crystals isolated from $SO_2$. They used the appropriate reacting proportions of $S_8$ and $AsF_5$ to obtain crystals supposedly containing $S_{16}^{2+}$ from the mixed solvent $SO_2$—$SO_2ClF$ at $-25°C$ and established crystallographically that the cation was in fact $S_{19}^{2+}$— two seven-membered rings of sulfur atoms joined by a chain of five sulfur atoms.[34]

It has become apparent over the last decade or so that the solution chemistry of sulfur cations in protonic superacids is very much more complex than that of the halogen cations. As stated above, the aspect of sulfur chemistry in strong acids which attracted attention nearly two centuries ago was the interaction of sulfur with acidic, oxidizing solvents to produce solutions exhibiting a range of intense colors. It now appears that these colors are not associated with the diamagnetic dispositive cationic species described earlier but with singly positively charged radical cations of the general formula $S_n^+$.

The existence of three such radical cations in various acidic media has been described in the general relevant literature, where they have been called $R_1$, $R_2$, and $R_3$. The natures of $R_2$ and $R_3$ remain unresolved at this stage, but $R_1$ has been identified by ESR spectroscopy as $S_5^+$ [35] and UV-visible bands have been assigned to it. One such band which occurs in a wide range of synthetic work involving oxidation of $S_8$ in superacids is located at 585 nm and frequently accounts for the blue color of such solutions. Burns et al. in their paper[34] on the characterization of the $S_{19}^{2+}$ cation have provided a very full account of the spectroscopic work on the radical cations and the implications for describing the spectroscopic properties of the dipositive cations of sulfur. For example, solutions containing $S_8^{2+}$ were believed for many years to owe their blue color to $S_8^{2+}$. Crystals such as

$S_8(AsF_6)_2$ were also blue, which lent credence to the reports that $S_8^{2+}$ was blue in solution. It is now believed that the solutions are blue because of small concentrations of $S_5^+$ in equilibrium with $S_8^{2+}$ and that $S_5^+$ is either doped into the crystals of $S_8^+$ or exists on the surface of these crystals. The red color of solutions containing the $S_{19}^{2+}$ cation and of the crystals derived from these solutions is also believed to be due to radical cation species.

Obviously, very complicated equilibria, which are dependent on the extent of oxidation of $S_8$ and which involve diamagnetic dipositive cations and paramagnetic singly charged radical cations, exist in superacidic media of differing acidity and basicity. $S_{19}^{2+}$ with its accompanying radicals, being more easily generated than, in turn $S_8^{2+}$ and $S_4^{2+}$ with their radicals, is able to tolerate somewhat higher levels of base than $S_8^{2+}$ and $S_4^{2+}$.

In their reviews[36] Gillespie and Passmore have delineated the limits of existence in the $H_2O$—$H_2SO_4$—$SO_3$ solvent system of the cations $S_4^{2+}$, $S_8^{2+}$, and the cation described by them at that stage as $S_{16}^{2+}$. It was shown above that later work from Gillespie's group indicated that the very large cation of small charge per S atom in solution was probably $S_{19}^{2+}$ and not $S_{16}^{2+}$. They say that $S_8$ dissolves slowly as such in 95–100% $H_2SO_4$ but that in 5% oleum, oxidation, presumably by $SO_3$, to "$S_{16}^{2+}$" is observed. In 10–15% oleum there is rapid oxidation to a mixture of "$S_{16}^{2+}$" and $S_8^{2+}$ which is then slowly oxidized to $SO_2$. The "$S_{16}^{2+}$" and $S_8^{2+}$ produced initially in 30% oleum are oxidized to $S_4^{2+}$ and finally to $SO_2$, whereas in 45% and 65% oleum $S_4^{2+}$, following initial generation of "$S_{16}^{2+}$" and $S_8^{2+}$, is rather stable and subsequent oxidation to $SO_2$ is very slow. This scheme seems to set the limits of acidity below which $S_8^{2+}$ and $S_4^{2+}$ will not be stable; but the system is very complex because $SO_2$ is the reduction product after $SO_3$ oxidizes to $S_8$ to a cation, or can be the end product of ultimate oxidation of $S_8$ right through to $SO_2$ or can be the higher-oxidation-state form of sulfur after disproportionation of cations in media which are not sufficiently acidic. Gillespie and Passmore[36] state the $S_8^{2+}$ disproportionates to "$S_{16}^{2+}$" and $SO_2$ in oleum containing less than 15% $SO_3$, and that $S_4^{2+}$ disproportionates to $S_8^{2+}$ and $SO_2$ below 40% $SO_3$. Even though $SO_3$ is a strong oxidant, $S_4^{2+}$ is more stable in 45 and 65% oleum than in 30% oleum. The higher acidity due to $SO_3$ prevents disproportionation to $SO_2$ and outweighs the oxidant effect of the trioxide.

The data in Table 4.3 suggest that in the oleum system $S_4^{2+}$ is stable for $H_0$ values more negative than $-14$, $S_8^{2+}$ at about $-13$, and "$S_{16}^{2+}$" for $H_0$ less negative than $-13$. The same order of stabilities with change in $H_0$ values is observed in the chemically simpler solvent system $HSO_3F$ in which the oxidant $S_2O_6F_2$ is reduced to the base of the solvent and in which there is a clearer differentiation between sulfur-containing oxidation products (cations) and products of disproportionation. Even though a smaller amount of the base $SO_3F^-$ will be available in the solvent $HSO_3F$ than of the base $HSO_4^-$ in the solvent $H_2SO_4$ because of the great difference in the self-ionization processes for the two solvents, the more electronegative $SO_3F^-$ would be expected to interact with cations more readily than the less electronegative $HSO_4^-$. On this basis, the absolute acidities in which different cations will be stabilized would be expected to differ from solvent to solvent.

This point has been emphasized by recent preliminary work carried out at Melbourne on generation and stability of polyatomic cations of sulfur in HF solutions of enhanced acidity.[37] The experimental approach was as in the study of stability of iodine cations in HF.[10] (See section 4.1.1.1.) $S_8$ in HF was oxidized by $F_2$ and a threshold was established at a Hammett function ($H_0$) of about $-18.5$ below which sulfur cations do not form; that is, any cations formed transiently would disproportionate to $S_8$ and $SF_4$. For $TaF_5$—HF and $AsF_5$—HF solutions with values of $H_0$ between $-18.7$ and $-20.1$, blue solutions containing red crystals were observed and the spectra recorded suggested that the radical cations $S_5^+$ and $R_2$ as well as $S_{19}^{2+}$ were the dominant species. For $H_0$ values between $-20.1$ and $-20.9$, $S_8^{2+}$ appears to be in equilibrium with $S_{19}^{2+}$ and with the radical cations $S_5^+$ and $R_2$. Spectra suggest that, in SbF—HF solutions more acidic than $H_0 = -22$, the major cationic species are $S_8^{2+}$ and $S_4^{2+}$. Study of sulfur cations in HF seems to be more difficult than in $HSO_3F$ and oleums. Even at very high acidities complex equilibria appear to exist in HF. Reasons will be suggested later in this chapter for the higher acidities required for cation stabilization in HF than in the other protonic superacids.

### 4.2.1.2 In Melts

There is very little definitive work on characterization of polyatomic cations of sulfur in melts and none on their isolation from melts. For example, the synthetic approach described in section 4.1.2 by which Corbett generated $I_3^+$ and $I_5^+$ in neutral melts, produced no compounds containing sulfur cations,[38] whereas compounds containing cations of selenium and tellurium can be synthesized in this way. These generalizations are consistent with the data in Tables 4.2 and 4.3 indicating that stabilization of sulfur cations in protonic superacids requires higher acidity levels than for iodine, selenium, or tellurium.

With their various co-workers, Bjerrum[38,39] and Mamantov[40] have applied spectroscopic and electrochemical techniques to try to identify the products obtained by chemical oxidation of $S_8$ with $Cl_2$ or by electrooxidation in the acidic eutectic melt 63% $AlCl_3$—37% NaCl. They describe very complex systems leading ultimately to S(II) and S(IV), probably as $SCl_2$ and $SCl_3^+$ in the acidic melts, but with formation of lower-oxidation-state cations such as $S_2^{2+}$, $S_4^{2+}$, $S_8^{2+}$, and $S_{16}^{2+}$ on the way—$S_{16}^{2+}$ was probably postulated on the basis of the earlier report of this species in $HSO_3F$ from the Gillespie group. In a subsequent study of the same chemical system, in which he used ESR spectroscopy, Bjerrum claimed the formation of the radical cations $S_4^+$ and $S_8^+$.[41] In that paper he stated that "it should, however, be noted that intermediate species produced on an electrode surface may not be those present at equilibrium."

Mamantov's group investigated electrochemical oxidation of $S_8$ in a restricted acidity-basicity range (AlCl$_3$:NaCl from 53:47 to 49.9:50.1 mol%, pCl from 5.4 to 1.5) and reported that $S_8^+$ and $S_8^{2+}$ were stable only when pCl was equal to or greater than 3.8.[42] This is consistent with disproportionation of the cations in the presence of significant concentrations of Cl$^-$ according to an equation:

$$4S_8^{2+} + 8Cl^- \rightleftharpoons 3S_8 + 4S_2Cl_2 \qquad\qquad 4.16$$

Under conditions of electrochemical oxidation, $S_8$ formed by disproportionation would be cyclically oxidized to $S_8^{2+}$, thence to $S_2Cl_2$, and even to $SCl_3^+$ in acidic solution.

Somewhat later Mamantov and co-workers used Raman spectroscopy and voltammetry to investigate generation of sulfur species in room-temperature chloroaluminates—mixtures of aluminium chloride and N-(n-butyl)pyridinium chloride.[43] In basic melts $S_8$ could be reduced to sulfide species but no oxidation to S(I) was observed, whereas this had been observed in basic AlCl$_3$—NaCl media, that is, at higher temperature. In acidic room temperatures melts $S_8$ could be oxidized to S(I), presumably $S_2Cl_2$, and through to S(IV), $SCl_3^+$; but this species was described as being stable in acidic melts for only a short time. The authors reported that "in acidic AlCl$_3$—BuPyCl melts, no evidence has been found for low-oxidation-state sulfur species that are important in determining the electrochemistry of sulfur in AlCl$_3$—NaCl melts."[43] Presumably, they are reporting that evidence for cations of the type $S_n^+$ or $S_n^{2+}$ was not found in the room-temperature melt electrochemistry.

## 4.2.2 Selenium Cations

The range of attainable oxidation states in the selenium cations $Se_4^{2+}$, $Se_8^{2+}$ and $S_{10}^{2+}$ is rather similar to that in the sulfur cations $S_4^{2+}$, $S_8^{2+}$, and $S_{19}^{2+}$, and structures are virtually identical for the cations of each element which have corresponding formulae. However, selenium cations are "easier" to generate in superacids than their sulfur counterparts—they can tolerate higher levels of basicity before disproportionating.

### 4.2.2.1 In Protonic Solvents

As indicated earlier, stable selenium cations can be generated at much lower acidities than the corresponding sulfur cations. Gillespie and Passmore have set out the conditions governing the stability of selenium cations in protonic superacidic media.[44] Cryoscopic and conductometric evidence has been used to show that the appropriate stoichiometries of Se and $S_2O_6F_2$ will produce $Se_8^{2+}$ and $Se_4^{2+}$ in $HSO_3F$, which is basic to the extent that it contains $SO_3F^-$ as the reduction product of $S_2O_6F_2$. Obviously, on reduction with Se, $Se_4^{2+}$, which is stable in slightly basic $HSO_3F$, can give a stable cation $Se_8^{2+}$

with a lower ratio of charge-to-element in the same medium; of course, $Se_8^{2+}$ would still be stable in a medium that was even more basic. Further, they report that Se can be oxidized by $H_2SO_4$ at 50–60°C to $Se_8^{2+}$. This cation, stable in $H_2SO_4$, which is basic because of $HSO_4^-$ produced in the redox reaction, can be oxidized by $SeO_2$ in that medium to $Se_4^{2+}$. In summary, $Se_8^{2+}$ and $Se_4^{2+}$ are both stable in slightly basic $HSO_3F$ and in the much more basic medium $H_2SO_4$. $Se_{10}^{2+}$, isolated from $SO_2$, as described in section 4.2.2.2, was reported[45] to be stable in 95.5% $H_2SO_4$ but was oxidized fairly rapidly to $Se_8^{2+}$ in the more acidic medium 100% $H_2SO_4$.

## 4.2.2.2 As Naked Cations in Weakly Basic Media

$Se_8(AsF_6)_2$,[46] $Se_8(Sb_2F_{11})_2$,[46] $Se_{10}(AsF_6)_2$,[45,46] and $Se_{10}(SbF_6)_2$[45,46] were prepared in $SO_2$ by oxidizing Se with $AsF_5$ or $SbF_5$, the pentafluorides forming the weakly basic anions $AsF_6^-$, $SbF_6^-$, and $Sb_2F_{11}^-$. Also, Se and $Se_8(AsF_6)_2$ in $SO_2$ in the ratio 2:1 gave $Se_{10}(AsF_6)_2$.[45] In this work the crystal structure of $Se_{10}^{2+}$ was determined. The stability of $Se_8(AsF_6)_2$ in $SO_2$ and the disproportionation of $Se_8(AlCl_4)_2$ (see section 4.2.2.3) in the same solvent[45] suggest that $AsF_6^-$ is more weakly basic than $AlCl_4^-$, at least in liquid $SO_2$.

## 4.2.2.3 In Melts

Investigation of selenium-containing species dissolved in melts is even more sketchy than for the corresponding sulfur case. Mamantov and Osteryoung[47] report that oxidation of selenium in acidic melts occurs at a potential ~0.1V less positive than that of sulfur and that, in basic melts, the species are anionic. Fehrmann and Bjerrum[48] have used a combination of potentiometry and spectrophotometry to study reacting mixtures of Se and $SeCl_4$ in the melt $AlCl_3$:NaCl, 63:37 at 150°C to identify the cations $Se_4^{2+}$ and $Se_8^{2+}$ and to propose the formation of three other low-oxidation-state cations— $Se_2^{2+}$, $Se_{12}^{2+}$, and $Se_{16}^{2+}$.

$Se_8(AlCl_4)_2$ and $Se_4(AlCl_4)_2$ have been prepared in neutral melts (i.e., those in which the ratio $Cl^-$:$AlCl_3$ is unity) by fusing appropriate stoichiometric proportions of Se, $SeCl_4$, and $AlCl_3$ and the structures of the compounds have been determined.[49] $Se_8(AlCl_4)_2$ is reported[45] to disproportionate in $SO_2$ to $Se_{10}(AlCl_4)_2$ and compounds such as $Se_2Cl_2$.

## 4.2.3 Tellurium Cations

The cation $Te_4^{2+}$ was first isolated from neutral chloroaluminate. Subsequent studies in protonic superacids and under "naked" conditions led to identification of $Te_6^{4+}$ and to study of the dependence of the stability of these ions on acidity of the medium.

## 4.2.3.1 In Protonic Solvents

Treatment of Te with $H_2SO_4$, weak oleums, and $HSO_3F$ gave rise to red solutions and the evolution of $SO_2$, suggesting reduction of the solvent with corresponding oxidation of Te to $Te_4^{2+}$.[50] The UV-visible spectra of the red solutions were identical with those in $AlCl_3$ melts where reaction stoichiometry indicated the presence of $Te_4^{2+}$ (see section 4.2.3.3) and gave similar spectra to the previously recorded Raman spectra for $Se_4^{2+}$. In stronger oleums (about 45%), the solution changed color from red to yellow-orange. Increase in $SO_3$ concentration was both increasing the acidity of the medium and providing stronger oxidizing conditions to produce $Te_6^{4+}$. This same color change occurred when $S_2O_6F_2$ and $S_2O_8^{2-}$ were used to oxidize $Te_4^{2+}$ in the acidic solvents $HSO_3F$ and $H_2SO_4$.[50] The oxidant $S_2O_8^{2-}$ is reduced initially in $H_2SO_4$ to $SO_4^{2-}$, an entity too basic to exist in $H_2SO_4$. It is protonated to $HSO_4^-$, that is, 1 mol of $S_2O_8^{2-}$ as an oxidant in $H_2SO_4$ would produce 4 mol of the base $HSO_4^-$. $Te_6^{4+}$, formed by oxidation, was observed to disproportionate with increasing time and temperature in 100% $H_2SO_4$, ultimately giving a precipitate of $TeO_2$.

## 4.2.3.2 As "Naked" Cations in Weakly Basic Media

From the solvent $SO_2$, Te oxidized with $S_2O_6F_2$, $SbF_5$, or $AsF_5$ gave $Te_4^{2+}$ associated with the very weakly basic anions $SO_3F^-$, $Sb_2F_{11}^-$, and $AsF_6^-$.[50] In appropriate reacting proportions, Te was reported[50] to react with $AsF_5$ in $SO_2$ to produce a residue, "$Te_3AsF_6$," considered more likely to be $Te_6^{2+}(AsF_6^-)_2$. Later Gillespie and co-workers demonstrated through structural determination the existence of a trigonal prismatic cation $Te_6^{4+}$ in the compounds $Te_6(AsF_6)_4 \cdot 2AsF_3$ and $Te_6(AsF_6)_4 \cdot 2SO_2$ which were isolated from the solvents $AsF_3$ and $SO_2$.[51] Their observations on the relative stabilities of $Te_6^{4+}$ and $Te_4^{2+}$ in $SO_2$, $AsF_3$, and in oleums are discussed in more detail in section 4.2.4 below.

## 4.2.3.3 In Melts

Bjerrum and Smith[52] reported purple melts resulting from dissolving Te and $TeCl_4$ in the 63–37% $AlCl_3$—NaCl eutectic melt and obtained a dark-purple solid by reacting Te, $TeCl_4$, and $AlCl_3$ in the mole ratio 7:1:4. They proposed formation of a cation $Te_{2n}^{n+}$, considered most likely to be $Te_4^{2+}$. By reacting appropriate proportions of Te, $TeCl_4$, and $AlCl_3$, Corbett and co-workers[53] isolated the solids $Te_4(AlCl_4)_2$ and $Te_4(Al_2Cl_7)_2$ and determined their structures. $Te_4^{2+}$ is square planar with $Al_2Cl_7^-$ bidentate to the square cation. Corbett commented that "no evidence has been found in the chloroaluminate system for a salt of the yellow ion $Te_n^{n+}$ ($n \geq 4$) which has been deduced in $HSO_3F$." Corbett's reaction mixtures were neutral. Enhanced acidity of the chloroaluminate system would be necessary to generate ions such as

$Te_6^{4+}$. $Te_4^{2+}$ is stabilized by the acid $AlCl_3$ which reacts with the base $Cl^-$ to form the anions $AlCl_4^-$ and $Al_2Cl_7^-$. In the presence of excess of the base $Cl^-$, $Te_4^{2+}$ disproportionates to Te and the covalent $TeCl_4$:

$$7Te + TeCl_4 \underset{Cl^-}{\overset{AlCl_3}{\rightleftharpoons}} 2Te_4^{2+} + 4Cl^- \qquad\qquad 4.17$$

Bjerrum and co-workers[54] used their combination of potentiometry and spectral observation and simulation to investigate reactions of $TeCl_4$ and Te in acidic chloroaluminates—the 63:37% eutectic of $AlCl_3$ and NaCl at 250°C—and reported identifying in solution the "solvated entities $Te_4^{2+}$, $Te_6^{2+}$ and $Te_8^{2+}$." However, there does not seem to have been any physical isolation of compounds with tellurium cations in oxidation states lower than in $Te_4^{2+}$.

## 4.2.4 Stabilities of Group VI Cations

Table 4.3 shows that the cation $Te_6^{4+}$ with a relatively high charge per tellurium atom has no counterpart in selenium or sulfur chemistry. Cations of low formal charge (i.e., below 0.5) per tellurium atom have not been isolated, although their existence in melts has been postulated. In proceeding through Se to S, cations of progressively lower charge per atom can be prepared relatively easily, the lowest reported being in $Se_{10}^{2+}$ and $S_{19}^{2+}$, respectively.

These same trends are observed in the acidities of the protonic media necessary to provide stable solutions of cations of S, Se, and Te. For all three elements, data in Table 4.3 are given for $H_2SO_4$-based solvent systems, that is, pure $H_2SO_4$ itself or the solvent made acidic with $SO_3$, or made slightly basic with $H_2O$ which disturbs the self-ionization equilibria of the solvent to generate large concentrations of $HSO_4^-$. Table 4.3 shows an obvious correlation between the formal oxidation state of an element in its cations and the acidity of the $H_2SO_4$-based medium in which those cations are stable.

Additionally, for sulfur cations, $H_0$ values are given for stable species in $HSO_3F$-based media. The correlation between charge-per-sulfur atom and acidity follows the same trend, but the absolute values of $H_0$ at which particular cations are stabilized differ for each solvent system. It should be recalled that a cation is destabilized by a disproportionation reaction which, in its final and simplest form, will lead to formation of both the element itself and of a *covalent* compound formed between the base of the solvent system and the element concerned, with the element in the compound in a higher oxidation state than in the original cation. In comparing the two solvent systems, it should be noted that $SO_3F^-$ is more electronegative than $HSO_4^-$ and will form covalent fluorosulfato- compounds more readily than bisulfato- compounds are formed. Therefore, a higher acidity, that is, a lower availability of base, is required to stabilize a sulfur cation in the presence of $SO_3F^-$ than with $HSO_4^-$ present.

**Table 4.3.** Stability of Homopolyatomic Cations of Chalcogens in Protonic Solvents

| General Formula | $X_6^{4+}$ | $X_4^{2+}$ | $X_8^{2+}$ | $X_n^{2+}(n>8)$ |
|---|---|---|---|---|
| Oxidation State | 0.67 | 0.5 | 0.25 | <0.25 |
| *Tellurium* | $Te_6^{4+}$ Stable in oleums >30% in $SO_3$ ($H_0 \cong -13.76$)[a] | $Te_4^{2+}$ Stable in $H_2SO_4$—$HSO_4^-$ ($H_o \approx -11$) | — | — |
| *Selenium* | — | $Se_4^{2+}$ Stable in 100% $H_2SO_4$ ($H_o \approx -11.9$) | $Se_8^{2+}$ Stable in 100% $H_2SO_4$ ($H_o \approx -11.9$) | $Se_{10}^{2+}$ Stable in 95.5% $H_2SO_4$ ($H_o \approx -10$) |
| *Sulfur* | — | (i) $S_4^{2+}$ Stable in oleums >40% in $SO_3$ ($H_o \approx -14.1$) (ii) Stable in $HSO_3F$—$SbF_5$[b] ($H_o \approx -18$) | $S_8^{2+}$ Stable in oleums >15% in $SO_3$ ($H_o \approx -13.2$) Marginally stable in $HSO_3F$— $SO_3F^-$ ($H_o \approx -13.8$) | $S_{19}^{2+}$ Stable in oleums >5% in $SO_3$ ($H_o \approx -12.7$) Stable in oleums $HSO_3F$— $SO_3F^-$ ($H_o \approx -13.8$) |

[a]Approximate values of Hammett Acidity Functions calculated on basis of footnote b of Table 4.2.
[b]For validity of comparisons, most media cited in Table 4.3 are based on $H_2SO_4$, either neutral (100% $H_2SO_4$) or made acidic by addition of $SO_3$ (oleums) or basic because the $HSO_4^-$ concentration has been enhanced either by reaction of $H_2SO_4$ acting as an oxidant for the appropriate chalcogen or by addition of $H_2O$ as in 95.5% $H_2SO_4$. In a particular medium, as for sulfur cations in $HSO_3F$, the general trends are the same, but dependence of stability on absolute $H_0$ values varies depending on the nature of chemical interaction of the cation with the base of the solvent system. (See section 4.2.4.)

A brief Melbourne study[37] of the dependence of stability of sulfur cations on acidity levels in HF shows that the $H_0$ values required for stabilization of $S_8^{2+}$ and $S_4^{2+}$ in HF are much more negative than those required, in turn, in media based on $HSO_3F$ and $H_2SO_4$. The bond energies and stability of a disproportionation product such as $SF_4$ would be greater than for corresponding initial products in $HSO_3F$ and $H_2SO_4$ and, again, a lower availability of the base ($F^-$) will be needed to initiate disproportionation of a sulfur cation than in the other superacidic media.

*A critical evaluation of the stability of chalcogen cations in superacid media based on the solvents $H_2SO_4$, $HSO_3F$, and HF provides a sharp warning against adoption of a simple-minded rationalization of cation stability based only on values of $H_o$ for the different media. Differing chemical interactions during disproportionations—that is, bond energies for the chalcogen concerned bound to the different bases of each system—must also be taken into account.*

In summary, no polyatomic cation of sulfur is stable in 100% $H_2SO_4$. Cations of "high" charge disproportionate through $S_8^{2+}$ and larger cations $S_n^{2+}$ to $S_8$ and $SO_2$ ultimately. $S_8^{2+}$ is stable in $HSO_3F$ and $S_4^{2+}$ can be produced in $HSO_3F$—$SbF_5$. By comparison, even $Se_4^{2+}$, with a "high" charge-to-element ratio, is stable to disproportionation in 100% $H_2SO_4$, as is $Se_8^{2+}$, and therefore, of course, both are stable in the more acidic $HSO_3F$. $S_{10}^{2+}$, with a smaller charge-to-selenium ratio, is stable in 95.5% $H_2SO_4$. $Te_4^{2+}$, with a high charge-to-atom ratio, has been reported as capable of existence in 98% $H_2SO_4$, a fairly basic medium in terms of superacidity studies. In their paper on the structure of the $Te_6^{4+}$ cation[51] the Gillespie group report that $Te_4^{2+}$ is "oxidized" to $Te_6^{4+}$ by increasing the $SO_3$ content of oleums. The role of $SO_3$ in increasing the acidity of the medium is probably much more important than its role as an oxidant—even in relatively dilute oleums there should be sufficient $SO_3$ to act solely as an oxidant considering the amount of solute present. This question of a single entity acting both as oxidant and as Lewis acid in a particular superacid system, and much of what follows in the next paragraph, will be dealt with in detail in chapter 9.

Gillespie and colleagues[51] say that $Te_4^{2+}$ and $Te_6^{4+}$ can exist in 100% $H_2SO_4$ but that a precipitate of $TeO_2$ is formed on prolonged standing, a process that is accelerated by heating. This observed reaction scheme probably begins with a disproportionation of the cations to Te and $TeO_2$ in this medium which is not strongly acidic. The Te, as formed, would be progressively reoxidized by the solvent to cations of fractional formal charge. Elemental Te, its cations, and $TeO_2$ are probably in equilibrium at this acidity, the position of equilibrium being disturbed as $TeO_2$ is formed. They report that in the solvents $SO_2$ and $AsF_3$, Te is oxidized initially to $Te_4^{2+}$ by $SbF_5$ and $AsF_5$. They say that further oxidation to $Te_6^{4+}$ occurs but that $Te_6^{4+}$ is in equilibrium with $Te_4^{2+}$ "even if a very large excess of oxidant is used." Both sets of observations are consistent with an interpretation that $Te_6^{4+}$ can be stabilized as the predominant cationic species in strong oleums but disproportionates, in part at least, to $Te_4^{2+}$ in the weakly basic solvents $SO_2$ and $AsF_3$.

Much of the electrochemical work on oxidation of sulfur, selenium, and telllurium in acidic melts is difficult to interpret, partly because of the intrinsic difficulty of the experimental procedures in melts, but also because the effect of an increasing voltage ramp is to produce in turn species like "$S_{16}^{2+}$", $S_8^{2+}$, $S_4^{2+}$, $(S_2^{2+})$, S(II), probably as $SCl_2$, and S(IV), probably as $SCl_3^+$ in acidic melts. Similar species have been proposed for Se and Te, although oxidation occurs at lower potentials. Under these conditions the applied potential can produce at the sensing electrode small quantities of species which would not necessarily be stable at the acidity or basicity of the bulk of the melt, as Bjerrum, a leading exponent of the technique, has acknowledged.[41]

## 4.3 Relative Stabilities of Homopolyatomic Cations of the Halogens and Chalcogens

Comparison of the elements of groups VI and VII shows that, in a general sense, the polychalcogen cations are much easier to isolate at lower acidities than polyhalogen cations. Comparing the cations of the lightest elements of the two groups (other than $O_2$ and $F_2$), all the known $S_n^{m+}$ cations can be synthesized in protonic solvents, whereas $Cl_2^+$ has not been prepared and $Cl_3^+$ exists only at low temperatures in the absence of solvent, which would provide a source of base. All the known cations of selenium can be isolated in 100% $H_2SO_4$, whereas $Br_3^+$ and $Br_2^+$ need, respectively, as solvents, $HSO_3F$ and $HSO_3F$—$SO_3$—$SbF_5$. While the tellurium cation $Te_4^{2+}$ is stable in media more basic than 100% $H_2SO_4$, $I_2^+$ and $I_4^{2+}$ require oleums, $HSO_3F$, or HF to prevent disproportionation.

## 4.4 Heteropolyatomic Cations of Nonmetallic Elements

As stated in the introductory section of this chapter, there is a vast range of heteropolyatomic cations of nonmetallic elements from the familiar $NH_4^+$ through to halogen- or chalcogen-based cations such as $ClO_2^+$, $IF_6^+$, and $SF_3^+$ which are entities with electronegative ligands bound to a central nonmetal in a relatively high formal oxidation state. These do not come within the purview of this book because they differ markedly from the homopolyatomic cations of the halogens and of the chalcogens which have been presented in considerable detail in this chapter.

The formal oxidation state of the appropriate element in all the homopolyatomic cations presented in this chapter has been less than unity. It has been shown that they can be generated in acidic solution or as "naked" cations, from solvents of very low basicity, such as $SO_2$, $SO_2ClF$, and $AsF_3$, and can be isolated as solids for characterization providing that the counteranions in the isolated compounds are themselves weakly basic anions, such as $Sb_2F_{11}^-$, $AsF_6^-$, $SO_3F^-$, and $HSO_4^-$, ions that increase in basicity in that order. However, while there has been some value, for the sake of complete coverage of their chemistry, in demonstrating that these homopolyatomic nonmetal cations can be isolated as "naked" cations in the absence of species sufficiently basic to cause disproportionation, the main purpose of this chapter has been to systematize their generation and stability *in solution* in superacidic media.

A very small number of heteropolyatomic cations based only on halogens in fractional oxidation states has been proposed. The existence of the cation $Cl_2F^+$ was first established by Christe and Sawodny[55] as a result of direct interaction at low temperature of a 2:1 mixture of ClF with a Lewis acid, $AsF_5$ or $BF_3$, to produce a solid such as $Cl_2F^+AsF_6^-$, stable at $-78°C$ but dissociating completely at room temperature. Gillespie and Morton[30] later used Raman spectroscopy to postulate that the bent triatomic cation is the asymmetrical $Cl—Cl—F^+$ rather than the symmetrical $Cl—F—Cl^+$, as proposed by Christe and Sawodny. A recent extensive computational and spectroscopic study[56] appears to confirm the asymmetrical structure. No solution chemistry of this cation has been reported. Aubke and colleagues have reported[23] that reaction of $[Br_2^+][Sb_3F_{16}^-]$ with a small excess of $Cl_2$ at room temperature resulted, after excess $Cl_2$ was allowed to escape, in formation of the residual solid compound $[Br_2Cl^+][Sb_3F_{16}^-]$ characterized by chemical analysis and Raman spectroscopy.

A wide range of heteropolyatomic cations involving the chalcogens has been reported. The Gillespie group at McMaster University has generated and characterized polyatomic cations incorporating both selenium and tellurium atoms, as well as sulfur—nitrogen cations. Passmore, at New Brunswick, has synthesized several sulfur—halogen and selenium—halogen cations.

In the majority of syntheses from both groups a fairly general chemical approach has been adopted. $AsF_5$ and $SbF_5$ have been used as oxidants of mixtures of chalcogens or of chalcogens and halogens and have also been the source of very weakly basic counteranions, such as $AsF_6^-$ and $SbF_6^-$. Occasionally specific oxidants have been used. $Te_6(AsF_6)_4$ oxidized $S_4N_4$ to $S_3N_2^+$, being itself reduced to $Te_4(AsF_6)_2$. $HSO_3F$, which dissociates to $SO_3$ and HF, oxidized $S_4N_4$ to $S_6N_4^{2+}$, and Gillespie used a clean oxidation by $Cl_2$ of $S_4N_4$ to $S_4N_4^{2+}$, with $AlCl_3$ providing an acidic medium and the counteranion $AlCl_4^-$.

In the main, the heterocations have been generated by oxidation in solvents of very low basicity, $SO_2$ and $AsF_3$. They have been generated as "naked" cations under conditions in which sufficient base is not present to cause disproportionation.

An appropriate experimental procedure has been described in detail by Passmore and colleagues.[57] The glass reaction vessel consists of two bulbs separated by a glass frit and fitted with a Teflon-stemmed valve. In a typical

**Table 4.4.** Synthesis of Heteropolyatomic Cations Containing Chalcogen Elements

| Cation | Reagents | Product | Reference |
|---|---|---|---|
| $Te_2Se_8^{2+}$ | $Se_8(AsF_6)_2$ + excess Te in liquid $SO_2$ | $Te_2Se_8(AsF_6)_2 \cdot SO_2$ | 58 |
| $Te_{3.7}Se_{6.3}^{2+}$ | Se–Te "alloy" + excess $AsF_5$ in liquid $SO_2$ | $Te_{3.7}Se_{6.3}(AsF_6)_2$ | 58 |
| $Te_3S_3^{2+}$ | Slight excess of 1:1 mixture of S and Te + $AsF_5$ in liquid $SO_2$ | $Te_3S_3(AsF_6)_2$ | 59 |
| $Te_2Se_4^{2+}$ | Excess of 1:1 mixture of Se and Te + $SbF_5$ in liquid $SO_2$ | $Te_2Se_4(SbF_6)_2$ | 59 |
| $Te_2Se_2^{2+}$ | Se–Te "alloy" (1:1) + excess $SbF_5$ in liquid $SO_2$ | $(Te_2Se_2)(Sb_3F_{14})(SbF_6)$ | 60 |
| $S_4N_4^{2+}$ | (a) $S_4N_4$ + excess $AsF_5$ in liquid $SO_2$ | $S_4N_4(AsF_6)_2 \cdot xSO_2$ | 61 |
|  | (b) $S_4N_4$ + ($AlCl_3$ : $Cl_2$, 1:1) in liquid $SO_2$ | $S_4N_4(AlCl_4)_2$ | 61 |
| $S_3N_2^+$ | $S_4N_4$ + $Te_6(AsF_6)_4$ (1:1) in liquid $SO_2$ | $S_3N_2(AsF_6)$ | 62 |
| $S_6N_4^{2+}$ | $S_4N_4$ + excess $HSO_3F$ in liquid $SO_2$ | $S_6N_4(S_2O_2F)_2$ | 62 |
| $S_7I^+$ | Excess of approx. 1:1 mixture of $S_8$ and $I_2$ + $AsF_5$ in liquid $AsF_3$ | $S_7I(AsF_6)$ | 63 |
| $[(S_7I)_2I]^{3+}$ | Approx. 1:1 mixture of $S_8$ and $I_2$ + excess $SbF_5$ in liquid $AsF_3$ | $[(S_7I)_2I](SbF_6)_3 \cdot 2AsF_3$ | 64 |
| $S_7Br^+$ | $AsF_5$ and excess $S_8$ dissolved in $AsF_3$; $Br_2$ added ($S_8$: $Br_2$, approx. 2:1) | $S_7Br(AsF_6)$ | 65 |
| $Se_6I_2^{2+}$ | Se : $I_2$ : $AsF_5$, 6:1:3 in liquid $SO_2$ | $Se_6I_2(AsF_6)_2 \cdot 2SO_2$ | 66 |

reaction, solid reactants would be introduced into one bulb and the solvent and oxidant would be condensed directly on the solids at $-196°C$ or the oxidant could be transferred in the solvent to the bulb maintained at that temperature. The reaction mixture is allowed to warm slowly to room temperature and maintained there for several hours or days. The frit allows filtration of impurities; crystals may be obtained from the reaction mixtures by suitable evaporation procedures.

Some representative examples of heteropolyatomic cations are presented in Table 4.4. They were all prepared using the general experimental approach outlined immediately above, although there were minor variations in procedure. For some of these, the literature describes several preparative reactions. The ones selected for Table 4.4 have been chosen to make them as general and as directly comparable as possible. Many of the reactions were not entirely specific for the product isolated. The use of $Cl_2$ and $AlCl_3$ as an alternative preparative route to $S_4N_4^{2+}$ is listed as a reaction of high specificity. In oxidizing $S_4N_4$ to $S_4N_4^{2+}$, $Cl_2$ is reduced to two $Cl^-$ ions that then react with the Lewis acid $AlCl_3$ to form the weakly basic counteranion $AlCl_4^-$. There is a nice example of specificity in the use of stoichiometric amounts of the reactants to produce $Se_6I_2(AsF_6)_2$. It appears to be fair to comment that most of these preparative methods lack the specificity of many of the reactions used in superacidic media for the preparation of homopolyatomic chalcogen cations, where, for example, $S_8$ was reacted with $S_2O_6F_2$ in $HSO_3F$ to produce $S_8^{2+}$ and $SO_3F^-$, the base of the solvent and the counteranion. (See section 4.2.1.1.)

There is no attempt here to offer a comprehensive list of these heteropolyatomic cations. It is deemed to be sufficient, within the context of this book, to show that they have been generated as "naked" cations under conditions where the reaction media and the counteranions were weakly basic. The principal interest in these compounds has been to use X-ray crystallography to determine the structures of the cations. The current literature does not report any studies on the stability and reactions of these heterocations in superacidic media and so they do not warrant further treatment in a book on inorganic systems in superacidic media.

# References

1. R.J. Gillespie, J. Passmore in *Advances in Inorganic Chemistry and Radiochemistry*, Vol. 17. H.J. Emeléus, A.G. Sharpe, eds. Academic Press, 1975, pp. 49–87.

2. R.J. Gillespie, J. Passmore, *Chem. Brit.*, **8**, 475 (1972).

3. R.J. Gillespie, J. Passmore in *M.T.P. Int. Rev. Sci. Inorg. Chem., Ser. 2*, Vol. 3, 1975, pp. 121–136.

4. A.A. Woolfe in *Advances in Inorganic Chemistry and Radiochemistry*, Vol. 9. H.J. Emeléus, A.G. Sharpe, eds. Academic Press, 1966, pp. 217–314.

5. R.J. Gillespie, M.J. Morton, *Quart. Rev.*, **25**, 553 (1971).

6. G. Adahmi, M. Herlem, *J. Electroanal. Chem.*, **26**, 363 (1970).

7. R.J. Gillespie, R. Kapoor, R. Faggiano, C.J.L. Lock, M. Murchie, J. Passmore, *J. Chem. Soc. Chem. Commun.*, *8* (1983).

8. R.J. Gillespie, T.E. Peel, *J. Am. Chem. Soc.*, *95*, 5173 (1973).

9. R.J. Gillespie, T.E. Peel, E.A. Robinson, *J. Am. Chem. Soc.*, *93*, 5083 (1971).

10. J. Besida, T.A. O'Donnell, *Inorg. Chem.*, *28*, 1669 (1989).

11. R.J. Gillespie, J. Liang, *J. Am. Chem. Soc.*, *110*, 6053 (1988).

12. R. Adrien, Ph.D. Thesis, University of Melbourne (1992).

13. D.J. Merryman, J.D. Corbett, P.A. Edwards, *Inorg. Chem.*, *14*, 428 (1975).

14. Z.J. Karpinski, R.A. Osteryoung, *Inorg. Chem.*, *23*, 4561 (1984).

15. Z.J. Karpinski, R.A. Osteryoung, *J. Electroanal. Chem.*, *164*, 281 (1984).

16. Z.J. Karpinski, R.A. Osteryoung, *J. Electroanal. Chem.*, *178*, 281 (1984).

17. R. Marassi, J.Q. Chambers, G. Mamantov, *J. Electroanal. Chem.*, *69*, 345 (1976).

18. K. Tanemoto, G. Mamantov, R. Marassi, *J. Inorg. Nucl. Chem.*, *43*, 1779 (1981).

19. R.D.W. Kemmitt, M. Murray, V.M. McRae, M.C.R. Symons, T.A. O'Donnell, *J. Chem. Soc., (A)*, 862 (1968).

20. C.G. Davies, R.J. Gillespie, P.R. Ireland, J.M. Sowa, *Can. J. Chem.*, *52*, 2048 (1974).

21. J. Passmore, G. Sutherland, P.S. White, *Inorg. Chem.*, *20*, 2169 (1981).

22. J. Passmore, P. Taylor, T. Whidden, P.S. White, *Can. J. Chem.*, *57*, 968 (1979).

23. W.W. Wilson, R.C. Thompson, F. Aubke, *Inorg. Chem.*, *19*, 1489 (1980).

24. V.M. McRae, Ph.D. Thesis, University of Melbourne (1966).

25. A.J. Edwards, G.R. Jones, R.J.C. Sills, *J. Chem. Soc., Chem. Commun.*, 1527 (1968).

26. K.C. Lee, F. Aubke, *Inorg. Chem.*, *19*, 119 (1980).

27. K.C. Lee, F. Aubke, *Inorg. Chem.*, *23*, 2124 (1984).

28. Ref. 5, pp. 559–562.

29. R.J. Gillespie, M.J. Morton, *Inorg. Chem.*, *11*, 586 (1972).

30. R.J. Gillespie, M.J. Morton, *Inorg. Chem.*, *9*, 811 (1970).

31. N. Bartlett, D.H. Lohman, *Proc. Chem. Soc.*, 115 (1962).

32. (a) Ref. 1, pp 63–65, (b) Ref. 2, p. 476.

33. (a) Ref. 1, p. 64, (b) Ref. 3, p. 123.

34. R.C. Burns, R.J. Gillespie, J.F. Sawyer, *Inorg. Chem.*, *19*, 1423 (1980).

35. H.S. Low, R.A. Baudet, *J. Amer. Chem. Soc.*, *98*, 3849 (1976).

36. (a) Ref. 1, p. 67, (b) Ref. 3, p. 126.

37. J. Besida, T.A. O'Donnell, unpublished observations.

38. R. Fehrmann, N.J. Bjerrum, F.N. Poulsen, *Inorg. Chem.*, *17*, 1195 (1978).

39. N.J. Bjerrum in *Characterization of Solutes in Non Aqueous Solvents*, G. Mamantov, ed. Plenum Press, New York, 1978, pp. 251–271.

40. R. Marassi, G. Mamantov, M. Matsunaga, S.E. Springer, J.P. Wiaux, *J. Electrochem. Soc.*, *126*, 231 (1979).

41. R. Fehrmann, N.J. Bjerrum, E. Pedersen, *Inorg. Chem.*, *21*, 1497 (1982).

42. K. Tanemoto, R. Marassi, C.B. Mamantov, Y. Ogata, M. Matsunaga, J.P. Wiaux, G. Mamantov, *J. Electrochem. Soc.*, *129*, 2237 (1982).

43. R. Marassi, T.M. Laher, D.S. Trimble, G. Mamantov, *J. Electrochem. Soc.*, *132*, 1639 (1985).

44. (a) Ref. 1, pp. 68–69, (b) Ref. 2, p. 477, (c) Ref. 3, p. 128.

45. R.C. Burns, W.-L. Chan, R.J. Gillespie, W.-C. Luk, J.F. Sawyer, D.R. Slim, *Inorg. Chem.*, *19*, 1432 (1980).

46. (a) Ref. 1, p. 70, (b) Ref. 2, p. 477, (c) Ref. 3, p. 128.

47. Ref. 39, p. 230.

48. R. Fehrmann, N.J. Bjerrum, *Inorg. Chem.*, *16*, 2089 (1977).

49. R.K. McMullan, D.J. Prince, J.D. Corbett, *Inorg. Chem.*, *10*, 1749 (1971).

50. Ref. 1, pp. 72–73; Ref. 3, p. 132.

51. R.C. Burns, R.J. Gillespie, W.-C. Luk, D. R. Slim, *Inorg. Chem.*, *18*, 3086 (1979).

52. N.J. Bjerrum, G.P. Smith, *J. Am. Chem. Soc.*, *90*, 4472 (1968).

53. T.W. Couch, D.A. Lokken, J.D. Corbett, *Inorg. Chem.*, *11*, 357 (1972).

54. R. Fehrmann, N.J. Bjerrum, H.A. Andreasen, *Inorg. Chem.*, *15*, 2187 (1976).

55. K.O. Christe, W. Sawodny, *Inorg. Chem.*, *8*, 212 (1969).

56. G. Frenking, W. Koch, *Inorg. Chem.*, *29*, 4513 (1990).

57. M.P. Murchie, J. Passmore, C.-M. Wong in *Inorganic Synthesis*, Vol. 27. A.P. Ginsberg, ed. John Wiley and Sons, 1990, pp. 332–339.

58. P. Boldrini, I.D. Brown, R.J. Gillespie, P.R. Ireland, W. Luk, D.R. Slim, J.E. Vekris, *Inorg. Chem.*, *15*, 765 (1976).

59. R.J. Gillespie, W. Luk, E. Maharajh,, D.R. Slim, *Inorg. Chem.*, *16*, 892 (1977).

60. P. Boldrini, I.D. Brown, M.J. Collins, R.J. Gillespie, E. Maharajh, D.R. Slim, J.F. Sawyer, *Inorg. Chem.*, *24*, 4302 (1985).

61. R.J. Gillespie, J.P. Kent, J.F. Sawyer, D.R. Slim, J.D. Tyrer, *Inorg. Chem.*, *20*, 3799 (1981).

62. R.J. Gillespie, J.P. Kent, J.F. Sawyer, *Inorg. Chem.*, *20*, 3784 (1981).

63. J. Passmore, G. Sutherland, P. Taylor, T.K. Whidden, P.S. White, *Inorg. Chem.*, *20*, 3839 (1981).

64. J. Passmore, G. Sutherland, P.S. White, *Inorg. Chem.*, *21*, 2717 (1982).

65. J. Passmore, G. Sutherland, T.K. Whidden, P.S. White, C.-M. Wong, *Can. J. Chem.*, *63*, 1209 (1985).

66. J. Passmore, P.S. White, C.-M. Wong, *Chem. Commun.*, 1178 (1985).

# CHAPTER 5

# Homopolyatomic Cations of Metallic Elements

For a long time it was assumed, if only implicity, that the element carbon was virtually unique in forming compounds containing chains, rings, and cages of identical or very similar atoms. Rings such as $S_8$, and cages such as $P_4$ and the structurally related $P_4O_6$ and $P_4O_{10}$ and the more complex $P_4S_4$, were established as neutral species, and the structural complexity of higher boranes and carboranes, either as neutral or anionic species, was determined subsequently. In the case of metallic elements, anionic homopolyatomic clusters such as $Sn_9^{4-}$, $Pb_7^{4-}$, $Sb_5^{3-}$, and $Bi_5^{3-}$ were formulated on the basis of analysis by Zintl in Germany in the 1930s. These will be discussed briefly in chapter 8 because of the acid-base properties of the media in which they are generated. They have been investigated more comprehensively than the homopolyatomic cations to be discussed in this chapter.

The existence and extent of heteropolyatomic cluster formation in transition metal chemistry and the very important part clusters play in structural chemistry have been extensively studied over the last three decades. Around the turn of the century it was reported that for the compound with the overall formula $Ta_6Br_{14}.7H_2O$ only one-seventh of the chloride could be precipitated from aqueous solutions of the compound and that the residual chloride must be present in polymeric form. Compounds such as this one have been shown by crystallographic techniques to be formulated as $[M_6X_{12}]X_2$, where M = Nb or Ta and X = Br, Cl, or F. Metal–metal bonding leads to an octahedron of six Nb or Ta atoms with each edge of the octahedron bridged by a halogen atom. The remaining two halogen atoms per formula bridge between the clusters and are therefore much more loosely bound and so can be precipitated from aqueous solution. "Molybdenum dichloride" is an aggregate of $[Mo_6Cl_8]^{4+}$ cluster cations with four discrete bridging chlorides per cluster. In this cluster each face of the octahedron of metal–metal bonded Mo atoms is capped by one Cl atom. Rhenium chemistry provides several examples of the great number of halogen-bridged metal cluster ions now characterized; for example, rhenium trichloride is an extended structure based on $Re_3Cl_9$ clusters, and the related anion, $[Re_3Cl_{12}]^{3-}$, has Cl atoms bridging $Re_3Cl_9$ clusters.

The halogen-bridged metal clusters provide only a small part of the general fund of transition metal cluster compounds. A few examples of carbonyl clusters are $[Co_4(CO)_{12}]$, $[Rh_6(CO)_{16}]$, $[Ni_5(CO)_{12}]^{2-}$, $[Ni_6(CO)_{12}]^{2-}$, and $[Pt_{19}(CO)_{22}]^{4-}$. Carbon, hydrogen, phosphorus, arsenic, sulfur, and other atoms and groups can be introduced into these carbonyl clusters.

The transition metal-halide cluster compounds often involve strongly reductive preparative methods and lead to compounds with the metals in low to very-low formal oxidation states. There do not appear to have been accounts of their preparation in protonic superacid media. Hussey[1] and Seddon[2] have investigated their redox behavior in basic room-temperature chloroaluminate melts. They are surveyed here, however briefly and inadequately, in order to contrast the vast number and range of these heteropolyatomic clusters with the relatively small number of currently known homopolyatomic *cations* of metallic elements. The homopolyatomic cations which have been characterized most adequately can be regarded as having been derived largely or exclusively from the posttransition metal main group elements, depending on whether members of the Zn—Cd—Hg group of elements, with their $d^{10}$ configuration, are regarded as true transition metals or not.

Accounts of preparation of homopolyatomic cations of metallic elements are rather fragmentary. With the exception of reliable characterization in the solvents $SO_2$ and $AsF_3$ of cations of mercury and some work in the same solvents on cations of bismuth, the postulated existence of most of the cations has been inferred from phase studies of melts or on the basis of species isolated from melts. It will be shown that the cationic species are produced and stable under conditions similar to those which have been shown in chapter 4 to be required for stabilization of polyatomic cations of elements of groups VI and VII; that is, they are usually generated as "naked" cations and depend for their preparation and stability on nonavailability of bases with which they would interact and disproportionate. Thus, research on homopolyatomic metallic cations differs in one respect from that on nonmetallic cations in that very little synthesis has been attempted in protonic superacids as such; however, relative Lewis acid strengths and levels of acidity and basicity in melts have been determinants of speciation in some of the syntheses.

In one very important way, the polyatomic cations of metals differ from those of nonmetals in that *lower* fractional formal oxidation states are stabilized with increasing acidity of the medium whereas, with the nonmetallic elements of groups VI and VII, higher fractional formal oxidation states, that is, higher charge-to-element ratios, were shown to be stabilized by oxidation in progressively more acidic media. Thus increase in acidity allowed formation of $I_2^+$ at the expense of $I_3^+$ and of $Br_2^+$ rather than $Br_3^+$. $S_{19}^{2+}$, $S_8^{2+}$, and $S_4^{2+}$ could be formed by oxidation in increasingly acidic solvents. We shall see that $Hg_2^{2+}$, stable in the basic solvent water, can be reduced to $Hg_3^{2+}$, which is marginally stable in $HSO_3F$, and that $Hg_4^{2+}$ can be formed only in equilibrium with $Hg_3^{2+}$ in this acidic medium.

This is consistent with the general chemistry of nonmetals and metals. Charged entities of nonmetals, unless they are complexed, are anionic—they exist in negative oxidation states. Nonmetals do occur in a wide range of compounds in formal high positive oxidation states, but as complexes—for example, $IF_6^-$ and $IO_4^-$. In chapter 4 it has been shown that "forcing" conditions—very high acidity or virtual absence of basic species—are necessary to form cationic species of uncomplexed nonmetals. The more acidic the medium the higher is the charge-to-element ratio. On the other hand, noncomplexed solvated metal *cations* occur in water and other protonic solvents and in acidic melts in "normal" oxidation states, $+1$ and more frequently $+2$ and $+3$ for transition metals and even $+4$ for actinides. For stable ions of metals in very high oxidation states, complexation must occur, as in $UO_2^{2+}$, $MnO_4^-$, and $AuF_6^-$. Metals can exist in very low formal oxidation states in complexes such as the carbonyls, but special conditions are required to stabilize metals in low oxidation states as *simple solvated or nonsolvated cations*. This concept will be developed further in chapter 7, which deals with monatomic cations of transition metals in very low oxidation states and also in this chapter in dealing with polyatomic cations of metallic elements.

## 5.1  Mercury Cations

The familiar cation $Hg_2^{2+}$ has been known since 1898, and it is the only homopolyatomic cation that can exist in aqueous solution. Its Raman spectrum was observed in aqueous $Hg_2(NO_3)_2$ solution in 1934. Justifiably, especially within the context of this and the preceding chapter, its stability in the basic solvent $H_2O$ has always been regarded as anomalous. However, it is consistent with the principles established here that addition of $NH_3$ or $OH^-$ to make the aqueous solution basic leads to disproportionation of $Hg_2^{2+}$ to Hg and nonionic basic compounds of Hg(II).

The disproportionation equilibrium

$$Hg_2^{2+} \rightleftharpoons Hg(0) + Hg(II) \qquad\qquad 5.1$$

can be disturbed in many other ways. For example addition of excess $I^-$ to aqueous $Hg_2^{2+}$ solution causes precipitation of Hg(0) because of formation of the strong complex $[HgI_4]^{2-}$. Hames and Plambeck[3] demonstrated, by electrochemical reduction of $Hg^{2+}$, the generation of $Hg_2^{2+}$ in the acidic eutectic $AlCl_3$:NaCl:KCl, 66:20:14, and showed it to be stabler in the acidic melt than in water, quoting values of $3.6 \times 10^4$ and $1.10 \times 10^2$ for the formation constant for $Hg_2^{2+}$—the reverse of equation 5.1—in the acidic melt and water, respectively.

Mamantov and colleagues[4] first identified $Hg_3^{2+}$ by obtaining a yellow solution with an absorption band at 325 nm in a highly acidic melt when they added $Hg_2Cl_2$ to an excess of both Hg and $AlCl_3$. Removal by distillation of

Hg and $AlCl_3$ yielded the compound $Hg_3(AlCl_4)_2$, which they reported as being "very sensitive to moisture and decomposes in water to give mercury metal and $Hg_2Cl_2$"; that is, it undergoes base-induced disproportionation. They noted that a preliminary X-ray crystallographic study showed an almost linear array of Hg atoms in their compound, with the angle for Hg—Hg—Hg being about 174°. Voltammetric reduction of $HgCl_2$ in acidic melt ($AlCl_3$:NaCl, 65:35) gave three reduction waves ($Hg^{2+} \rightarrow Hg_2^{2+} \rightarrow Hg_3^{2+} \rightarrow Hg$) with diffusion currents in the ratio 3:1:2, this ratio being consistent with the reductions proposed earlier. In basic melt, $AlCl_3$ saturated with NaCl, only two waves of equal height were observed, indicating that $Hg_3^{2+}$ is stable only in acidic melts.

Gillespie and co-workers[5] oxidized Hg with $AsF_5$ and $SbF_5$ in solution in $SO_2$ and obtained yellow solids $Hg_3(AsF_6)_2$ and $Hg_3(Sb_2F_{11})_2$, stable in the presence of the solvent and of $AsF_3$ and $SbF_3$ (the products of reduction of $AsF_5$ and $SbF_5$); all three compounds $SO_2$, $AsF_3$, and $SbF_3$ are very weakly basic. Solutions of these solids in $HSO_3F$ gave the characteristic absorption at 325 nm observed by Mamantov.[4]

It is highly significant in the context of this and the preceding chapter that in $HSO_3F$, the compounds were reported to form $Hg_2^{2+}$ slowly.[5] It was postulated earlier (section 4.1.5) that $Br_2^+$ and $I_2^+$ disproportionate in the presence of the base $SO_3F^-$ but are stabilized by the more acidic anions $AsF_6^-$ and $Sb_2F_{11}^-$. One can see a similar situation here. $Hg_3^{2+}$, stable in compounds containing $AsF_6^-$ and $Sb_2F_{11}^-$, could be disproportionating in "neat" $HSO_3F$, that is, a medium in which a reasonable concentration of the more basic anion $SO_3F^-$ is available. Hg would be forming in the disproportionation, but it had been reported as early as 1932 that Hg dissolves very slowly in $HSO_3F$, presumably through oxidation by $SO_3$. Therefore Hg from the disproportionation would be oxidized to polyatomic cations that would in turn disproportionate and be reoxidized. These authors also report that $Hg_3^{2+}$ disproportionates rapidly and completely to Hg and Hg(II) in the presence of "water and other basic substances."

The difference in disproportionation products for $Hg_3^{2+}$ as reported by Mamantov and Gillespie groups is consistent with the known chemistry of mercurous compounds. The insoluble $Hg_2Cl_2$ is stable in aqueous solution but $Hg_2^{2+}$ compounds disproportionate further in aqueous $F^-$ solutions.

Oxidative reaction of Hg with $AsF_5$ in the ratio 4:3, designed to yield $Hg_4(AsF_6)_2$ and the reduction product $AsF_3$, gave a mixture of yellow crystals of $Hg_3(AsF_6)_2$ and red-black $Hg_4(AsF_6)_2$, the crystal structure of which was determined[6] and shown to be nearly linear. $Hg_4^{2+}$ is unstable in the weakly basic solvent $SO_2$, disproportionating to $Hg_3^{2+}$ and the formal species $Hg^{0.35+}$ in the compound $Hg_{2.86}AsF_6$. In a review[7] on cations of Zn, Cd, and Hg, Gillespie and colleagues discuss the preparation and structural determination of this very unusual and interesting compound which is based on discrete close-packed $AsF_6^-$ anions with infinite cationic chains of Hg atoms running through channels in the solid, the average charge per Hg atom being

0.35. However, this compound does not have any observable solubility in superacidic media and therefore is marginally relevant in this book only because, as a solid, it appears to occur in equilibrium with both $Hg_3^{2+}$ and $Hg_4^{2+}$ in $SO_2$.

In summary, the dimeric cation $Hg_2^{2+}$ is unique in being stable in aqueous solution, although it disproportionates to Hg and nonionic Hg(II) compounds on addition of bases of the aqueous system or species such as $I^-$ that complex strongly with Hg(II). $Hg_3^{2+}$ can be prepared relatively easily but disproportionates slowly even in the acidic solvent $HSO_3F$. $Hg_4^{2+}$ disproportionates to $Hg_3^{2+}$ and $Hg^{0.35+}$ under the attempted conditions of preparation, that is, in the presence of the weak base $AsF_3$ or in solution in the weakly basic solvent $SO_2$. It is interesting to speculate that $Hg_4^{2+}$ may be isolable from very acidic media such as the solvents $CF_3SO_3H$, $HSO_3F$ or HF to which appropriate Lewis acids would be added.

## 5.2 Cadmium Cations

Cutforth et al.[7] give a brief review of the many experiments performed since 1890 in which metallic cadmium has been shown to dissolve in molten $CdCl_2$ to produce intensely colored, even black, solutions. Explanations of this coloration have been based on the effect of charge-transfer bands associated with species containing both Cd(I) and Cd(II) joined by halogen bridges. They report that Corbett and co-workers showed that this intense coloration was discharged by the addition of $AlCl_3$. Corbett demonstrated[8] that reduction of $CdCl_2$ by Cd to $Cd_2^{2+}$ was at a maximum when the ratio $AlCl_3$:$CdCl_2$ in the melt was approximately 3:1 or greater; that is, as the melt was made sufficiently acidic. Corbett was able to isolate $Cd_2(AlCl_4)_2$ from such melts. Cutforth et al. quote Corbett[8] in saying that "it seems that the charge-transfer complexes between $Cd^I$ and $Cd^{II}$, which are held together by a chloride bridge, are destroyed when chloride is replaced by the much less basic $AlCl_4^-$ ion." It seems much simpler to state that discrete $Cd_2^{2+}$ ions are formed at the expense of all other formally positively charged species when the system is made sufficiently acidic with $AlCl_3$ under reducing conditions.

Corbett postulated weak Cd—Cd bonding in $Cd_2^{2+}$ on the basis of a strong Raman band for the melt at 183 cm$^{-1}$ which can be related to a similar Raman band at 173 cm$^{-1}$ in the solid compound with formula $CdAlCl_4$. Corbett also[9] reports absorptions at 289 and 218 nm in the electronic spectrum for $Cd_2^{2+}$ in $NaAlCl_4$. He says that, when these melts are made 60% or more in $Cl^-$, the spectra are characteristic of those of $Cd^{2+}$—$Cl^-$ systems. All of these observations are consistent with the reversible equilibrium:

$$Cd + CdCl_2 \underset{Cl^-}{\overset{AlCl_3}{\rightleftharpoons}} Cd_2^{2+} + 2Cl^- \quad (2AlCl_4^-)$$

5.2

It is not surprising that "CdAlCl$_4$" gave a dark precipitate of metallic Cd resulting from disproportionation on contact with basic solvents such as water, dioxane, and ethanol.

In this author's opinion, Corbett complicates the discussion of stable compounds containing polyatomic cations by concerning himself too much with the relative lattice energies of compounds containing Cd$_2^{2+}$ and Cd$^{2+}$ (or of polyatomic cations of other metals and the "normal" cations of those metals) in compounds with AlCl$_4^-$ or other anions such as Cl$^-$ as the counterions.[8,10] It seems to be sufficient to postulate the stabilization in acidic melts of polyatomic cations and, additionally, their base-induced disproportionation. Their isolation as solid compounds is a separate issue. In chapter 10 it will be suggested that there are anionic types which are probably superior to AlCl$_4^-$ for producing stable solids containing polyatomic cations such as Cd$_2^{2+}$.

Cutforth et al.[7] recount attempts to isolate Cd$_2^{2+}$ and more complex homopolyatomic cations of Cd in media other than conventional melts. They report that Cd, interacting with AsF$_5$ or SbF$_5$ dissolved in AsF$_3$, does not form Cd$_2$(AsF$_6$)$_2$ but yields Cd(AsF$_6$)$_2$ and a gray compound. They say that the gray compound was formed by reaction of equimolar amounts of Cd and AsF$_5$ in AsF$_3$ at room temperature and had the formula Cd$_3$(AsF$_6$)$_2$. Not surprisingly it gave a white product and metallic Cd on reaction with atmospheric moisture. Excess Cd reacting with AsF$_5$ in AsF$_3$ gave Cd$_4$(AsF$_6$)$_2$. They state that relatively sharp Raman bands at 75 cm$^{-1}$ and 112 cm$^{-1}$ may reasonably be assigned to the two symmetrical stretching modes of a symmetrical linear Cd$_4^{2+}$ cation. Understandably, they point out the oddity of Cd$_2^{2+}$ disproportionating in AsF$_3$ whereas Cd$_3$(AsF$_6$)$_2$ and Cd$_4$(AsF$_6$)$_2$ can be isolated from this medium and suggest that this might be due to the insolubility of Cd$_3$(AsF$_6$)$_2$ in AsF$_3$. This insolubility would "drive" the disproportionation equilibrium in favor of formation of Cd$_3^{2+}$:

$$2Cd_2^{2+} \rightleftharpoons Cd_3^{2+} + Cd^{2+} \qquad\qquad\qquad 5.3$$

## 5.3 Zinc Cations

To date the only polyatomic cation of zinc that is known is the diatomic species Zn$_2^{2+}$ and that has proved difficult to generate and has low stability. In their 1968 application of a range of electrochemical techniques to the study of Zn, Cd, and Hg ions in the acidic eutectic melt AlCl$_3$:NaCl:KCl, 66:20:14, Hames and Plambeck[3] reported evidence for the formation of Hg$_2^{2+}$ and Cd$_2^{2+}$, but not for Zn$_2^{2+}$.

Kerridge and Tariq[11] obtained very good evidence for the generation of Zn$_2^{2+}$ when they obtained a yellow glass by rapidly chilling molten ZnCl$_2$ saturated with metallic Zn from 600°C. The glass was stable indefinitely in a dry atmosphere and dissolved in water to deposit gray zinc particles. It gave a Raman signal at 175 cm$^{-1}$, consistent with a Zn-Zn bond by comparison with Corbett's reported bands[9] at $183cm^{-1}$ and $173cm^{-1}$ for $Cd_2^{2+}$ in a melt and in the solid chloroaluminate. The calculated force constant for the Zn—Zn bond of $0.6$ mdyne/A° can be compared with those of $1.1$ and $2.5$ for $Cd_2^{2+}$ and $Hg_2^{2+}$. The Raman band for the Zn—Zn bond was observed in a solution of the glass in dry ether in which the cation was stable for 5–10 minutes after which the yellow color faded and zinc was deposited.

Kerridge and Tariq reported further[11] that addition of the chloride acceptor (i.e., Lewis acid) $CeCl_3$ to Zn—$ZnCl_2$ melts resulted in a steadily increasing percentage of formation of lower-oxidation-state zinc, reaching a limit of about 16–17% conversion, that is, about four times that occurring in a pure $ZnCl_2$ melt, when the $CeCl_3$:$ZnCl_2$ molar ratio was about 65%, that is, when effectively complete formation of $CeCl_4^-$ had occurred. Free $Cl^-$, which could cause disproportionation of $Zn_2^{2+}$, had been reduced to a minimum by the reaction

$$Zn + ZnCl_2 + 2CeCl_3 \rightarrow Zn_2^{2+} + 2CeCl_4^- \qquad 5.4$$

Triatomic and tetratomic cations similar to those in cadmium and mercury chemistry have not been characterized or isolated for zinc.

## 5.4 Bismuth Cations

Virtually all of the work on polyatomic cations of bismuth has been done in melts, initially simply by studying the dissolution of metallic Bi in $BiCl_3$. Later Bjerrum and Smith adopted a procedure that might well have been expected to prove more profitable than proved to be the case—they dissolved Bi and $BiCl_3$ in the acidic eutectics 63 mol% $AlCl_3$—37 mol% NaCl and $ZnCl_2$—NaCl (72:28) and obtained spectrophotometric evidence for the cations $Bi^+$ and $Bi_5^{3+}$.[12] They observed complete reaction in the more acidic melt $AlCl_3$—NaCl; that is, reaction proceeded until all the Bi or $BiCl_3$ was consumed, but, in the less-acidic medium $ZnCl_2$—NaCl, reaction reached equilibrium with both unreacted metal and unreacted $BiCl_3$ still present. When they reacted a large excess of Bi with dilute $BiCl_3$ in the 63–37% eutectic $AlCl_3$—NaCl they identified $Bi_8^{2+}$.[13]

Corbett[14] crystallized $Bi_5(AlCl_4)_3$ from $NaAlCl_4$ after reacting stoichiometric amounts of Bi and $BiCl_3$. He reported isolation of $Bi_4(AlCl_4)$, presumably $Bi_8(AlCl_4)_2$, which was characterized by chemical analysis, after reaction of excess Bi and $BiCl_3$. The analytical figures suggest that $Bi_8(AlCl_4)_2$ when isolated from $NaAlCl_4$ is not particularly pure. The analyses for the compound isolated from a melt containing excess $AlCl_3$ are much better, and he reported that he did not obtain direct reduction to "$Bi_4(AlCl_4)$" unless excess $AlCl_3$ was used. In much later work Krebs and co-workers isolated $Bi_5(AlCl_4)_3$ and $Bi_8(AlCl_4)_2$ from $NaAlCl_4$ and showed crystallographically that $Bi_5^{3+}$ is a trigonal bipyramidal cluster cation[15] and that $Bi_8^{2+}$ is a square antiprism.[16]

It is highly significant within the framework of the acid-base equilibria discussed in this book that the Krebs group reported that, in order to get "good" crystals of $Bi_5(AlCl_4)_3$, "there should be a slight but significant acidity of the melt by an excess of about 0.1 mol $AlCl_3$ [about 10 mol % $AlCl_3$] over the neutral composition given"[15] and that they prepared "pure" $Bi_8(AlCl_4)_2$ single crystals "by reaction of Bi with $BiCl_3$ + $4AlCl_3$ (33 mol% excess of $AlCl_3$ over a 'neutral' melt)."[16] Krebs's observations support the thesis that the formation of progressively lower fractional oxidation states for metal cluster cations requires progressively higher acidity of the media.

Hershaft and Corbett[17] isolated from Bi—$BiCl_3$ the compound believed from some earlier studies to have been BiCl. They assigned the formula $Bi_{12}Cl_{14}$ on the basis of their X-ray structural analysis from which they reported that the compound was the ionic aggregate $Bi_9^{5+}(Bi^{III}Cl_5^{2-})_2$ $(Bi_2^{III}Cl_8^{2-})_{0.5}$. Krebs and co-workers have given a preliminary mention of a refinement of the $Bi_{12}Cl_{14}$ structure in their paper on the structure of $Bi_5^{3+}$.[15] They suggest that the anionic part of the structure might be "described as polymeric instead of the idealized $Bi_2Cl_8^{2-}$ and $BiCl_5^{2-}$"; but this discussion of the nature of the anion does not have relevance in this chapter, where stabilities of cationic clusters are being rationalized in acid-base terms.

In describing the formation of $Bi_{12}Cl_{14}$, Corbett states that while small additions of KCl to the melt increased the yield—an effect that he puts down to effective production of chlorobismuthate(III) anions—larger additions caused "the amount of reduction possible with excess metal to diminish rapidly because an excess of the $BiCl_3$ is converted into complex anions." It seems much more likely that polybismuth cations of fractional charge disproportionated with increasing basicity of the medium. Formation of anions such as $BiCl_4^-$ in the basic melt would then be incidental to the formation of $BiCl_3$ by disproportionation, although formation of the anions would help to "drive" the disproportionation in favor of $BiCl_3$ buildup in the disproportionation equilibrium.

Later Corbett isolated the compound $Bi^+Bi_9^{5+}(HfCl_6^{2-})_3$, which contains the tricapped trigonal prismatic cluster cation by reducing Bi(III) in a 3:2 mixture of $HfCl_4$ and $BiCl_3$ with Bi.[18] This is still a neutral melt, but he provided a good isolating anion, the octahedral $HfCl_6^{2-}$—a more favorable combination of reactants than simple admixture of Bi and $BiCl_3$. This and other refinements of synthetic procedures in melts will be discussed in more detail in chapter 10.

There has been little work reported on generation of polyatomic cations of Bi other than that in melts. Burns et al.[19] investigated the possible oxidation of Bi to complex cations by $PF_5$, $AsF_5$, $SbF_5$, $SbCl_5$, $HSO_3F$, and $HSO_3Cl$ in the solvent $SO_2$. Only the pentafluorides oxidized Bi. $PF_5$ did so very slowly and incompletely; but both $AsF_5$ and $SbF_5$ generated $Bi_5^{3+}$ after initial formation of $Bi_8^{2+}$. The medium was probably too basic to allow formation of a stable product of $Bi_8^{2+}$ which would disproportionate to $Bi_5^{3+}$ in the first instance. A stable compound, $Bi_5(AsF_6)_3 \cdot 2SO_2$, was isolated and characterized. These authors made the following comment: "Interestingly, the cations $Bi_8^{2+}$ and $Bi_5^{3+}$ were only observed when the counteranions were derived from the group 5 pentafluorides. It is somewhat surprising that these cations, or perhaps others, were not formed as $SbCl_6^-$, $SO_3F^-$ and $SO_3Cl^-$ salts."[19] They report that when $SbCl_5$, $HSO_3F$, and $HSO_3Cl$ were used as potential oxidants, the products were white and crystalline, admixed with unreacted metal, depending on reaction stoichiometry. It would appear that the anions $SbCl_6^-$, $SO_3F^-$, and $SO_3Cl^-$, in the presence of $SO_2$, would be sufficiently basic to cause disproportionation of any polyatomic cations that might form and that cyclic oxidation and disproportionation would eventually lead to the formation of white compounds of Bi(III).

Burns et al.[19] reported that "the $Bi_5^{3+}$ cation was found to be slowly oxidized in 100% $H_2SO_4$. This was indicated by a complete loss of color after some hours, together with the concomitant formation of a white precipitate—containing only trivalent bismuth." This behavior can be compared with the months'-long stability in HF—$BF_3$ of the yellow species $Bi_5^{3+}$,[20] characterized as having the same electronic spectrum as a fresh solution of $Bi_5(AsF_6)_3 \cdot 2SO_2$ in 100% $H_2SO_4$. It can be argued that, even though HF—$BF_3$ is weakly acidic within the HF solvent system, it provides an environment of sufficient acidity for $Bi_5^{3+}$ to be stable indefinitely whereas, in the much more basic solvent 100% $H_2SO_4$, disproportionation to Bi(III) and Bi(0) would occur, the latter requiring only gentle oxidation to $Bi_5^{3+}$. A cycle of disproportionation reactions would give Bi(III) as the final product. It is not necessary to postulate that S(VI) in $H_2SO_4$ can cause direct oxidation of Bi(0) to Bi(III) even though a band assigned to $SO_2$ grew in intensity as the solution was losing color. $SO_2$ would be the reduction product of oxidation by $SO_3$ of Bi, formed progressively from disproportionation, to $Bi_5^{3+}$ or other fractional oxidation state cations.

## 5.5  Polyatomic Cations of Other Metals

While many metallic polyatomic cations other than those given above have been reported, the evidence for their existence is sketchy and fragmentary. It usually arises from simple studies of solubilities of metals in their halides or other salts with resultant mutual oxidation and reduction—much less frequently from electrochemical, magnetic, or spectroscopic (including electron spin resonance (ESR)) studies.

In a potentiometric study in weakly acidic chloroaluminate (5% $AlCl_3$—95% $NaAlCl_4$). Munday and Corbett[21] identified $Cd_2^{2+}$, previously characterized by that research group (section 5.2), and provided evidence for lead in an oxidation states between II and 0 and for silver between I and 0. Anders and Plambeck,[22] using potentiometry and voltammetry in a more acidic melt, ($AlCl_3$:NaCl:KCl, 66:20:14), not surprisingly identified Cu(I), Au(I), Ga(I), In(I), and Tl(I) but, more interestingly, obtained electrochemical evidence for Ag in an oxidation state between I and 0. As Hames and

Plambeck had shown for the cadmium and related systems,[3] equilibrium constants for most of the reactions—

$$M(II) + M(0) \rightleftharpoons M(I) \qquad\qquad 5.5a$$
$$\text{or} \quad M(III) + M(0) \rightleftharpoons M(I) \qquad\qquad 5.5b$$

were much greater in acidic chloroaluminate than in molten chloride or water. As Anders and Plambeck concluded, "[acidic] chloroaluminate melts favour the existence of lower-valence ionic species more strongly than water or chloride melts."[22]

In a review on polyatomic cations of group VI elements, Gillespie and Passmore[23] make brief reference to $Zn_2^{2+}$, $Pb_2^{2+}$, $Mg_2^{2+}$, $Ca_2^{2+}$, $Sr_2^{2+}$, $Ba_2^{2+}$, $Ag_2^+$, $Ag_4^+$, and $Ag_4^{3+}$. Many of these are proposed on the basis of reduction of salts of metals by the metals themselves. Again it is significant that often a Lewis acid is reported as markedly increasing the formation of lower oxidation states, for example, the extent of the unfavorable direct reduction of $ZnCl_2$ by Zn was increased when the Lewis acid $CeCl_3$ was added to the binary mixture $Zn-ZnCl_2$.[11] (See section 5.3.)

## 5.6 Summary

Despite the limited evidence that is available for the existence of individual homopolyatomic cations of metallic elements, there are some common general features associated with their synthesis, well illustrated for Hg, Cd, and Bi but less so for many others. In melts, the existence of these cations is favored as the melt is made more acidic and, where the evidence exists, inhibited as the melt is made more basic.

As was the case for generation of compounds containing polyatomic halogen cations from melts, for example, $I_3^+$ and $I_5^+$ as the neutral chloroaluminates by reaction of stoichiometric amounts of $I_2$, ICl, and $AlCl_3$, much of the work on synthesis of polyatomic cations of metallic elements from melts has been limited by using neutral media. For example, $Bi^+$ and $Bi_9^{5+}$ were isolated together under neutral conditions (section 5.4). The authors[18] speculate that their compound $Bi^+Bi_9^{5+}(HfCl_6^{2-})_3$ might result from disproportionation of $Bi_5^{3+}$ in the hypothetical compound $(Bi_5^{3+})_2(HfCl_6^{2-})_3$ to $Bi^+$ and $Bi_9^{5+}$. The single cations $Bi_5^{3+}$ and $Bi_8^{2+}$—of low charge per bismuth atom—required progressively more acidic melts for their isolation as pure compounds (section 5.4).

Burns et al.[19] studied possible oxidation of Bi to polyatomic cations, investigating $PF_5$, $AsF_5$, $SbF_5$, $SbCl_5$, $HSO_3F$, and $HSO_3Cl$ as potential oxidants in the solvent $SO_2$. Again, acid-base factors seem to have governed the synthesis of stable compounds of polyatomic bismuth cations. These appear to be stable when the counteranion is the hexafluoroanion of P, As, or Sb, but are unable to be formed when the potential counteranion is $SO_3F^-$, $SO_3Cl^-$, or $SbCl_6^-$. Fragmentary though it might be, the available evidence suggests that these latter anions are more basic than the hexafluoroanions and so would cause disproportionation of any bismuth cations formed transiently.

The Gillespie group has used its experimental approach effectively to produce "naked" cations, in the case of Hg and Cd, which had not been postulated from melt studies. They used an oxidant, for example, a pentafluoride, which also produces a very weakly basic fluoroanion. The medium for their studies was usually a very weakly basic solvent such as $AsF_3$ or $SO_2$.

As in the case of the polyatomic nonmetal cations, trends in stabilities of polyatomic metal cations within groups of the periodic classification can be observed, although they are not as well documented at this stage. It becomes progressively easier to produce "low-charge" cations of greater complexity in proceeding down a group. Thus $Zn_2^{2+}$ is generated with great difficulty and in low yield and has not been isolated in a separate compound. $Cd_2^{2+}$ is much easier to characterize and isolate from melts than $Zn_2^{2+}$ and $Cd_3^{2+}$ and $Cd_4^{2+}$ can be synthesized as "naked" cations. $Hg_2^{2+}$ is stable even in water and, again, the triatomic and tetratomic cations can be isolated. The force constants for the $M_2^{2+}$ cations, for M = Zn, Cd, and Hg, as quoted in section 5.3, support this stability trend within the group.

Much more limited information is available for trends within other groups. There is a wide range of polyatomic cations of Bi (section 5.4). Corbett[10] reports that the reduction of $SbCl_3$ by Sb is increased dramatically on addition of $AlCl_3$, but there do not appear to be reports of arsenic in formal positive oxidation states lower than III.

# References

1. (a) C.L. Hussey, *Pure and Appl. Chem., 60,* 1763 (1988). (b) P.A. Barnard, I-Wen Sun, C.L. Hussey, *Inorg. Chem., 29,* 3670 (1990).

2. K.R. Seddon in *Molten Salt Chemistry,* G. Mamantov, R. Marassi, eds. D. Reidel, 1987, pp. 377–379.

3. D.A. Hames, J.A. Plambeck, *Can. J. Chem., 46,* 1727 (1968).

4. G. Torsi, K.W. Fung, G.M. Begun, G. Mamantov, *Inorg. Chem., 10,* 2285 (1971).

5. B.D. Cuthforth, C.G. Davies, P.A.W. Dean, R.J. Gillespie, P.R. Ireland, P.K. Ummat, *Inorg. Chem., 12,* 1343 (1973).

6. B.D. Cutforth, R.J. Gillespie, P. Ireland, J.F. Sawyer, P.K. Ummat, *Inorg. Chem., 22,* 1344 (1983).

7. B.D. Cutforth, R.J. Gillespie, P.K. Ummat, *Revue de Chimie Minérale, 13,* 119 (1976).

8. J.D. Corbett, W.J. Burkhard, L.F. Druding, *J. Am. Chem. Soc., 83,* 76 (1971).

9. J.D. Corbett, *Inorg. Chem., 1,* 700 (1962).

10. J.D. Corbett in *Progress in Inorganic Chemistry,* Vol. 21. S.J. Lippard, ed. John Wiley, New York, 1976, pp. 129–158.

11. D.H. Kerridge, S.A. Tariq, *J. Chem. Soc. (A),* 1122 (1967).

12. N.J. Bjerrum, C.R. Boston, G.P. Smith, *Inorg. Chem., 6,* 1162 (1967).

13. N.J. Bjerrum, G.P. Smith, *Inorg. Chem., 6,* 1968 (1967).

14. J.D. Corbett, *Inorg. Chem.*, *7*, 198 (1968).

15. B. Krebs, M. Mummert, C. Brendel, *J. Less-Common Metals, 116*, 159 (1986).

16. B. Krebs, M. Hucke, C.J. Brendel, *Angew Chem., Int. Ed. Engl., 21*, 445 (1982).

17. A. Hershaft, J.D. Corbett, *Inorg. Chem., 2*, 979 (1963).

18. R.M. Friedman, J.D. Corbett, *Inorg. Chem., 12*, 1134 (1973).

19. R.C. Burns, R.J. Gillespie, W.-C. Luk, *Inorg. Chem., 17*, 3596 (1978).

20. J. Besida, T.A. O'Donnell, unpublished observations.

21. T.C.F. Munday, J.D. Corbett, *Inorg. Chem., 5*, 1263 (1966).

22. U. Anders, J.A. Plambeck, *Can. J. Chem., 47*, 3055 (1969).

23. R.J. Gillespie, J. Passmore, *Acc. Chem. Res., 4*, 413 (1971).

# CHAPTER 6

# Transition Element Cations in "Normal" Oxidation States in Protonic Superacids and Acidic Melts

In considering "acidic" and "basic" nonaqueous solvents in chapter 1, it was convenient to take as a reference point, in discussing acidity and basicity, the familiar solvent of inorganic chemistry, water. Elements of groups I and II of the periodic classification form compounds that are essentially ionic in nature, and to the extent that these compounds are soluble in water, they dissolve to form solvated—that is, hydrated—cations and anions derived from the solute. Compounds of the heavier members of groups III–V, such as Tl, Sn, Pb, and Bi, can dissolve in water to form solvated cations. Compounds of other main group elements may undergo hydrolysis, but will not form simple hydrated cations. It has been seen in chapters 4 and 5 that many other main-group elements form polyatomic cations, but these cations are hydrolyzed on contact with water. These elements cannot be cationic in *aqueous* solution. They react with the bases of the aquosolvent system and disproportionate to form oxo- and hydroxoanions even in very strongly acidic aqueous solutions.

When transition elements of the first row of the periodic classification form simple compounds in which the elements are in oxidation state II or III, these compounds can dissolve in an aqueous solution that is sufficiently acidic and free of complexing agents and can form aquocations of the type $M(OH_2)_n^{2+}$ and $M(OH_2)_n^{3+}$ where $n$ is usually 6.

Even for cations of metals in these low oxidation states, stable in aqueous solution, the higher the charge on a cation the greater the interaction of the cation with the base or bases of the solvent or with any other bases present. Thus $CrOH^{2+}$ and $CrX^{2+}$ (for X = a uninegative base other than $OH^-$) form to a greater extent than $CrOH^+$ and $CrX^+$ under similar solution conditions. Similarly, $Cr^{3+}$ (and $Fe^{3+}$, etc.) form polymeric species such as $Cr_2(OH)_4^{2+}$ and $Cr_3(OH)_4^{5+}$ to a greater extent than $Cr^{2+}$ and analogous systems. The formation of species of the general type $Cr_x(OH)_y^{(3x-y)+}$ by interaction of the cation with the base $OH^-$ provides further opportunity for reducing the ratio of positive-charge–to–metal. In aqueous solutions which are approximately neutral, so-called hydroxides precipitate. These are certainly not species like $Cr(OH)_3$ or $Fe(OH)_3$, but very complex neutral polymeric compounds—and therefore of very low solubility—containing many $Cr^{3+}$ (or $Fe^{3+}$) formal cations with $H_2O$, $OH^-$, and $O^{2-}$ as ligands. Their stoichiometry is quite indefinite, depending on pH and metal concentration of the solution. In moderately basic solutions Cr(III) species are no longer cationic but anionic, for example, $Cr(OH)_4^-$.

It follows that with higher formal charge on a $d$-transition metal, the discrete aquocation as such will not be observed for most elements. The ratio of charge to ionic volume is too great for the formal aquocation $Ti(OH_2)_6^{4+}$ to exist in aqueous solution, even under strongly acidic conditions. The aquocation can be considered to undergo deprotonation or to interact with $OH^-$, the base of the solvent system, to form $Ti(OH)_2(OH_2)_4^{2+}$ or even the trihydroxospecies. These are the forms in which cationic titanium(IV) exists in acidic aqueous solution.[1a] Similarly, vanadium(IV) and vanadium(V) can be considered as being in solution as $V(OH)_2(OH_2)_4^{2+}$ and $V(OH)_4(OH_2)_2^+$, each of which can lose one or more water molecules from one or more pairs of $OH^-$ ligands to form $VO(OH_2)_5^{2+}$ and $VO_2(OH_2)_4^+$, frequently represented simply as $VO^{2+}$ and $VO_2^+$.[1b]

A well-established feature of the $d$-transition elements of the second and third row of the periodic classification is that they form compounds in which they are in much higher oxidation states than for the corresponding first-row elements, for example, elements like Mo and W form anionic molybdates and tungstates and related compounds in aqueous solution where the metals are in oxidation state VI. They do not form simple aquocations like $Cr(OH_2)_6^{2+}$ and $Cr(OH_2)_6^{3+}$. Even for oxidation state IV, the Zr(IV) cation has been described in aqueous solution as $[Zr_4(OH)_8]^{8+}$. For metals with larger cationic radii, for example, for $Th^{4+}$ and for actinides with atomic numbers greater than that of Th, the aquocation of charge $+4$ can be observed in aqueous solutions which are sufficiently acidic, that is, which have very low concentrations of $OH^-$, and are free from other bases which can form complexes. For $f$-transition metals with formal charges $+5$ and $+6$, cations in solution, and in the solids derived from aqueous solution, are based on the $MO_2^+$ and $MO_2^{2+}$ entities. These dioxocations are monomeric in aqueous solution only at low pH values.[1b] As the pH is gradually increased, they accept bridging base anions to form polymeric cations in which, with increasing pH, the charge-to-metal ratio is gradually decreased still further.

Some $d$-transition elements of the first row can exist in acidic aqueous solution in high oxidation states, for example, Cr(VI) and Mn(VII), but they then exist as *anionic* species, the familiar $CrO_4^{2-}$ and $MnO_4^-$ being examples. $CrO_4^-$ can be considered as being derived from $Cr(OH_2)_6^{6+}$, which is much too highly charged to exist as such even in highly acidic aqueous solutions. $Cr(OH_2)_6^{6+}$ could be considered as interacting with the base of the solvent $H_2O$ to form $Cr(OH)_6$ in the first instance. Elimination of two water molecules from four hydroxyl ligands would give $(HO)_2CrO_2$, the strong acid origin of $HCrO_4^-$ and $CrO_4^{2-}$. In an interesting main-group analogy S(VI) and Se(VI) exist in water as $SO_4^{2-}$ and $SeO_4^{2-}$, both being protonated, depending on the level of acidity of the solution. The parent acids are $(HO)_2SO_2$ and $(HO)_2SeO_2$. The larger Te(VI) can accommodate more ligands in aqueous solution than S(VI) and Se(VI) and exists in water as the acid $Te(OH)_6$.

Until a little over a decade ago, virtually nothing was known about existence and speciation of monatomic transition metal cations in protonic superacid media and acidic melts, although there already existed a good body of investigational work on the more complex polyatomic nonmetal and metal cations in both types of media and of haloanions of transition metals in basic melts.

The bulk of the work on transition-metal cation formation in superacids has been done for the HF solvent system. Much more recently there has been a very limited amount of investigation of speciation in the superacids $HSO_3F$, $CF_3SO_3H$, and $H_2SO_4$. What has emerged is very small by comparison with the vast amount of work on the nature of ions derived from metallic elements in aqueous solution, but the general principles applying to formation of cations, anions, and neutral insoluble species appear to be the same in water, in protonic superacids, and in melts. There is nothing remarkable about formation of solvated cations in superacids, but because it is a field of inorganic solution chemistry opened up quite recently, it is presented in this chapter, where cations in "normal" oxidation states are discussed, and in the following chapter where it will be shown that cations in "low" oxidation states can be used to demonstrate some new chemistry in superacidic media of controlled acidity and basicity.

Classification of cations as to whether they are in "normal" or "low" oxidation states is necessarily arbitrary. Complexation with ligands of the solvent occurs when metallic elements in high oxidation states occur in aqueous solution. This can be seen with stable water-soluble ions such as $Cr_2O_7^{2-}$, $MnO_4^-$, $VO_2^+$, and $UO_2^{2+}$. Compounds in exceptionally high oxidation states for a particular element would oxidize water to $O_2$ and other oxidation products in any attempt to form an aqueous solution. $K_2Ni^{IV}F_6$ is an example of such a compound.

Within this framework, a compound or cation is considered to contain the metal in a "low" oxidation state if it would cause reduction of water. The classification does not depend on the numerical magnitude of the charge on the cation. Thus $U^{3+}$ reduces water while $Ni^{2+}$ does not. Clear-cut examples of low-oxidation-state cations to be discussed in chapter 7 are $Ti^{2+}$ and $Sm^{2+}$, which are both stable in HF but reduce $H_2O$. Inevitably there will be some overlap in an arbitrary classification between "low" and "normal" oxidation states. $V(OH_2)_6^{2+}$ would reduce water. $Cr(OH_2)_6^{2+}$ is stable in acidic aqueous solution in the absence of air, but these will all be presented, with others, in chapter 7.

## 6.1  Cations of Transition Elements in Nonaqueous Media

As stated earlier, most of the spectroscopic and other investigations of the generation, stability, and solvation of transition-metal cations in superacids have been carried out using HF as a solvent, despite the fact that rather specialized experimental procedures must be adopted for HF solutions, as outlined in section 1.1.3.5. Little has yet been published in the open literature on cationic speciation in $HSO_3F$, $CF_3SO_3H$, and $H_2SO_4$ even though solutions based on these solvents can be studied in conventional glass apparatus. Most of the work in these three superacids is included in postgraduate theses from the University of Melbourne, and it has been summarized in a short review on cations of $d$- and $f$-transition elements in acidic nonaqueous media published in 1988.[2]

Some spectra for transition-metal compounds in acidic chloroaluminates have been available for nearly 30 years, but there has been little speculation on speciation in these melts. It will be argued here that, by comparing these spectra with those for corresponding elements in water and in protonic superacids, particularly HF, a strong case can be made for postulating that these species are cationic in acidic melts, as in $H_2O$ and HF.

In the following sections, spectra in nonaqueous acidic media of $d$- and $f$-transition elements will be compared with corresponding spectra in water to show that solvated cations exist in these media.

## 6.1.1  Cations of $d$-Transition Elements

In the sections below, the general proposition will be seen to hold that, for both $d$- and $f$-transition elements, the fluorides, bisulfates, fluorosulfates, triflates, and chlorides are not sufficiently soluble in neutral media, that is, in the appropriate neutral protonic superacids or chloroaluminates, to provide observable spectra in those media. However increase of the acidity of each of the media by addition of Lewis acids leads to formation of solutions with observable spectra which then provide information about the existence of solvated cations in those media. For $d$- and $f$-elements, there are sufficient significant differences in the ease of generation of solutions and in the observed spectra to warrant separate presentation.

### 6.1.1.1  Solutions in Anhydrous Hydrogen Fluoride

Reference to Table 1.2 of chapter 1 shows that solid fluorides containing singly positively charged cations have quite high solubilities in HF at ambient temperature. The largely ionic alkaline earth difluorides are moderately soluble. By comparison, the much-less-ionic solids, the $d$-transition element difluorides, are very much less soluble (about $1–4 \times 10^{-3}$ molal) and the trifluorides are generally less soluble again. Fluorides of transition elements in very high oxidation states, for example, $VF_5$, $ReF_7$, and the many hexafluorides of the actinides and of the second- and third-row transition metals, have relatively high solubilities in HF (0.5–3.3 molal)[3] but are present in solution as molecular, not cationic, species. For most of these higher fluorides the electronic configuration of the metal is $d^0$ and their HF solutions present very little of interest for electronic spectroscopic studies. The cations of transition metals in low or intermediate oxidation states, that is, those with partially filled $d$-orbitals, are the ones of interest to spectroscopists and those that will provide information on solvation of the cations in HF and other nonaqueous solvents.

Prior to 1975 there had been no spectroscopic investigations of transition-metal cations in anhydrous hydrogen fluoride because of the low solubilities of the di-, tri-, and tetrafluorides and because all anions other than complex fluoro anions are solvolyzed, in part at least, in anhydrous HF to fluoride.[3] The implication of the latter point is that if there were to be an attempt to dissolve a transition metal compound containing an anion other than a fluoroanion, the fluoride produced by solvolysis of the anion would cause precipitation of the transition-metal cation as the fluoride. In that year Cockman and O'Donnell[4] deliberately enhanced the acidity of HF by addition of fluorides that are Lewis acids of the solvent system and produced solutions sufficiently concentrated to allow the use of UV-visible, Raman, and ESR spectroscopy to study the nature and extent of solvation by HF molecules of the cations in solution. Contemporaneously, Court and Dove reported[5] a value of the wavelength for the major peak and its shoulder in a solution of $Co^{2+}$ in HF acidified with both $BF_3$ and $AsF_5$ as part of an investigation of the solution chemistry of Co(II), Co(III), and Co(IV) in HF.

In supposedly neutral HF that, if it were to contain no base-producing impurities, would be slightly basic because of the generation of solvated fluoride ions through dissolution of the solid fluoride $MF_n$, the equilibrium

$$MF_n + (m + n)HF \rightleftharpoons M^{n+}_{(FH)_m} + nHF_2^-$$  6.1

lies overwhelmingly to the left as written. Reaction of the Lewis base $F^-$ (written here in the solvated form $HF_2^-$) with a Lewis acid, for example,

$$BF_3 + HF_2^- \rightleftharpoons BF_4^- + HF$$  6.2

displaces the equilibrium in equation 6.1 to an extent dependent on the strength of the Lewis acid. The weak Lewis acid $BF_3$ produces solutions in HF of dipositive $d$-transition metals and tripositive $f$-transition metals of sufficient strength for spectroscopy.[4] Stronger Lewis acids were required to give solutions of tripositive $d$-transition metal cations[6] or tetrapositive actinide ions[7] because the concentration of the base $F^-$ initially in equilibrium with the solid fluoride must be reduced still further in order to dissolve solid fluorides with more highly positively charged cations, that is, with higher lattice energies. In all of this work the spectra recorded for the solutions support the postulation[4,6,7] that each solution contains cations of $d$- or $f$-transition metals solvated by HF molecules in a fashion exactly analogous to the coordination by water molecules in the case of the corresponding aquocations.

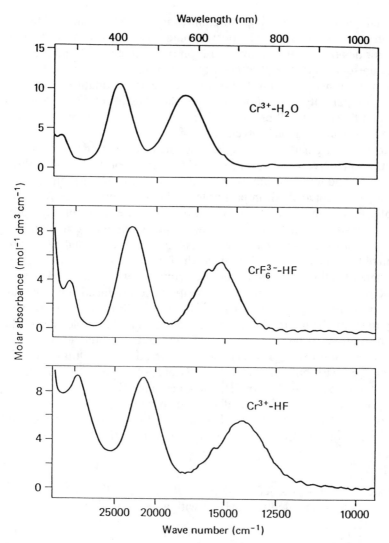

**Figure 6.1.** UV-Visible Spectra of Chromium(III) as $Cr(OH_2)_6^{3+}$ in $H_2O$, $CrF_6^{3-}$ in HF, and $Cr(FH)_6^{3+}$ in HF.

In their initial report on electronic spectra of cations in HF, Barraclough et al.[4] presented spectra for $Ni^{2+}$, $Co^{2+}$, $Pr^{3+}$, and $Nd^{3+}$ and discussed them in terms of HF solvation of the cations by comparing spectra for the corresponding aquocations. In a subsequent paper[6] on the spectra of $Cr^{3+}$ and $Mn^{2+}$ they provided more detailed evidence for hexacoordination by solvent molecules by studying a wider range of analogues for the entity postulated as $Cr(FH)_6^{3+}$ and by recording the ESR spectrum for $Mn^{2+}$. Most emphasis will be placed on these two systems in this presentation.

Figure 6.1 shows the familiar spectrum for the hexaaquochromium(III) cation which is well established as the stable form of chromium(III) in aqueous solutions which are acidic and contain no strong complexing species. Hydroxospecies form as the acidity is reduced, and the spectrum is modified if even simple complexing species such as chloride ion are present. Figure 6.1 also shows the very similar spectrum for chromium(III) in HF, the acidity of which has been very much increased by use of the strong Lewis acid $SbF_5$.[6] The similarity of these two spectra indicates hexacoordination of Cr(III) in acidic HF, and the postulated entity is $Cr(FH)_6^{3+}$. The proposed hexacoordination of Cr(III) in HF was supported by preparing $(NH_4)_3CrF_6$ and recording spectra of the stable anion in water and in HF made strongly basic with $NH_4F$ which was added to prevent dissociation of $CrF_6^{3-}$. The spectra were virtually identical in each medium.[6] Data derived from the spectra in Figure 6.1 are given in Table 6.1. Also included in Table 6.1 are data from the spectrum of a solution prepared by dissolving $CrF_3$ in HF which was 0.5 M in $NH_4F$.[8] This shows that $CrF_3$ is amphoteric in HF, giving $CrF_6^{3-}$ as the anionic species.

Two general features of *all* spectra recorded for *d*-transition metal cations in $H_2O$ and in HF can be seen from Figure 6.1. First, molar absorbances are less in HF than in $H_2O$, usually about half the aqueous case. Second, corresponding peaks are shifted to lower energy, the shifts in the case of Cr(III) spectra being of the order of about 50 to about 150 nm, depending on band positions. The three spectra in Figure 6.1 indicate that the ligand strengths of the three ligands $H_2O$, $F^-$, and HF as members of the spectrochemical series decrease in that order.

The shifts were also visually apparent as the colors of the solutions were blue-violet, green, and yellow-green, respectively. The similarities in peak position, profiles, and relative intensities of the bands in the three media quite clearly indicate that the *d-d* transitions of the octahedrally coordinated chromium(III) ion are being observed in each case. Three spin-allowed transitions were observed together with a number of low-intensity spin-forbidden bands which appear as shoulders on the lowest-energy spin-allowed band.

Good evidence for hexacoordination of $Mn^{2+}$ by HF solvent molecules was provided by the ESR spectrum for a solution of $Mn^{2+}$ in HF—$AsF_5$,[6]

**Table 6.1.** Peak Maxima of Principal Bands in Chromium(III) Spectra

| Solution | Peak maxima (nm) | | |
|---|---|---|---|
| $Cr(OH_2)_6^{3+}$ in 1 M $HClO_4$ | 265 | 407 | 575 |
| $(NH_4)_3CrF_6$ in $H_2O$ | 291 | 441 | 671 |
| $(NH_4)_3CrF_6$ in 1 M NaF—HF | 289 | 448 | 671 |
| $CrF_3$ in 5 M $SbF_5$—HF | 308 | 476 | 714 |
| $CrF_3$ in 0.5 M $NH_4F$—HF | 286 | 436 | 673 |

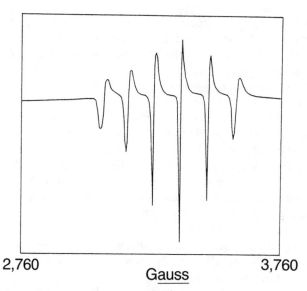

2,760                                                     3,760

Gauss

**Figure 6.2.** ESR Spectrum of Manganese(II) in $AsF_5$—HF.

shown in Figure 6.2. It is very similar to that for $Mn(OH_2)_6^{2+}$ in water show-ing the characteristic six lines due to the interaction of the unpaired electrons with the nuclear spin of 5/2 on the manganese.

As stated earlier, the first publication[4] in this program of spectroscopy of transition-metal cations in HF contained spectra for Ni(II) and Co(II) which showed that the cations in solution were $Ni(FH)_6^{2+}$ and $Co(FH)_6^{2+}$. Cockman[9] recorded similar HF solution spectra for $Cu^{2+}$ and $Fe^{2+}$.

In the Melbourne program there were several unsuccessful attempts to obtain from acidified HF crystals which might contain solvated cations, as can be done with aquocations. All attempts resulted in the formation of an-solvates. For example, good crystals obtained from a solution of $Mn^{2+}$ in HF—$BF_3$ were shown to be the ansolvate $Mn(BF_4)_2$[10]—an interesting struc-ture in which the Mn(II) atoms are seven-coordinate and are bridged through F atoms of $BF_4^-$ anions in an arrangement in which four of the $BF_4^-$ anions bridge four $Mn^{2+}$ atoms and the remaining three bridge to three $Mn^{2+}$ atoms; that is, the formula can be written as $Mn(BF_4)_{4/4}(BF_4)_{3/3}$.

## 6.1.1.2 Solutions in Other Protonic Superacids

Chapter 4 in particular contains extensive accounts of the work done by Gillespie and his research group on generation and spectroscopic character-ization of polyatomic cations of nonmetallic elements in $H_2SO_4$, oleums, and $HSO_3F$. Until very recently nothing has been reported on the generation of cations of transition metals in these acids and the related "triflic" acid $CF_3SO_3H$.

The transition-metal compounds $MX_n$ (where X represents $HSO_4^-$, $SO_3F^-$, or $CF_3SO_3^-$, the bases of the solvent systems $H_2SO_4$, $HSO_3F$, or $CF_3SO_3H$) are sparingly soluble in the "neat" or nearly neutral solvents as is the case for transition-metal fluorides $(MF_n)$ in "neat" HF (6.1.1.1).

Following the experience with HF solutions, the expectation would be that, as the concentration of the base $X^-$ is reduced by addition of Lewis acids of the solvent system HX, the concentration of the transition metal cation $M^{n+}$ would increase sufficiently to allow electronic and other spectra to be recorded so that information on the solvation of these cations in the different superacids could be gathered. Since the mid-80s an initial survey of cationic speciation in oleums and acidified $HSO_3F$ and $CF_3SO_3H$ has been undertaken at the University of Melbourne.

In the Melbourne-based program involving recording of electronic spectra of transition-metal cations in superacid systems, and thence studies of coordination by solvent molecules, the preferred preliminary system has been Co(II) because spectra provide information on the environment in solution of Co(II). Hexacoordinated Co(II) gives spectra similar to the well-known spectrum for $Co(OH_2)_6^{2+}$—a broad peak above 500 nm with a shoulder 50–70 nm to higher energy.[11] Tetracoordination of Co(II) provides more intense, more complex spectra with a cluster of three well-defined peaks. These differences can be seen in Figure 6.4 of section 6.1.1.3 for Co(II) in basic and acidic melts, and the differences in coordination will be discussed in somewhat greater detail there.

The hydrogensulfates, fluorosulfates, and triflates of cobalt(II) were prepared and shown to be sparingly soluble in their nearly neutral parent acids. Addition of appropriate Lewis acids increased the concentration of cobalt(II) in each case. For example, the solubility of $Co(HSO_4)_2$ in 100% $H_2SO_4$ was found to be approximately $3 \times 10^{-5}$ mol $kg^{-1}$. The solubility increased with addition of $SO_3$, reaching a maximum of about $7 \times 10^{-4}$ mol $kg^{-1}$ at 20% $SO_3$.[12] The spectrum in this medium is shown in Figure 6.3.[12] It is very similar to those for $Cr(OH_2)_6^{2+}$ in aqueous solution[11] and for $Co(FH)_6^{2+}$ in acidified HF,[4] indicating hexacoordination of Co(II) in solution.

The cationic nature of Co(II) in 20% $SO_3$—$H_2SO_4$ was demonstrated in an electrolysis experiment by using a three-compartment cell with sintered glass membranes separating each compartment and with platinum electrodes in the anode and cathode compartments, each of which contained 20% $SO_3$—$H_2SO_4$ initially. A solution of Co(II) in 20% $SO_3$—$H_2SO_4$ was introduced into the middle compartment and a DC potential was applied. After many hours, the anode compartment was found to contain no Co(II) detectable by analytical spectrophotometry. Co(II) had moved to the cathode compartment, and, spectrophotometrically, it was shown that the total amount of Co(II) in the central and cathode compartments at the cessation of the experiment was equal to the amount of Co(II) in the central compartment initially.[12] The experimental indication of the existence of the hexacoordi-

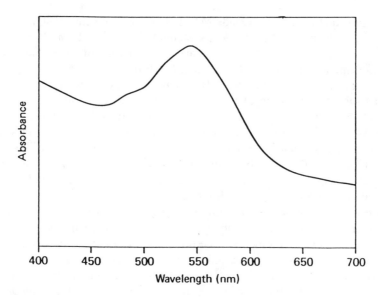

**Figure 6.3.** UV-Visible Spectrum of $Co^{2+}$ in 20% $SO_3$—$H_2SO_4$.

nated cation in oleum solutions leads to the proposal that $Co^{2+}$ in oleums is solvated by three bidentate $H_2SO_4$ molecules as ligands.

Similar spectra were recorded when $Co(SO_3F)_2$ was dissolved in $HSO_3F$ made acidic with $SbF_5$,[13] and when cobalt triflate, $Co(CF_3SO_3)_2$, was dissolved in $CF_3SO_3H$ made more acidic by the addition of the Lewis acid boron triflate, $B(CF_3SO_3)_3$.[14] The spectrum in each case consists of a broad peak arising from a *d-d* electronic transition with a shoulder assumed to be due to splitting by spin-orbit coupling, as for the aquocation. Peak and shoulder positions are given for each solvent in Table 6.2. For $H_2O$ the molar absorption coefficients ($mol^{-1}$ $dm^3$ $cm^{-1}$) are 2.8 for the shoulder and 4.5 for the main peak, while the corresponding values for HF are 1.3 and 2.8.[4]

The similarity of the spectra for Co(II) in aqueous solution where the cation is accepted as being $Co(OH_2)_6^{2+}$, in HF where the evidence for the solvated cation $Co(FH)_6^{2+}$ is good, and in $H_2SO_4$, $HSO_3F$, and $CF_3SO_3H$ leads to the proposal, already made for $H_2SO_4$, that in these superacids the dipositive cation is solvated by three bidentate $HSO_3F$ or $CF_3SO_3H$ molecules.

For the reasons given, most of the Melbourne work has been directed toward generation and spectral characterization of cations of cobalt(II) in the four superacids of major interest. More recently Adrien[15] has recorded

**Table 6.2.** Peak Positions for Cobalt(II) in Water and in Superacids

| Solvent | $H_2O$ | HF | $HSO_3F$ | $CF_3SO_3H$ | $SO_3$—$H_2SO_4$ |
|---|---|---|---|---|---|
| Peak positions | 458sh[a] | 471sh | 477sh | 482sh | 483sh |
| (nm) | 510 | 532 | 545 | 555 | 540 |

[a]sh = shoulder.

spectra for Ni(II) and Cr(III) in $CF_3SO_3H$ acidified with $B(OSO_2CF_3)_3$ and found them to be similar to those in acidified HF and $H_2O$, pointing again to a cation that is hexacoordinated by solvent molecules. He observed Cr(III) spectra for $CF_3SO_3H$ made basic with $CF_3SO_3^-$, indicating that, as in HF, the insoluble binary compound is amphoteric. Most of Adrien's work on transition metals in $CF_3SO_3H$ has been directed toward generation of cations in unusually low oxidation states in this solvent and is presented in chapter 7.

## 6.1.1.3 Speciation in Chloroaluminate Melts

Most of the definitive work on coordination of transition-metal species in melts has been done in basic melts, for example, in alkali metal halides. A classic early example was the work of Gruen and McBeth,[16] who dissolved 3d-transition-metal di- and trichlorides in alkali-metal-chloride eutectics and in neutral melts such as $CsGaCl_4$ and $Cs_2ZnCl_4$. They postulated the formation of anionic species such as $CoCl_4^{2-}$, $NiCl_4^{2-}$, $FeCl_4^{2-}$, and $CrCl_6^{3-}$ in all melts, with V(II), V(III), and Ti(III) exhibiting equilibria between the hexachloro- and the tetrachloroanions, depending on the pCl of the melt.

A little later Øye and Gruen[17] recorded spectra for 3d-transition-metal dichlorides in molten $AlCl_3$—an acidic medium—and interpreted all of the spectra simply "on the basis of octahedral configurations of *chlorides* about the central transition metal ions . . . in contradistinction to the situation in alkali chloride melts where, with the exception of $V^{2+}$, the dipositive 3d ions display fourfold coordination." They did speculate that in $AlCl_3$ the dipositive cations $M^{2+}$ could be coordinated to two tridentate $Al_2Cl_7^-$ anions.

The general description "hexacoordination by chlorides" has been incorrectly extended by several authors to infer that the species in solutions of $MCl_2$ in molten $AlCl_3$ are $MCl_6^{4-}$ *anions*. For example, Lever presents the spectrum for $VCl_2$ in $AlCl_3$ as being that for the $VCl_6^{4-}$ anion[18a] and describes the spectrum for $TiCl_2$ in $AlCl_3$ as being "presumably derived from the $[TiCl_6]^{4-}$ ion."[18b] The question of speciation when transition-metal dichlorides are in the highly acidic solvent $AlCl_3$ will be discussed in greater detail below when spectra for $CoCl_2$ in both basic and acidic chloride melts will be compared.

As stated above, Gruen and McBeth[16] studied several 3d-transition metal chlorides in the eutectic LiCl—KCl, a basic melt of the chloroaluminate system, at temperatures between 400 and 1,000°C. Their observed spectrum for $CoCl_2$ (Figure 6.4a) was virtually identical to that for an aqueous solution of $[Ph_3MeAs]_2CoCl_4$ containing excess $Cl^-$ to stabilize the $CoCl_4^{2-}$ anion,[19] and they postulated the existence in the basic melt of the tetrahedral anionic species $CoCl_4^{2-}$.

Subsequently Øye and Gruen recorded spectra at 227°C of dichlorides of 3d-transition metals in pure $AlCl_3$, the strongest acid of the same solvent system.[17] Their spectrum for Co(II), is characteristic for that of the cation

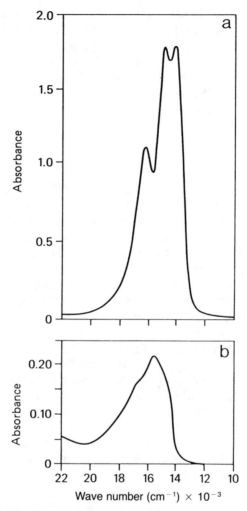

**Figure 6.4.** UV-Visible Spectra Cobalt(II) (a) as $CoCl_4^{2-}$ in Basic Melt and (b) as $Co^{2+}$ in Acidic Melt.

in an octahedral environment. The broad peak with a shoulder is very similar to that for $Co^{2+}$ in water and in protonic superacids (Section 6.1.1.2) and is part of a simpler spectrum than for four-coordinate Co(II). As expected, the intensities of bands for the octahedral environment are significantly less than for those arising from the tetrahedral arrangement of ligands. Table 6.3 shows that, for similar spectra, peak maxima are shifted to much lower energies in acidic melts than in protonic acidic solvents.

In their first joint publication in the area, Øye and Gruen stated simply that Co(II) in the acidic medium $AlCl_3$, like the other $3d$-metals cations, was in an octahedral environment of chlorides.[17] They did suggest the possibility that the cation $M^{2+}$ was coordinated to two $Al_2Cl_7^-$ anions, each of which

**Table 6.3.** Peak Maxima (nm) For Co(II) and Ni(II) in Protonic Solvents and Acidic Melts

|       | $H_2O$ | $HF^4$ | $CF_3SO_3H^{15}$ | $AlCl_3{}^7$ |
|-------|--------|--------|------------------|--------------|
| $Co^{2+}$ | 458sh  | 471sh  | 482sh            | 590sh        |
|       | 510    | 532    | 555              | 633          |
| $Ni^{2+}$ | 393    | 424    | 435              | 476          |
|       | 654    | —      | 585              | —            |
|       | 722    | 850    | 850              | 926          |
|       | 1176   | —      | —                | 1560         |

was tridentate to provide the octahedral arrangement of ligands. In a subsequent, more detailed spectroscopic study[20] of cobalt(II) in chloroaluminates of varying acidity and basicity, Øye and Gruen demonstrated octahedral coordination of Co(II) in a moderately acidic melt (AlCl₃:KCl, 64.5:35.5) and tetrahedral coordination, presumably $CoCl_4^{2-}$, in a melt that was extremely weakly acidic (AlCl₃:KCl, 50.1:49.9).

They showed that between 42% KCl and 49.9% KCl, that is, in melts which were weakening progressively in acidity, the spectra changed from that for octahedral Co(II) to one typical of tetrahedral Co(II). In this paper they postulated the existence of electrically neutral $Co(Al_2Cl_7)_2$ in acidic melts, with the two $Al_2Cl_7^-$ ligands being tridentate. They argued that as the basicity of the melt increases, that is, the amount of available $Cl^-$ increases, $Al_2Cl_7^-$ is replaced by $AlCl_4^-$ and $Co^{2+}$ is coordinated in the 40–50% KCl range by both $AlCl_4^-$ and $AlCl_4^-$ in differing proportions, causing gross distortion of the octahedral environment of the $Co^{2+}$ present in more acidic melts.

Øye and Gruen reported[20] that addition of KCl to a melt containing 49.3% KCl led to partial formation of a precipitate shown to be CoCl₂. Of this observation they said that "instead of CoCl₂, one might have expected $Co(AlCl_4)_2$ to precipitate, but this compound has a melting point 225°." They reported further that "addition of KCl to a melt whose KCl:AlCl₃ ratio was exactly equal to 1 caused the CoCl₂ precipitate to redissolve completely with only a few times more than the stoichiometric amount of KCl needed to form $CoCl_4^{2-}$. By 50.3% KCl, the melt was showing the spectrum for $CoCl_4^{2-}$."

In fact CoCl₂ is the expected precipitate from neutral chloroaluminate, just as "hydroxides" precipitate from H₂O, binary fluorides precipitate from HF, and the binary compounds of transition metals with bases of the other protonic superacids are sparingly soluble in the neutral parent acids. Furthermore, CoCl₂ is expected to redissolve in excess base, just as amphoteric "hydroxides" dissolve in excess $OH^-$. The generalized behavior of forming soluble cationic and anionic species and precipitates of the compounds of transition-metal cations with appropriate bases, regardless of whether the solution medium is water, a superacid, or a chloroaluminate—or indeed ammonia—is discussed in chapter 9 in a summary of the systems presented in this book.

As stated earlier in this section, others have taken the simple statement about the octahedral arrangement of chlorides and extended it to state or infer that the species in solution in the acidic melt would be $CoCl_6^{4-}$. It would be unrealistic to postulate that, in passing from an alkali chloride medium, where virtually all of the chloride is available for complexing, to $AlCl_3$, where the chloride-ion concentration would be vanishingly small, the Co(II) entity would change from $CoCl_4^{2-}$ to $CoCl_6^{4-}$. Essentially all of the chloride resulting from the dissolution of a very small amount ($3 \times 10^{-3}$ mol kg$^{-1}$) of metal dichloride in pure $AlCl_3$ would form $Al_2Cl_7^-$ or $Al_3Cl_{10}^-$. However, the concentration of the chloroaluminate anion would be very much less than that of dimeric $Al_2Cl_6$, the predominant species in molten $AlCl_3$, and complete 1:2 interaction of the cations and anions in very dilute solution seems very unlikely.

It seems much easier to postulate that the dissolution and solvation process is like that of a metal dichloride in water where the aquocation is formed with chloride acting simply as a counterion. In pure $AlCl_3$ three dimeric $Al_2Cl_6$ molecules could act as bidentate ligands to give the solvated cation, with $Al_2Cl_7^-$ as the counterions. There is strong chemical and spectroscopic evidence that, when $CoF_2$ dissolves in HF—$BF_3$, the cation is solvated as $Co(FH)_6^{2+}$ with $BF_4^-$ as counterions. The electrolysis experiment in oleum demonstrated beyond doubt that Co(II) in that medium is cationic and not neutral, as it would be in $AlCl_3$ if, as suggested by Øye and Gruen, $Co^{2+}$ was hexacoordinated in the melt by two $AlCl_7^-$ ligands. The solvation is obviously more complex in weakly acidic melts.

Hussey and Laher[21] recorded spectra for Co(II) in the room-temperature melt system aluminium trichloride—N-n-butyl pyridinium chloride ($AlCl_3$—BuPyCl), varying the melt component ratios to study Co(II) in basic and acidic melts. Absorption spectroscopy indicated that Co(II) was tetrahedrally coordinated as the $CoCl_4^{2-}$ anion in the basic melt for which the $AlCl_3$:BuPyCl ratio was 0.8:1. Potentiometric measurements on Co(II)/Co(0) in the basic melt showed a fourth-power dependence of the potential of the couple on the logarithm of chloride ion mole fraction, consistent with the formation of $CoCl_4^{2-}$.

In an acidic melt for which the $AlCl_3$:BuPyCl ratio was 2:1, the Co(II) spectrum indicated an octahedrally coordinated species, with bands much less intense than in the spectrum for tetrahedral $CoCl_4^{2-}$. Hussey and Laher report that in "slightly acidic" melt ($AlCl_3$:BuPyCl, 1.33:1) voltammetry suggests coexistence of "two somewhat different Co(II) species" and that $CoCl_2$ precipitates from the neutral melt ($AlCl_3$:BuPyCl, 1:1). The latter observation is normal and expected, as discussed earlier. The two Co(II) species in slightly acidic chloroaluminate may be $Co^{2+}$ and $CoCl^+$ with complex solvation.

Osteryoung and co-workers[22] reported spectroscopic and electrochemical behavior for Ni(II) similar to that of Co(II) in the room-temperature chloroaluminate, $AlCl_3$—BuPyCl, of varying acidity and basicity. They described "sea green/blue" solutions in basic melts ($AlCl_3$:BuPyCl, 0.8:1.0), with electronic spectral peaks at 705 and 658 nm and a shoulder at about 617 nm. These peak positions can be compared with those for $(Et_4N)_2NiCl_4$ in $CH_3NO_2$, 700, 660, and 610(sh) nm.[19] Potentiometry indicated the existence of $NiCl_4^{2-}$ in the basic melt by showing a fourth-power dependence of the potential for the Ni(II)—Ni(0) couple on chloride ion concentration. When $NiCl_2$ was dissolved in acidic melts ($AlCl_3$:BuPyCl, 1.5:1), the resulting solutions were described as "sandy" colored and the only feature of the spectrum in the 400–800-nm region was a weak band at about 547 nm.

Reference to Table 6.3 shows this band to be at a very different position from that in the 400–800-nm region ascribed to solvated $Ni^{2+}$ in the pure acid of the system, $AlCl_3$. Crystalline $NiCl_2$ is yellow and its room-temperature spectrum has a band at 455 nm, close to that for Ni(II) in $AlCl_3$. It is possible that, instead of the formal $Ni^{2+}$ in crystalline $NiCl_2$ and in pure $AlCl_3$ at 227°C, a species such as solvated $NiCl^+$ is the predominant one in the weakly acidic solvent $AlCl_3$:BuPyCl, 1.5:1. Osteryoung and colleagues reported irreversibility of Ni(II)—Ni(0) reduction in the acidic melt. Strong complexation of $Ni^{2+}$ by $Cl^-$, as distinct from simple solvation, could account for this. They reported no reduction wave at vitreous carbon electrodes within the accessible solvent potential limits for $NiCl_4^{2-}$ in basic melts, indicating formation of a very stable complex of Ni(II).

Very good evidence for drastic differences in metal speciation in basic and acidic melts comes from Osteryoung's voltammetric investigation[23] of the Fe(III)/Fe(II) system in the room-temperature melt $AlCl_3$—BuPyCl as the melt is changed from being basic to moderately acidic. The reduction potential (vs. $Al^{III}$/Al) for Fe(III)/Fe(II) is about 1.8 V more positive in basic melt ($AlCl_3$:BuPyCl, 0.75:1.00) than in acidic melt ($AlCl_3$:BuPyCl, 2.00:1.00), suggesting the presence in basic melts of anionic species that are far more stable toward electrochemical reduction than the species in acidic melts, proposed here as being cationic. Osteryoung and colleagues showed potentiometrically[23] that the anionic species were $FeCl_4^-$ and $FeCl_4^{2-}$, with the former stabler toward dissociation to $Cl^-$ than the latter. As appears to be quite general behavior for transition-metal chlorides in chloroaluminate melts, they reported that precipitation of the electroactive solute $FeCl_2$ occurs near neutrality, commencing near a melt ratio of $AlCl_3$:BuPyCl, 0.99:1, with redissolution occurring in the 1.1:1 acidic melt.

Reference to Figure 6.5 shows that the redox processes are close to being reversible in basic melts. A value of 62 mV for the difference, $\Delta E_p$, between the cathodic peak potential $E_p^c$ and the anodic peak potential $E_p^a$ in cyclic voltammetry is observed for a perfectly reversible one-electron process. The greater the departure from this number for experimentally observed peak

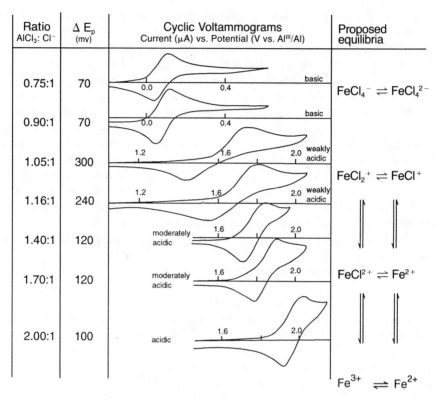

**Figure 6.5.** Cyclic Voltammograms for Fe(III)/Fe(II) in Chloroaluminates of Varying Acidity and Basicity.

potential differences the greater is the degree of irreversibility of the redox process. Osteryoung observed a value of about 65mV for $\Delta E_p$ in basic melts, indicating a high degree of reversibility for the process:

$$\text{Fe(III)} + 1\varepsilon \rightleftharpoons \text{Fe(II)} \qquad\qquad 6.3$$

An important factor in electrochemical reversibility is that the oxidized and reduced species of the couple should have similar or identical stereochemistry. In the light of all the evidence the most likely redox equilibrium for this system in basic melts is

$$\text{FeCl}_4^- + 1\varepsilon \rightleftharpoons \text{FeCl}_4^{2-} \qquad\qquad 6.4$$

The region of melt neutrality could not be studied electrochemically because of $\text{FeCl}_2$ insolubility, but, on proceeding to a slightly acidic melt ($\text{AlCl}_3$:BuPyCl, 1.05:1), two effects were observed. There was a massive change of about 1.6 V in the position of the half-wave potential, $E_{1/2}$, for reduction of Fe(III), and the value of $\Delta E_p$ increased from 70 to 300 mV, indicating a high degree of irreversibility and therefore of speciation for the components of the redox couples in basic and weakly acidic melts. The very large shift in the value of $E_{1/2}$ is consistent with a change from anionic to cationic Fe(II) and Fe(III). The irreversibility suggests a difference in stereochemistry between Fe(II) and Fe(III) in the weakly acidic melt.

As the melt was made progressively more acidic there was a steady shift in values of $E_p^c$ and $E_p^a$ to more positive potentials, but more significantly, $\Delta E_p$, the measure of reversibility, decreased from 300 mV to 100 mV for the melt in which the acid-to-base ratio was 2.00:1. This suggests a simplifying of cationic speciation with increasing melt acidity. It would be consistent with the observed chemistry for Co(II) and Ni(II) in pure $AlCl_3$, the acid of the chloroaluminate system, to suggest, as indicated in the last column of Figure 6.5, that in pure $AlCl_3$ the redox couple in the iron system would be $Fe^{3+}/Fe^{2+}$, appropriately solvated.

Both Fe(III) and Fe(II) would be expected to undergo chlorocomplexation in melts of weak or moderate acidity, the extent of complexation for Fe(III) being greater than for Fe(II). Therefore it is suggested in Figure 6.5 that the species in the weakly acidic melt could be the stereochemically dissimilar $FeCl_2^+$ and $FeCl^+$ and that, as the availability of $Cl^-$ was reduced, these cations could be gradually replaced by $FeCl^{2+}$ and $Fe^{2+}$, in a series of chloride-dependent equilibria. The difference in the extent and strength of chlorocomplexation of Fe(II) and Fe(III) at any $Cl^-$ level, expected on chemical grounds and indicated potentiometrically by Osteryoung as being complete in basic melts, could introduce sufficient stereochemical difference for the two components of the redox couple to account for the observed reversibility trends in melts of varying acidity.

## 6.1.2 Cations of $f$-Transition Elements

Most of the study of spectra of $f$-transition metal cations in normal oxidation states in superacids has been carried out in HF solutions where, as for the $d$-transition metal fluorides, dissolution of solid fluorides, which are insoluble in "neat" HF, resulted from addition of appropriate Lewis acids. Dissolution of metallic $f$-transition elements in "neat" or acidified HF has yielded cations in low oxidation states, as demonstrated in chapter 7. It has been demonstrated that spectra can be recorded in other protonic superacids by adopting similar experimental approaches, but fewer examples are available. For the lanthanides, molten salt spectra are restricted to cations in low oxidation states. The technological importance of the actinides has led to a reasonable amount of study of these cations in acidic molten salts, but most of the work has been in basic melts, where the absorbing species are anionic.

### 6.1.2.1 Solutions in Hydrogen Fluoride

The first published work on electronic spectra in anhydrous HF of transition metal cations[4] included spectra of $Pr^{3+}$ and $Nd^{3+}$. These were very similar to the aqueous spectra, the results indicating similar coordination of the cations in each solvent. Subsequently Cockman[9] recorded spectra for all lanthanide(III) elements except Pm; these again closely resembled aqueous spectra. The only significant differences between the two sets of spectra are that lanthanide peaks are shifted 3–5 nm to higher energy in HF than in $H_2O$ and that absorbances are less in HF, usually about half of those in water.

These features are all easily discernible in the spectra for Nd(III) presented in Figure 6.6.[4,9] Typical peak maxima for Nd(III) in $H_2O$ are 345, 575, 740, and 794 nm compared with peaks in HF at 344, 572, 735, and 791 nm. The large broad band at about 280 nm in the HF—$BF_3$ spectrum for Nd(III) is an impurity band, believed to be due to protonated unsaturated low-molecular-weight fractions of KelF leached from the polymer container by acidic HF. (See section 2.4.1.) It could be quenched with minute amounts of diluted $F_2$, as was done for the solution which gave the HF—$SbF_5$ spectrum.

Another important aspect of the Nd(III) spectra is that the sharpness of the bands resulting from *f-f* transitions allows band broadening to be observed very much more readily than for *d-d* bands. The bands in weakly acidic HF, that is, as acidified with $BF_3$, are broader than those in the very

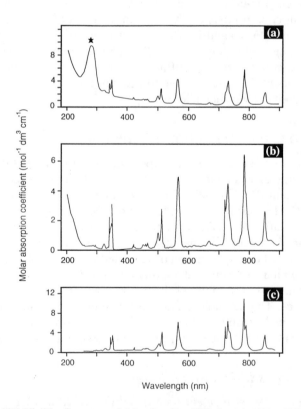

**Figure 6.6.** UV-Visible Spectra of Neodymium(III) in (a) $BF_3$—HF, ★-Impurity Band (see text), (b) 5 M $SbF_5$—HF, and (c) 1 M $HClO_4$—$H_2O$.

strongly acidic 5 M $SbF_5$—HF, suggesting that there might be some degree of fluorocomplexation to species such as $NdF^{2+}$ or even $NdF_2^+$ in $BF_3$—HF whereas the formation of $Nd(FH)_n^{3+}$ would be expected in $SbF_5$—HF.

From a synthetic point of view, it is worth recording that lanthanide tri-fluorides dissolved readily in HF—$BF_3$,[4] whereas $d$-transition element tri-fluorides such as $CrF_3$ were virtually insoluble in HF—$BF_3$ and solutions in HF—$SbF_5$ were prepared with difficulty.[6] This probably is a consequence of the greater size of the tripositive lanthanide cations than for the $d$-transition metal cations and the greater ionic character of the solid lanthanide tri-fluorides. HF solution spectra for Np and Pu in oxidation states III and IV will be presented before the corresponding U spectra because $Np_{aq}^{3+}$, $Np_{aq}^{4+}$, $Pu_{aq}^{3+}$, and $Pu_{aq}^{4+}$ are stable species in acidified aqueous solutions and in HF, whereas U(III) compounds reduce water—so $U^{3+}$ spectra will be dealt with in chapter 7—and U(IV) spectra indicate the existence of fluorocomplexed cations in HF while U(IV) exists in strongly acidic aqueous solutions as the simple solvated cation $U_{aq}^{4+}$.

Figure 6.7 shows the marked similarity of the Np(III) spectrum in $BF_3$—HF[7] to that in an aqueous $HClO_4$ solution. As is the case for all HF spectra of $f$-transition metals relative to the corresponding aqueous spectra, peak maxima are shifted to higher energies by a small number of nanometers, usually 20–30 for actinides. Major peaks for $Np^{3+}$ in HF occur at 544, 760, 966, and 1,338 nm, compared with corresponding peaks in aqueous solution at 552, 786, 991, and 1,363 nm.[24]

In Figure 6.8 the presence of the cation $Np(FH)_n^{4+}$ in $AsF_5$—HF[7] is indicated by the close similarity of the spectrum for that solution to that for Np(IV) in nomcomplexing acidified aqueous solution[24] where the existence of $Np_{aq}^{4+}$ is generally accepted. The major peaks for the aqueous spectrum are at 723 and 960 nm and these correspond with peaks at 718 and 949 nm in HF—$AsF_5$. Two peaks in the HF spectrum at 793 and 819 nm can be compared with a peak reported at 820 nm as part of a general area of low-intensity absorbance in the aqueous spectrum. The order of intensities of the two major peaks are reversed between $H_2O$ and HF. This HF spectrum is one example of the many in which spectral resolution is significantly better for cations in HF solution than in analogous aqueous solutions. This could indicate that for the tetrapositive cation hydroxocomplexation to form species like $NpOH^{3+}$ is stronger than fluorocomplexation to $NpF^{3+}$ in HF. This is discussed in the summary to this chapter (section 6.2).

Work reported by Avens and colleagues from the Los Alamos National Laboratory in 1989[25] provides evidence for the existence of the solvated cations $Pu^{3+}$ and $Pu^{4+}$ in superacidic media. They dissolved $PuCl_3$ in 5% $SbF_5$—HF and in 25% $SbF_5$—$HSO_3F$ and recorded peak positions in the region 300–800 nm for the $HSO_3F$ solutions virtually identical with those for $Pu^{3+}$ in acidified water, with the equivalent peaks shifted to slightly higher energy in HF, just as the Melbourne group has observed for actinides generally. Pu metal in HF—$SbF_5$ gave, as would be expected, a spectrum con-

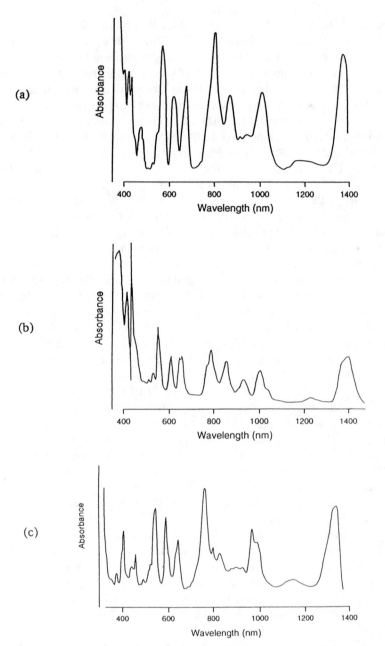

**Figure 6.7.** Neptunium(III) Spectra in (a) 2 M HClO$_4$—H$_2$O, (b) AlCl$_3$—BuPyCl, 2:1, and (c) BF$_3$—HF.

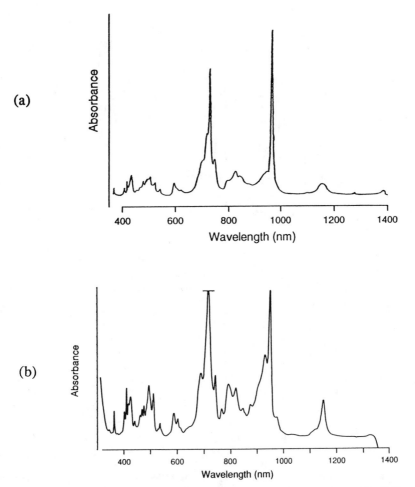

(a)

(b)

**Figure 6.8.** Neptunium(IV) Spectra in (a) 2 M $HClO_4$—$H_2O$ and (b) $AsF_5$—HF.

siderably better resolved than the published spectrum for $Pu^{3+}$ in acidified water, but with peaks shifted to somewhat lower wavelengths by about 20 nm. Dissolution of $PuO_2$ in 5% $SbF_5$—HF and in 25% $SbF_5$—$HSO_3F$ provided solutions in which the 300–850-nm spectra were very similar to that for $Pu^{4+}$ in acidified water. The $HSO_3F$ solution peaks were at wavelengths very close to those for aqueous $Pu^{4+}$ and the HF solution peaks were at slightly shorter wavelengths.[25]

On the basis of the close similarities of the spectra that have been observed for solvated cations in aqueous and HF solutions of Np(III), Np(IV), Pu(III), and Pu(IV), it might be expected that the spectrum obtained by dissolving $UF_4$ in HF—$AsF_5$ or HF—$SbF_5$ would be very similar to that ascribed to the aquocation $U^{4+}$ in noncomplexing, acidic water. Figure 6.9 shows that the sharp peaks in the 400–600-nm region associated with the

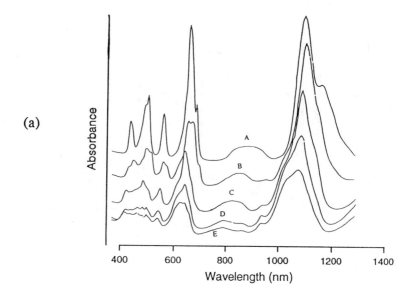

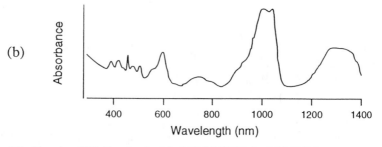

**Figure 6.9.** Uranium(IV) Spectra in (a) 0.03 M U(IV) in 1 M $HClO_4$—$H_2O$ with HF added at the following concentrations: A = 0.00, B = 0.020, C = 0.033, D = 0.053, E = 0.080 M; and in (b) $AsF_5$—HF.

spectrum $U_{aq}^{4+}$ are not observed in the HF spectrum and that the very strong peaks at about 650 and 1070 nm for the aqueous spectrum are replaced by weaker, broader bands at about 600 nm and in the region of 1,020 nm.

Preliminary information on the nature of the U(IV) cation in HF came from comparison with the work of Stein and his colleagues[26] at the Argonne National Laboratory, who in 1966 reproduced the aqueous spectrum for U(IV) in 1 M $DClO_4$—$D_2O$ (spectrum A of Figure 6.9a) recorded earlier by their colleagues Cohen and Carnall.[27] Stein et al. added aqueous HF progressively to their 0.03 M U(IV) solution until the HF concentration was 0.08 M; that is, HF was in two- to threefold excess over U(IV), and they recorded the spectra B to E of Figure 6.9a. They postulated the formation of $UF_2^{2+}$ in their final solution and in a separate experiment they crystallized

$UF_2.SO_4.2H_2O$ from an aqueous solution initially 6 M in $H_2SO_4$ and 0.9 M in HF. The marked similarity of the spectrum for U(IV) in HF—$AsF_5$ (Figure 6.9b)[7] to that in Figure 6.9a E leads to the postulation that the cationic species in the HF solution is $UF_2^{2+}$.

There are small shifts of about 20 nm to higher energies for the proposed $UF_2^{2+}$—HF—$AsF_5$ spectrum relative to Stein's aqueous spectrum (E of Figure 6.9a) that are consistent with the shifts observed for simpler cationic actinide species in going from $H_2O$ solutions to HF, but otherwise the spectra for U(IV) in HF—$AsF_5$ and in acidified aqueous solution to which HF has been added are very similar. Stein et al. report[26] "that bands at 430, 485, 495, and 549 nm (Figure 6.9a A) drop very rapidly as HF is added to the $U_{aq}^{4+}$ solution to be replaced by vibrational fine structure in the region 400 to 650 nm" (Figure 9.9a E). Similar vibrational structure is observed in the U(IV)—HF—$AsF_5$ spectrum (Figure 6.9b), and its detail is compared with that recorded by Stein et al. in the paper describing the U(IV)—HF—$AsF_5$ system.[7] As indicated above, the two intense peaks of the spectrum for $U_{aq}^{4+}$ are replaced in the $H_2O$—$HClO_4$—HF and in the HF—$AsF_5$ spectra by less-intense, broader bands at about 600 and 1020 nm.

Since the original U(IV)—HF—$AsF_5$ spectrum recorded in 1976,[7] spectra obtained in later work have indicated that U(IV) in acidic media containing available fluoride ions is not a simple solvated cation, but is more likely to be a fluorocomplex. Avens and colleagues at the Los Alamos National Laboratory dissolved $UF_4$ in 5% $SbF_5$—HF and $UO_2$ in 25% $SbF_5$—$HSO_3F$ and recorded spectra between 320 and 800 nm[25] which are very similar to the $UF_4$—$AsF_5$ spectra in that wavelength range. It will be seen in section 6.1.2.2 that the spectrum now associated with $UF_2^{2+}$ has been observed for U(IV) in solution in both $HSO_3F$ and $CF_3SO_3H$ to which $F^-$ has been added.

## 6.1.2.2 Solutions in Other Superacids

By comparison with the amount of work involving spectroscopic investigation of lanthanides and actinides in HF, the extent of investigation of these elements in solution in the other protonic superacids is quite small.

In a Melbourne program of investigation of metal and nonmetal cations in superacid media it was found that $Pr_2(SO_4)_3$, treated with 100% $H_2SO_4$, gave little or no discernible absorption characteristic of $Pr^{3+}$. However, when 20% $SO_3$ was added to $H_2SO_4$—that is, the acidity of the medium, as shown by previously measured Hammett Acidity Functions,[28] was increased by a factor greater than 10—a spectrum for $Pr^{3+}$ was observed with peak positions virtually identical with those in aqueous solution.[29]

For convenience of presentation, the generation of solutions of Pu(III), Pu(IV), and U(IV) in 25% $SbF_5$—$HSO_3F$ and recording of their spectra by the Los Alamos group were reported in section 6.1.2.1 with the directly parallel studies in 5% $SbF_5$—HF.

Recently at Melbourne Adrien has studied U(III) and U(IV) solutions in HSO$_3$F and CF$_3$SO$_3$H of differing acidities.[15] Because the interpretation of his work requires a prior discussion of the behavior of U(III) in HF solutions of varying acidity, all of his work will be presented in chapter 7.

### 6.1.2.3 Solutions in Chloroaluminates

There appears to have been very little investigation of lanthanides in their normal oxidation state III in molten halides. Russian workers[30] recorded electronic spectra for Nd(III) and Er(III), as well as for several actinides, in alkali metal chloride melts and under these basic conditions, interpreted their spectra in terms of anionic species such as NdCl$_6^{3-}$ and ErCl$_6^{3-}$.

Lipsztajn and Osteryoung[31] conducted a voltammetric investigation of Nd(III) in the room-temperature melt aluminium chloride—1-methyl-3-ethylimidazolium chloride (AlCl$_3$—IMEC) and found that NdCl$_3$ "dissolves in the basic melt to form bluish nonelectroactive chloride complexes." After monitoring the decrease of the limiting current for Cl$^-$ oxidation that occurred on addition of NdCl$_3$ to weakly basic AlCl$_3$—IMEC they said that "the stoichiometry of the complex formed was NdCl$_6^{3-}$." Their voltammetric study as they added NdCl$_3$ to weakly acidic melts is of greater interest relative to this chapter. They reported that the observed decrease of the Al$_2$Cl$_7^-$ reduction current to form AlCl$_4^-$ corresponded almost exactly to the amount of Cl$^-$ added to the melt as NdCl$_3$. They say that "this indicates that no 'free' chloride remains in the Nd(III) complex in acidic melts and suggests that Nd(III) is solvated by either AlCl$_4^-$ or Al$_2$Cl$_7^-$." It is postulated here that Osteryoung's observation suggests the formation of Nd$^{3+}$, however that entity might be solvated. Lipsztajn and Osteryoung[31] conclude their report on Nd(III) in AlCl$_3$—IMEC by saying that addition of NdCl$_3$ to weakly acidic melts gave cyclic voltammograms showing the presence of a new redox system at potentials more negative than that for reduction of Al$_2$Cl$_7^-$, but that the low solubility of the Nd(III) "complex" in the weakly acidic melt precluded further studies at that stage. It was likely that their work was restricted by the low solubility of NdCl$_3$ in the weakly acidic melt.

As stated earlier, the technological importance of uranium and the lighter transuranic elements is such that there has been a reasonable body of study of these elements in molten halides, but most of it has been done in basic melts, where anionic rather than cationic species will be formed.

Russian workers,[30] quoted earlier, recorded electronic spectra for chloroanions of Np(III), Np(IV), Np(V), Pu(III), Pu(IV), Pu(VI), and Am(III) in molten alkali metal chorides. Earlier, Young had reported spectra for U(III) and U(IV) in molten fluorides,[32] and Morrey[33] conducted a spectrophotometric study of solutions of UCl$_4$ in basic and acidic KCl—AlCl$_3$ mixtures. In the melt KCl:AlCl$_3$, 1.047:1 at 400°C, the observed spectrum was quite similar to that for [R$_3$NH]$_2$UCl$_6$ in xylene at 25°C,[34] with a large sharp peak at a little over 2,000 nm and a similar distribution of peaks with similar relative intensities at wavelengths to 400 nm. For $UCl_4$ in the acidic melt $KCl:AlCl_3$, 0.806:1 at 350°C, the sharp feature at about 2,000 nm was replaced by a less-intense, very-much-broader band in the same region, a band similar to that observed for $U_{aq}^{4+}$ in 1 M $DClO_4$ at 25°C.[27] Other bands down to about 450 nm were in similar positions for U(IV) in the acidic melt and $U_{aq}^{4+}$ in acidic aqueous solution. These spectra suggest very strongly that U(IV) is anionic in basic chloroaluminates, existing probably as $UCl_6^{2-}$, and cationic in acidic melts. However, the tetrapositive cation $U^{4+}$, because of its high charge, would probably be chlorocomplexed to cationic species such as $UCl^{3+}$, $UCl_2^{2+}$, and $UCl_3^+$, depending on the availability of $Cl^-$ in the relatively weakly acidic melt.

Much more recently, Heerman et al.[35] recorded a spectrum for a solution of $UCl_4$ in the basic room-temperature melt $AlCl_3:BuPyCl$, 0.9:1, and although very much better resolved, it was strikingly like that obtained by Morrey in the high-temperature basic melt[33] and for $UCl_6^{2-}$ in xylene.[34]

The same authors[36,37] have carried out electrochemical and spectroscopic studies of U(IV) in *acidic* $AlCl_3$—BuPyCl. In the 2:1 melt they reported that reduction of U(IV) to U(III) on glassy carbon electrodes is irreversible. This probably reflects differences in stereochemistry—chlorocomplexation—for the U(IV) and U(III) species, since they say[36] that U(III) exists as $U^{3+}$ while U(IV) is present as cationic chlorocomplexes, $UCl_x^{(4-x)+}$ with $3 \geq x \geq 1$, the major species in the 2:1 melt being $UCl^{3+}$. In a subsequent paper,[37] this Belgian group used potentiometry and spectroscopy to study U(III) and U(IV) in $AlCl_3$—BuPyCl melts of varying acidity. They reported that $UCl_4$ precipitated when the melt ratio approached 1:1 and they observed that the spectrum for $UCl_4$ "in a slightly acidic $AlCl_3$—BuPyCl melt is identical with the spectrum of uranium(IV)[33] in slightly acidic $AlCl_3$—KCl melts at 250°C." They continued: "the species in these melts was represented by Morrey[33] as $U(AlCl_4)_6^{2-}$ but such a formulation is obviously in contradiction with the potentiometric results of this work," where chlorocations of U(IV) are postulated.

The work of De Waele et al. on uranium(III) in acidic $AlCl_3$—BuPyCl[37] is reported in chapter 7, because, by the convention adopted in this book, oxidation state (III) is not regarded as a "normal" oxidation state for uranium. There is some reinterpretation in chapter 7 of some minor experimental observations of these workers in terms of U(III) disproportionation, which is presented there as an example of interesting chemistry in nonaqueous media of elements in low oxidation states.

Whereas uranium(III), in reducing aqueous solutions, is regarded here as being in an unusually low oxidation state, neptunium(III) is stable in aqueous solution, as indicated by the aqueous spectrum in Figure 6.7. Its molten salt chemistry, together with that of Np(IV), is presented in this chapter.

Schoebrechts and Gilbert[38] carried out voltammetric, potentiometric, and spectroscopic investigations of Np(III) and Np(IV) in neutral, acidic, and basic AlCl$_3$—BuPyCl melts at 40°C. They found that NpCl$_3$ and NpCl$_4$ were insoluble in slightly acidic melts, redissolving as the AlCl$_3$ content of the melts reached 55 and 51%, respectively. Reduction of Np(IV) to Np(III) was quasi-reversible at a glassy carbon electrode in both acidic and basic melts. Potentiometry and spectroscopy were used to suggest that the species in basic melts are NpCl$_6^{3-}$ and NpCl$_6^{2-}$. Potentiometry suggests that speciation of Np(IV) in acidic melts is like that of U(IV), namely, NpCl$_x^{(4-x)+}$ (with 3 $\geq$ x $\geq$ 1). Their spectrum for Np(III) in the acidic 66.6 mol% AlCl$_3$ melt (see Figure 6.7b) leads them to suggest "that Np(III) exists as the solvated Np$^{3+}$ cation over the whole acidic composition range." They say "the solvation is of course different from that in perchloric solutions." In this they appear to be differentiating between the situation in aqueous HClO$_4$ solution where there is strong aquocomplexation and that in melts, where they are "led to conclude that the chloroaluminate melts are among the least solvating media."

## 6.2 Summary

Study of ionic speciation for $d$- and $f$-transition metals shows, even at this fairly early stage, that for protonic superacids and for chloroaluminate melts there is a common pattern of generation of cationic, anionic, and neutral species that is dependent on the acidity, basicity, and neutrality of the medium, regardless of the chemical nature of the medium or of its temperature domain. In neutral media a binary compound of the metal and the base of the solvent system is sparingly soluble, for example, "M(OH)$_n$" for water, MX$_n$ for the protonic superacids HX, and MCl$_n$ for chloroaluminates. In basic media anions are formed with the base of the solvent system as the ligand, for example, M(OH)$_{n+m}^{m-}$, MX$_{n+m}^{m-}$, and MCl$_{n+m}^{m-}$, as has been well established for a considerable period of time. Study of acidic media has been much more recent, with the obvious exception of acidic aqueous systems. There is now strong evidence that cations, solvated or complexed to varying degrees, are formed in all the acidic media. This generalization is presented in somewhat more detail in chapter 9 of this book.

Transition-metal speciation has been studied much more extensively in hydrogen fluoride than in the other superacids. Because of the structural simplicity of the two solvating molecular species, H$_2$O and HF, and of the ligands OH$^-$ and F$^-$, comparison of electronic spectra in these two media readily provides information on the nature of species in HF solutions of differing acidity and basicity. The Cr(III)—HF system is a good example. CrF$_3$ is very insoluble in HF solutions, unless the acidity is very high, for example, in SbF$_5$—HF. Then dissolution of the trifluoride occurs and, as discussed above, the spectrum for the species soluble in acidic HF indicates, by comparison with those for hexacoordinated Cr(OH$_2$)$_6^{3+}$ and CrF$_6^{3-}$, that it is Cr(FH)$_6^{3+}$. (See Figure 6.1.) The peak maxima listed in Table 6.1 for a solution of $CrF_3$ in 0.5 M $NH_4F$—HF again indicate hexacoordination of Cr(III). By comparison with the data for the synthesized compound $(NH_4)_3CrF_6$ dissolved in strongly basic HF to minimize the chance of dissociation to $F^-$ of the anion $CrF_6^{3-}$, it can be seen that Cr(III) exists in basic HF as $CrF_6^{3-}$.

Spectra for Co(II), Ni(II), Mn(II), Cu(II), and Fe(II) in acidic HF indicated, by comparison with aqueous spectra, the existence of $M(FH)_6^{2+}$ in each case, with complementary ESR spectroscopic evidence in the case of Mn(II). General features of the HF spectra of $d$-transition metal cations when compared with aqueous spectral peaks were that for HF-solvated cations spectral peaks were shifted 50–150 nm to lower energies than for corresponding peaks for aquocations and that intensities for the peaks in HF were about one-half of those in aqueous solutions. Weaker solvation by HF than by $H_2O$ is indicated by the fact that all attempts to prepare crystals containing HF-solvated cations of $d$-transition metals using methods analogous to those which yielded crystals containing the corresponding aquocations have proved unsuccessful. The case of the attempted crystallization of $Mn(FH)_6(BF_4)_2$ has been described above. The crystals isolated were $Mn(BF_4)_2$.

$4f$-element cation peaks were shifted to slightly higher energies, by 2–5 nm, than for the very similar aqueous spectra, while the shift to higher energy was by 20–30 nm for $5f$ elements. Band intensities for $f$-element cations in HF were also about one-half of those in aqueous solution and, while there was a very high degree of correspondence between peak positions, some bands changed a little in relative intensity in moving from $H_2O$ to HF.

A noticeable feature of HF solution spectra of cations of $4f$ and $5f$ elements was that much better resolution was obtained than for corresponding aqueous spectra, reflecting less complexation of cations in HF than in $H_2O$.

There are several reasons for the simplicity and better resolution of spectra in HF and they have been touched on in an earlier review.[3] First, potential ligands such as ions like halides, sulfates, and carboxylates or molecules with oxygen, nitrogen, or other donor atoms are protonated or solvolyzed in HF and are not available to act as ligands. Second, HF can lose only one proton to give a ligand formally analogous to $OH^-$. $H_2O$ can lose a second proton and the $O^{2-}$ ion is a powerful ligand, leading frequently to solute polymerization in aqueous solution. The only ligand other than molecular HF that can be derived from the solvent HF is the fluoride ion, and some electrochemical evidence is cited in that review[3] to show that $F^-$ in HF appears to be a weaker complexing ligand than $OH^-$ in $H_2O$.

That some measure of fluorocomplexation of metal cations does occur in weakly acidic HF has been shown in Figure 6.6, where absorption bands for Nd(III) in HF solutions were significantly broader in $BF_3$—HF than in $SbF_5$—HF, where optimum conditions obtain for the existence of the simple solvated cation $Nd(FH)_x^{3+}$. Figure 6.6 also shows, as for much other work in HF spectroscopy, that band positions are independent of whether the anion derived in HF solution from the Lewis acid is $BF_4^-$, $AsF_6^-$, or $SbF_6^-$; that is, there is no spectroscopic evidence that these ions are coordinating to the metal cations in solution.

Much less investigation of cation generation and speciation has been done in the three superacids $H_2SO_4$, $CF_3SO_3H$, and $HSO_3F$ than in HF. The Melbourne group decided to investigate Co(II) in these superacids because of the strong indication of tetra- or hexacoordination that was available from electronic spectra (Figure 6.4). Co(II) hydrogensulfates, triflates, and fluorosulfates were all sparingly soluble in the "neat" parent acid. Addition of Lewis acids gave spectra in each case indicative of six-coordinate Co(II) (Table 6.2), but the strongest evidence for cationic speciation was the electromigration experiment described in section 6.1.1.2, in which Co(II) in 20% $SO_3$—$H_2SO_4$ migrated to the cathode and not to the anode.

Addition of the bases $HSO_4^-$, $CF_3SO_3^-$, and $SO_3F^-$ as the potassium salts to suspensions of $Co(HSO_4)_2$, $Co(SO_3CF_3)_2$, and $Co(SO_3F)_2$ in the "neat" parent acids $H_2SO_4$, $CF_3SO_3H$, and $HSO_3F$ caused four- to fivefold increase of the Co(II) concentration as the concentration of base reached 2 M in each case.[39] The spectrum for Co(II) in basic $H_2SO_4$ was clearly that for Co(II) in an octahedral environment.[39] There is no hard evidence on the nature of the ligands in either the anionic or cationic Co(II) species. An attractive possibility is that three bidentate $HSO_4^-$ ions are involved in the formation of an anion $[Co(HSO_4)_3]^-$ with similar complexation to form $[Co(SO_3CF_3)_3]^-$ and $[Co(SO_3F)_3]^-$. The cationic species in acidified $H_2SO_4$ could involve coordination by three bidentate $H_2SO_4$ molecules and similar cationic species involving three bidentate solvent molecules could be imagined for Co(II) in acidified $CF_3SO_3H$ and $HSO_3F$.

Two general types of chloroaluminate melt have been studied—"high-temperature" melts which are usually $AlCl_3$—KCl or $AlCl_3$—NaCl mixtures and "room-temperature" melts, mixtures of $AlCl_3$ and either N-butylpyridinium chloride (BuPyCl) or 1-methyl-3-ethylimidazololium chloride (IMEC).

These appear to differ in two significant respects. Studies can be made in the high-temperature regime in pure $AlCl_3$, where virtually all of the solvent particles are $Al_2Cl_6$ molecules, or in $AlCl_3$-rich melts where, depending on the melt ratio, $Al_2Cl_6$, $Al_2Cl_7^-$, and $AlCl_4^-$ can all coexist in equilibrium. Boxall et al.[40] have used electrochemical measurements to show that in the $AlCl_3$:NaCl, 2:1 melt $Al_2Cl_6$, $Al_2Cl_7^-$, and $AlCl_4^-$ are all present in comparable concentrations at typical working temperatures, namely, 300–350°C. A room-temperature melt having the 2:1 ratio contains virtually no $Al_2Cl_6$, being composed mainly of $Al_3Cl_{10}^-$, $Al_2Cl_7^-$, and $AlCl_4^-$. Furthermore pCl is reported to change by about 2.2 units in the region of neutrality for high-temperature melts whereas the change is given as 8–10 units for room temperature melts. (See section 2.5.)

Reports for a wide range of studies of d- and f-transition metal chlorides in all melts suggest that the binary chlorides $MCl_n$ are sparingly soluble in the immediate region of neutrality and that melts ranging from those that are slightly basic to the pure chlorides (KCl or BuPyCl) contain chloroanions, usually $MCl_4^{2-}$ for first-row $d$-transition elements or $MCl_6^{(6-n)-}$ for $f$-transition elements.

There has been far less agreement in the molten salt literature on speciation in acidic melts. The extreme in unacceptability has been the proposal referred to in section 6.1.1.3—that $VCl_2$, $TiCl_2$, $CoCl_2$, and other dichlorides dissolved in pure $AlCl_3$ are present as $VCl_6^{4-}$, $TiCl_6^{4-}$ and $CoCl_6^{4-}$. As stated there, it is almost impossible to accept the proposition that Co(II) is present as $CoCl_4^{2-}$ in basic melt where $Cl^-$ is in large concentration, that it precipitates as $CoCl_2$ in neutral melt, and that in pure $AlCl_3$, where the chloride concentration is vanishingly small, it exists as $CoCl_6^{4-}$.

One fact is certain. Speciation in acidic melts must be very different from that in basic melts. Irreversible reduction of Ni(II) to Ni(0) was observed in acidic $AlCl_3$:BuPyCl (2:1) at about $+0.3$ V vs. an $Al/AlCl_3/BuPyCl$ (2:1) reference electrode while "no wave for $NiCl_4^{2-}$ anion reduction appeared within the accessible solvent potential limits."[22] The reduction potentials for Fe(III)/Fe(II) in basic and in acidic $AlCl_3$—BuPyCl differed by about 1.8 V.[23] While there would be general agreement that in basic melt the redox couple would be $FeCl_4^-/FeCl_4^{2-}$, there is no fixed position amongst authors on speciation of Fe(III) and Fe(II) in acidic melts.

The most widely accepted position in the older literature appears to have been that, in acidic melts, metal cations are solvated by $Al_2Cl_7^-$ ions to give neutral species. This implies that when a transition metal dichloride is dissolved in an acidic melt the $Al_2Cl_7^-$ units are tridentate ligands leading to the formation of $M(Al_2Cl_7)_2$, but if the initial solute is a trichloride, the formation of a neutral $M(Al_2Cl_7)_3$ implies different; bidentate coordination on the part of $Al_2Cl_7^-$ ligands.

A further complication is introduced by considering simple chlorocomplexation of cations in chloroaluminate media. As the formal charge increases and as the value of pCl for the acidic medium decreases, cations will become progressively more complexed by chlorides to give $MCl_m^{(n-m)+}$ species.

In a recent EXAFS study[41] of solutions of $MnCl_2$, $CoCl_2$, and $NiCl_2$ in acidic and basic $AlCl_3$—MEIC media, Seddon and co-workers indicate the expected existence of $MCl_4^{2-}$ anions for each metal in basic melts, but interpret their results as showing that each metal ion ($M^{2+}$) is "coordinated by a first shell of six chlorine centres and a second shell of three aluminium centres. These data support a model in which the metal centre is coordinated by three bidentate $[AlCl_4]^-$ ligands and do not fit any model involving chelating $[Al_2Cl_7]^-$." If formal coordination of $M^{2+}$ cations by anions is accepted, this model requires that, in acidic chloroaluminates, the transition metal is present as *anionic* $[M(AlCl_4)_3]^-$ units, as Seddon states explicitly.

The Belgian molten salt chemists[36,38] report their electrochemical and spectroscopic investigations of uranium and neptunium in acidic melts as being interpreted in terms of solvated $U^{3+}$ and $Np^{3+}$, with the tetrapositive metals in solution as $UCl_x^{(4-x)+}$ and $Np_x^{(4-x)+}$ where $3 \geq x \geq 1$.

As stated at the close of section 6.1.2.3, Schoebrechts and Gilbert[38] wrote that they were "led to conclude that the chloroaluminate melts are among the least solvating media." Until firmer electrochemical or spectroscopic evidence becomes available on the nature (or absence) of complexation of cations of low charge in chloroaluminate melts, it may be better to regard those cations simply as occupying octahedral sites in the melts.

Seddon's EXAFS study[41] of $CoCl_2$, $MnCl_2$, and $NiCl_2$ in acidic and basic $AlCl_3$—MEIC media can be used to provide some support for this simple concept. He gives a coordination number of 4 and Co—Cl distances in the basic melt as 2.28 A°, compared with 2.22, 2.33, and 2.31 A° in crystalline $Cs_2CoCl_4$,[42] indicating that Cl's are bonded to Co in the melt as $CoCl_4^{2-}$ anions. As stated above, Seddon reports a coordination of $Co^{2+}$ by six Cl's in the acidic melt with a Co—Cl distance of 2.44 A°. In a crystal structure determination of $Co(AlCl_4)_2$, Ibers[43] measured a Co—Cl distance of 2.47 A°. It could be argued that in the acidic melt $Co^{2+}$ occupies an octahedral site in an environment of $AlCl_4^-$ anions in much the same way as it does in crystalline $Co(AlCl_4)_2$. Seddon comments that coordination of $Co^{2+}$ by three bidentate $AlCl_4^-$ ligands instead of three $Al_2Cl_7^-$ units "is a somewhat surprising result, in view of the expected steric strain within such a complex." This strain would not be present if, as in the solid, $Co^{2+}$ cations occupy octahedral sites in a host of $AlCl_4^-$ anions.

# References

1. J. Burgess, *Metal Ions in Solution*, Ellis Horwood, Chichester, 1978, (a) pp. 25–49, (b) 269–270.

2. T.A. O'Donnell, *Aust. J. Chem.*, *41*, 1433 (1988).

3. T.A. O'Donnell, *J. Fluorine Chem.*, *11*, 467 (1978).

4. C.G. Barraclough, R.W. Cockman, T.A. O'Donnell, *Inorg. Chem.*, *16*, 673 (1977).

5. T.L. Court, M.F.A. Dove, *J. Fluorine Chem.*, *6*, 491 (1975).

6. C.G. Barraclough, R.W. Cockman, T.A. O'Donnell, M. Snare, *Inorg. Chem.*, *27*, 4504 (1988).

7. M. Baluka, N. Edelstein, T.A. O'Donnell, *Inorg. Chem.*, *20*, 3279 (1981).

8. J. Besida, T.A. O'Donnell, unpublished observations.

9. R.W. Cockman, Ph.D. Thesis, University of Melbourne (1983).

10. R.W. Cockman, B.F. Hoskins, M.J. McCormick, T.A. O'Donnell, *Inorg. Chem.*, *27*, 2742 (1988).

11. A.B.P. Lever, *Inorganic Electronic Spectroscopy*, 2nd ed. Elsevier, Amsterdam, 1984, pp. 480–481.

12. K. Milkins and T.A. O'Donnell, unpublished data, University of Melbourne (1986).

13. A. Dohrmann and T.A. O'Donnell, unpublished data, University of Melbourne (1984).

14. S. Hay and T.A. O'Donnell, unpublished data, B.Sc. (Honors) Report, University of Melbourne (1987).

15. R. Adrien, Ph.D. Thesis, University of Melbourne (1992).

16. D.M. Gruen, R.L. McBeth, *Pure Appl. Chem.*, *6*, 23 (1963).

17. H.A. Øye, D.M. Gruen, *Inorg. Chem.*, *3*, 836 (1964).

18. Reference 11, (a) p. 416, (b) p. 399.

19. N.S. Gill, R.S. Nyholm, *J. Chem. Soc.*, 3997 (1959).

20. H.A. Øye, D.M. Gruen, *Inorg. Chem.*, *4*, 1173 (1965).

21. C.L. Hussey, T.M. Laher, *Inorg. Chem.*, *20*, 4201 (1981).

22. R.J. Gable, B. Gilbert, R.A. Osteryoung, *Inorg. Chem.*, *4*, 1173 (1965).

23. C. Nanjundiah, K. Shimizu, R.A. Osteryoung, *J. Electrochem. Soc.*, *129*, 2474 (1982).

24. P.C. Hagan, J.M. Cleveland, *J. Inorg. Nucl. Chem.*, *28*, 2905 (1966).

25. K.D. Abney, P.G. Eller, L.R. Avens, W.H. Smith, *Proc. A.C.S. Winter Fluorine Chemistry Symposium*, St. Petersburg, Florida (1989).

26. L. Stein, C.W. Williams, I. Fox, E. Gebert, *Inorg. Chem.*, *5*, 662 (1966).

27. D. Cohen, W.T. Carnall, *J. Phys. Chem.*, *64*, 1933 (1960).

28. R.J. Gillespie, T.E. Peel, E.A. Robinson, *J. Am. Chem. Soc.*, *93*, 5083 (1960).

29. K. Milkins and T.A. O'Donnell, unpublished data, University of Melbourne (1986).

30. Yu. A. Barbanel, V.R. Klokman, *Radiokhimiya*, *18*, 699 (1970).

31. M. Lipsztajn, R.A. Osteryoung, *Inorg. Chem.*, *24*, 716 (1985).

32. J.P. Young, *Inorg. Chem.*, *6*, 1486 (1967).

33. J.R. Morrey, *Inorg. Chem.*, *2*, 163 (1963).

34. J.L. Ryan, *Inorg. Chem.*, *3*, 211 (1964).

35. L. Heerman, R. De Waele, W. D'Olieslager, *J. Electroanal. Chem.*, *193*, 289 (1985).

36. R. De Waele, L. Heerman, W. D'Olieslager, *J. Electroanal. Chem.*, *142*, 137 (1982).

37. R. De Waele, L. Heerman, W. D'Olieslager, *J. Less-Common Metals*, *122*, 319 (1986).

38. J.P. Schoebrechts, B. Gilbert, *Inorg. Chem.*, *24*, 2105 (1985).

39. T.A. O'Donnell, P.B. Smith, unpublished data, University of Melbourne (1988).

40. L.G. Boxall, H.L. Jones, R.A. Osteryoung, *J. Electrochem. Soc.*, *120*, 233 (1973).

41. A.J. Dent, K.R. Seddon, T. Welton, *Chem. Commun.*, 315 (1990).

42. R.W.G. Wyckoff, *Crystal Structures*, Vol. 3, 2nd ed. Interscience (1965), p. 100.

43. J.A. Ibers, *Acta Crystallogr.*, *15*, 967 (1962).

# CHAPTER 7

# Cations of Transition Elements in Low Oxidation States

In a fundamental sense, there is nothing remarkable about the generation of cations of $d$- and $f$-transition elements in normal oxidation states in protonic superacidic media and in acidic melts. As discussed in chapter 6, the insolubility of binary fluorides, bisulfates, fluorosulfates, and triflates in the neutral parent acids parallels the insolubility of "hydroxides" in aqueous solutions. Similarly, the generation of cationic species in the appropriate acidic nonaqueous media and of anionic species in basic media corresponds to formation of aquocations in acidic aqueous solutions and of hydroxoanions in aqueous base.

One justification for the presentation of the work on characterization of cationic species in chapter 6 lies in the fact that most of the work is relatively new and is still very restricted in coverage. Since the first reports of solvated cations of transition metals in acidic HF in 1977,[1] there has been a reasonable amount of work, almost exclusively from Melbourne, on generation of metallic cations in HF. Only a very small amount of corresponding work on *metallic* cations in $H_2SO_4$, $HSO_3F$, and $CF_3SO_3H$ has been done. Explicit postulation of the existence of discrete cationic species of $f$-transition elements in chloroaluminates has come from Belgian investigators only within the last decade.[2,3] Much of chapter 6 is devoted to an attempt to show that many other observations in chloroaluminates could be interpreted in terms of formation in acidic melts of cations of $d$-transition elements in "normal" oxidation states, for example, $Co^{2+}$, $Ni^{2+}$, $Fe^{2+}$, and $Fe^{3+}$.

The material presented in chapter 6 provides a background for studies of species which can exist only in highly acidic media, namely, cations of transition elements in oxidation states that are so low that they would reduce water—and in many cases have been demonstrated to have done so. Of course, transition metals can exist in extremely low oxidation states in a very wide range of compounds, such as the carbonyls and their derivatives. However, in this chapter, and throughout this book, we are concerned only with discrete solvated cations, in solution, and containing only the metal (or nonmetal) regardless of whether the cation is monatomic or polyatomic. It will be shown that controlled, high levels of superacidity are required for generation and stabilization of some of these cations, e.g. $Ti^{2+}$, and that even when they exist in a superacid, addition of the base of the solvent causes them to be destabilized by disproportionation reactions. Studies of the stabilization of low-oxidation-state cations in protonic superacids and their disproportionation as the basicity of the superacidic media is increased will be related to generation and stability of low-oxidation-state transition metal cations in chloroaluminate melts of varying acidity and basicity.

Chronologically, the first system to show enhanced stability of low oxidation states in protonic superacids was the U(III)—U(IV) system. In 1976 metallic uranium was shown to react with HF acidified with $BF_3$ to produce $H_2$ and a solution of $U^{3+}$ which was stable indefinitely. Subsequently $U^{3+}$ was shown to disproportionate to U(0) and $UF_4$ when $F^-$ was added to the acidic solution. A similar general experimental procedure was then used to generate $Cr^{2+}$, $V^{2+}$, $Ti^{2+}$, $Zr^{3+}$, $Hf^{3+}$, $Eu^{2+}$, $Yb^{2+}$, and $Sm^{2+}$. The differing experimental conditions for preparation of each of these solutions and for characterization of the cationic species and references to the original work are given throughout this chapter. In chapter 6 $d$-transition metal cations are presented before $f$-transition metal species and that same order will be preserved in this chapter despite the chronology and the importance of the U(III)—U(IV)—HF system in these studies.

## 7.1 Cations of $d$-Transition Metals in Low Oxidation States

Characterization of cations in HF has preceded the small amount of similar work in other protonic superacids and has led to a considerable amount of reinterpretation of related and often earlier investigations in chloroaluminate media. Therefore HF systems will be presented first.

## 7.1.1 Solution Preparation and Characterization in Hydrogen Fluoride

Solutions containing $Cr^{2+}$, $V^{2+}$, and $Ti^{2+}$ as HF-solvated cations were prepared by reaction of the metal in each case with HF which was acidified by an appropriate Lewis acid.[4] Experience at this stage showed that if the sample of metal used was extensively oxidized, low-oxidation-state cations were not formed or were in equilibrium with cations in higher oxidation states. Combined oxygen on the metal would be converted by the superacid to $H_2O$, which would then be protonated to $H_3O^+$. This could act as an oxidant to produce $H_2$ and $H_2O$, which would be reprotonated, leading to a cyclic oxidation process. Use of metallic powder or sponge was avoided because of the large surface available for oxidation. Where possible crystalline metal was used. Any surface oxide was removed with a very acidic solution, for example, $AsF_5$—HF, which was discarded, and this process was repeated several times.

Treatment of cleaned metallic Cr and V with nonoxidizing, weakly acidic $BF_3$—HF gave solutions similar in color to that of the cation $Cr(OH_2)_6^{2+}$, stable in acidic aqueous solution, and to the reported color for $V(OH_2)_6^{2+}$ prepared in aqueous solution by electrolytic reduction. $V^{2+}$ slowly reduces $H_2O$.

There was no apparent reaction between metallic Ti and $BF_3$—HF. The more acidic, but potentially oxidizing, solvent $AsF_5$—HF gave blue or green solutions with complex spectra, suggesting the presence of Ti(III). The still-more-acidic solvent 3 M $SbF_5$—HF caused evolution of $H_2$ and gave, surprisingly, a stable orange solution. Bluish solutions of Ti(III) had been expected. As shown below by comparison with spectra of known Ti(II) systems, the spectrum of the orange solution indicated the presence in solution of $Ti(FH)_6^{2+}$. It was established that there was a narrow "window" of $SbF_5$—HF acidity in which $Ti^{2+}$ was stable. At concentrations of $SbF_5$ in HF above 3 M, $Ti^{2+}$ was oxidized. High concentrations of $AsF_5$ or $SbF_5$, in which significant amounts of the molecular pentafluorides have been shown to exist in HF in equilibrium with the fluoroarsentates or the fluoroantimonates,[5] gave blue solutions or precipitates, indicating oxidation of Ti(II) to Ti(III). The molecular pentafluorides are much stronger oxidants than the fluoroanions. $Ti(FH)_6^{2+}$ was not stable in $SbF_5$—HF solutions which were less concentrated than about 2 M in $SbF_5$. It is proposed that, even in this highly acidic medium, there is sufficient concentration of $F^-$ present to cause disproportionation of Ti(II), as described in section 7.1.2.

The recorded UV-visible spectra[4] for $Cr^{2+}$, $V^{2+}$, and $Ti^{2+}$ in HF are reproduced in Figure 7.1 and positions of peak maxima in HF spectra are compared in Table 7.1 with those in aqueous spectra, where available[6], and with spectra for Cr(II), V(II), and Ti(II) in molten $AlCl_3$.[7] The latter are presented in detail in section 7.1.4. As reported in chapter 6 for $d$-transition metal cations in "normal" oxidation states, peak positions in HF spectra occur at lower energies in HF than in $H_2O$ and are lower again in molten chloride media.

Octahedral coordination of $Cr^{2+}$ in HF can be postulated by comparison of spectra in HF and in $H_2O$, where the $Cr(OH_2)_6^{2+}$ is the recognized species. Øye and Gruen[7] had postulated an octahedral environment of chlorines for Cr(II) in molten $AlCl_3$, although they were not particularly explicit about the entities to which these chlorines were bonded. This matter was discussed in section 6.1.1.3. Solvation to form $V(FH)_6^{2+}$ was postulated[4] on the basis of comparison of the V(II)—HF spectrum with that for $V(OH_2)_6^{2+}$ in water[6] and for this aquocation in crystalline alums. Øye and Gruen[7] interpreted their spectrum of $VCl_2$ in solution in molten $AlCl_3$ in terms of six-coordination of the V(II). Table 7.1 shows the similarities in spectra for Cr(II) and V(II) in $H_2O$, HF, and molten $AlCl_3$.

Comparison with aqueous spectra is impossible for the Ti(II)—HF system—solid $TiCl_2$ is reported to react violently with water in reducing it. $Ti^{2+}$ was shown to be the absorbing species by comparison of its UV-visible spec-

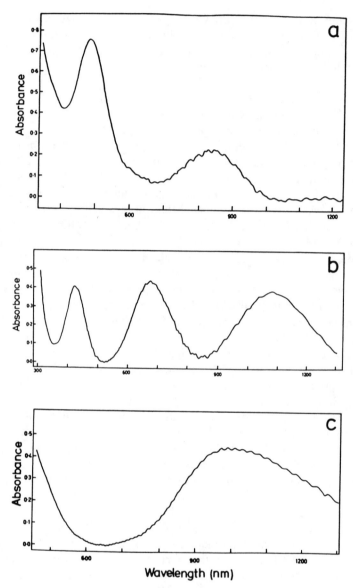

**Figure 7.1.** UV-Visible Spectra in Acidified HF: (a) $Ti^{2+}$; (b) $V^{2+}$; (c) $Cr^{2+}$.

trum reproduced in Figure 7.2a with two other sets of spectra. As they had done for Cr(II) and V(II), Øye and Gruen[7] interpreted their spectrum for a solution of $TiCl_2$ in molten $AlCl_3$ (Figure 7.2b) in terms of hexacoordination of Ti(II). The presence of $Ti^{2+}$ in anhydrous HF at room temperature as the six-coordinate species $Ti(FH)_6^{2+}$ is supported even more strongly by the comparison of its spectrum with that of $Ti^{2+}$ in a host crystal of NaCl (Figure 7.2c). Smith and co-workers[8] added stoichiometric amounts of metallic Ti

**Table 7.1.** Peak Maxima (nm) for Transition Metal Cations in Oxidation State (II)

| | $H_2O^6$ | $HF^4$ | $AlCl_3^7$ |
|---|---|---|---|
| Cr(II) | 710 | 1,010 | 1,190 |
| V(II) | 358, 540, 810 | 424, 674, 1,085 | 493, 826, 1,250 |
| Ti(II) | — | 480, 840 | 690, 1,316 |

and $CdCl_2$ at the 1% level to molten NaCl. After halogen exchange, volatile Cd was dispelled from solution and, after crystallization of the NaCl, $Ti^{2+}$ remained "doped" into the host crystal in which the coordination number for ions is six and the spectrum for the solid was recorded.

An extinction coefficient of about 4.5 $mol^{-1} dm^3 cm^{-1}$ for the 480-nm band of the $Ti^{2+}$ spectrum, the most novel of the spectra presented to date in this

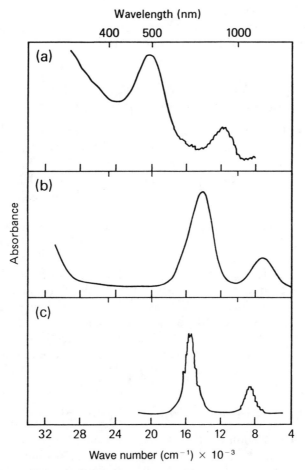

**Figure 7.2.** Spectra of Titanium(II) in (a) $SbF_5$—HF, (b) Molten $AlCl_3$ at 500 K, and (c) Crystalline NaCl at 10 K (Doped at 1% Level).

section, is typical of the magnitude expected for a $d$-transition metal cation in solution in HF.

Metallic Zr, treated with HF—$SbF_5$, yielded a pale-yellow solution with a major absorption band at 300 nm and a shoulder at approximately 370 nm.[9] Treatment of this solution with successive minute amounts of $F_2$ caused reductions in the intensities of these bands until they were eliminated. This indicates a low-oxidation-state compound in solution and the peak positions ascribed to $Zr^{3+}$ fit into a sensible series with those for Zr(111) in molten $AlCl_3$, $AlBr_3$, and $AlI_3$[10] in an analogous fashion to the correspondence between spectra for Ti(11) and V(II) in HF and in molten $AlCl_3$; that is, there was a shift to lower energies of about 300 nm for the higher-energy bands as the medium was changed from HF at room temperature to molten $AlCl_3$ at 500 K. Larsen and colleagues[10] reported much smaller shifts in peak positions (60–70 nm) for the solvents molten $AlCl_3$, $AlBr_3$, and $AlI_3$. A corresponding experiment in which metallic Hf was reacted with HF—$SbF_5$[9] provided a spectrum with a very strong band centered at 325 nm, assumed to be due to $Hf^{3+}$. No comparison with aqueous spectra is possible in the case of solutions of $Zr^{3+}$ and $Hf^{3+}$ in acidified HF.

## 7.1.2 Disproportionation in Hydrogen Fluoride of Low-Oxidation-State $d$-Transition Metal Cations

As stated in the introductory section of this chapter, the first investigation of base-induced disproportionation of a low-oxidation-state cation in HF was in relation to the U(III) system. (See section 7.2.2.) Subsequently Besida and O'Donnell[11] studied the addition of solutions of NaF in HF to the yellow-orange solutions of $Ti^{2+}$ in $SbF_5$—HF. Mixed precipitates of blue $TiF_3$, white $TiF_4$, and black Ti metal were observed and the supernatant solution was colorless. Disproportionation of Ti(II) to the element and to both Ti(III) and Ti(IV) were observed simultaneously because of the insolubility of $TiF_3$ in HF.

In section 7.1.1, it was stated that $Ti(FH)_6^{2+}$ cannot be generated in HF which is less than about 2 m in $SbF_5$. Besida and O'Donnell[11] established that reaction of metallic Ti with dilute $SbF_5$—HF solutions produced colorless solutions and $TiF_3$ or $TiF_3$ with $TiF_4$, depending on acid strength, indicating that if Ti(11) were to have transient existence in such solutions, it would disproportionate to Ti(0) and Ti(III) or Ti(IV). The finely divided metallic Ti would redissolve to give Ti(III) or Ti(IV).

## 7.1.3 Solutions in other Protonic Superacids

It was stated in section 6.1 that, by comparison with HF solution studies, very few transition metal cations in normal oxidation states had been generated in $H_2SO_4$, $HSO_3F$, and $CF_3SO_3H$. It is hardly surprising that even fewer studies have been made in these superacids of cations in unusually low oxidation states.

Aubke and co-workers at UBC, Vancouver, have very recently adduced ESR spectroscopic evidence for the existence of solvated $Au^{2+}$ in $HSO_3F$.[12] No stable aquocation $Au^+$ exists in water. The aqueous chemistry of Au(I) is very largely limited to anionic species such as $Au(CN)_2^-$ and $AuCl_2^-$. The halogen chemistry of Au(I) is exhibited as the chain polymeric solids $(AuX)_\infty$ where $X = Cl$, Br, or I. The chemistry of gold(II) is also very limited. There appear to have been no reports of the existence of discrete, solvated $Au^{2+}$ cations in solution. Solid compounds that formally contain Au(II) are either compounds with Au—Au bonds, mixed-oxidation-state diamagnetic solids such as $Au_4Cl_8$, which contains Au(I) and Au(III), or anionic species complexed with ligands such as dithiolene.

Aubke and colleagues[12] reduced $Au(SO_3F)_3$ with CO in $HSO_3F$ to produce $Au(SO_3F)_2$ and $Au(CO)SO_3F$ as solid products. More importantly, within the context of this chapter, they conproportionated metallic Au and $Au(SO_3F)_3$ in $HSO_3F$ for 22 days at room temperature to produce the solid precipitate $Au(SO_3F)_2$, formulated as a mixed-valency-state compound, Au(I) Au(III) $(SO_3F)_4$. ESR spectroscopy of the red-black supernatant solution was interpreted as indicating the presence of the discrete paramagnetic solvated cation $Au^{2+}$, which appears to be stable because of the high level of acidity of the medium in which it was generated.

At Melbourne, Adrien[13] has studied the interaction of metallic Cr with $CF_3SO_3H$ acidified with $B(OSO_2CF_3)_3$. The initial spectrum indicated that the solution contained both Cr(II) and Cr(III). After a standing period of several days, the solution spectrum showed the presence of Cr(III) only. There is now a considerable body of experimental evidence to indicate that solutions of $B(OSO_2CF_3)_3$ in $CF_3SO_3H$ are thermally unstable at room temperature, especially if they are reasonably concentrated. An interpretation of Adrien's observations would be that initially the solution was sufficiently acidic to stabilize Cr(II), in part at least, but that, with decomposition of the Lewis acid, the medium was reduced in acidity to the point where Cr(II) disproportionated to Cr(III) and Cr(0), the latter species being continuously reoxidized by the medium.

## 7.1.4 Solutions in Acidic Chloroaluminate Melts

Relatively little of the study in acidic melts of $d$-transition metal halides in low oxidation states has been directed toward identification of cations, as such, in the melts. Indeed, this comment is valid as it relates to all melt studies, regardless of the "normalcy" or otherwise of the oxidation states.

As shown in chapter 6, there is general acceptance that, in neutral melts, precipitation of binary halides occurs to a greater or less extent and that anionic species, usually formulated as hexahalometallates or tetrahalometallates, exist in basic melts. Gruen and McBeth[14] used electronic spectroscopy to characterize such anionic species for tripositive Ti, V, and Cr and for dispositive V, Cr, Mn, Fe, Co, Ni, and Cu in neutral and basic chloride melts. They found that for all the dipositive metals except V, the anionic species was $MCl_4^{2-}$ in neutral and in basic melts. V(II) and the tripositive metals provided both tetrachloro- and hexachlorometallates, equilibria favoring the latter as the basicity was increased. In the later studies in high-temperature and room-temperature chloroaluminates, reported in chapter 6, the anionic nature of transition metals in basic melts has been either established or assumed.

Soon after the 1963 report[14] of the spectra of the eight first-row transition metals, Ti to Cu, in neutral and basic molten chloride media, Øye and Gruen[7] published the spectra of the dichlorides of the same elements in the highly acidic medium, molten $AlCl_3$. They used directly as solutes the dichlorides of the elements Mn to Cu, which are stable from a redox point of view in this and in most other media. It is of interest here that they conproportionated Cr and $CrCl_3$ in $AlCl_3$ to form Cr(II). $VCl_2$ was prepared by reaction of metallic V with HCl gas at 950°C. $TiCl_2$ was prepared in situ in the sealed spectral cell by a metathetical reaction between $PbCl_2$ and Ti metal in molten $AlCl_3$ at about 230°C.

As discussed for the spectrum of $CoCl_2$ in $AlCl_3$ (section 6.1.1.3), Øye and Gruen said[7] that spectra in $AlCl_3$ for all eight of the dipositive metals could be interpreted "on the basis of octahedral configurations of chlorides about the central transition metal ions." They might have been better to postulate an octahedral environment of six "chlorines," since some authors have been led to postulate the existence of anionic species $MCl_6^{4-}$ in acidic $AlCl_3$, a proposition rejected in section 6.2. They did suggest the possibility of coordination of $M^{2+}$ by two tridentate $Al_2Cl_7^-$ anions. An alternative, proposed in section 6.1.1.3, is weak coordination by three bidentate $Al_2Cl_6$ molecules, overwhelmingly the major species in the solvent. $Al_2Cl_7^-$ would be present to a very small extent, namely, in an amount stoichiometrically equal to the very low $MCl_2$ concentration, usually $10^{-1}$ to $10^{-2}$ M. It might be sufficient to interpret the spectra in terms of discrete $M^{2+}$ cations in octahedral sites in the melts.

The spectra of Øye and Gruen for Cr(II), V(II), and Ti(II) in $AlCl_3$ were used in section 7.1.1, as illustrated in Figure 7.2, to support the proposal that in superacids such as HF the cations $Cr^{2+}$, $V^{2+}$, and $Ti^{2+}$ are in an octahedral environment; that is, they exist as the solvated species $M(FH)_6^{2+}$.

The recording of spectra of $d$-transition-metal dichlorides in $AlCl_3$[7] and of the dichlorides and trichlorides in neutral and basic chloride media[14] represents the most comprehensive and wide-reaching survey of metal speciation in molten salts. Others have mounted more detailed electrochemical studies of differing oxidation states for individual $d$-transition elements in melts of differing acidity and basicity. It is probably fair comment to say that these studies have been directed less to an attempt to characterize ionic species in solution than to the task of establishing electrochemical parameters for redox behavior in terms of melt composition. An underlying, perhaps fundamental, quest seems to have been to determine the conditions for electrodeposition of the so-called refractory metals.

Fung and Mamantov[15] studied electrochemical oxidation of $Ti(AlCl_4)_2$ in $AlCl_3$—NaCl melts from quite strongly acidic (65% $AlCl_3$) to weakly acidic (52% $AlCl_3$) and reported stepwise oxidation in the acidic melts of Ti(II) to Ti(III) and Ti(IV), with both processes being electrochemically reversible at platinum electrodes. They commented that "the lower oxidation states were found to be more stable at lower temperatures and in more acidic melts." Osteryoung and colleagues[16] used as medium the acidic room-temperature chloroaluminate melt ($AlCl_3$:1-ethyl-3-methylimidazolium chloride, 1.5:1) in an electrochemical investigation of the reduction of Ti(IV). They reported reduction to Ti(III) and Ti(II) in two one-electron steps, each of which exhibited slow electron-transfer kinetics. The recent electrochemical investigation[17] of reduction of Ti(IV) through Ti(III) and Ti(II) to electrodeposited Ti(0) was conducted in the basic LiCl—KCl eutectic and so lies outside this discussion of cationic species in acidic melts.

Sørlie and Øye[18] have conducted a very elegant spectroscopic study of $TiCl_2$ and $TiCl_3$ in a range of chloroaluminates of varying acidity and basicity from the pure acid $AlCl_3$, through differing mixtures of $AlCl_3$ and KCl, to the strongly basic eutectic LiCl—KCl. Dependence of speciation of Ti(III) on acidity and basicity of the medium will be discussed immediately below, but the Ti(II) system will be presented in section 7.1.5 because it provides a nice example of base-induced disproportionation.

Sørlie and Øye[18] dissolved $TiCl_3$ in molten $AlCl_3$ and obtained the spectrum in Figure 7.3a with a main peak at 13,800 $cm^{-1}$ and a well-defined shoulder, or another peak, at 10,700 $cm^{-1}$. It has been shown several times in this book (Tables 6.3 and 7.1) that, for spectra of the same species, there is a large shift of the major peak positions to lower energies as the medium is changed from water to chloroaluminates. It is instructive to note that the spectrum for $Ti(OH_2)_6^{3+}$ is characterized by a broad peak with a maximum at 20,100 $cm^{-1}$ and a shoulder at 17,400 $cm^{-1}$, a splitting of 2,700 $cm^{-1}$. The spectra for $TiCl_6^{3-}$ and $Ti(OH_2)_4Cl_2^+$—distorted octahedral entities related to the aquocation and to the species to be discussed here—have major peaks at 13,000 and 19,220 $cm^{-1}$, respectively, with splittings of 3,000 and 4,250 $cm^{-1}$.[19]

In separate experiments they made $AlCl_3$—KCl mixtures of decreasing acidity by increasing the KCl concentration relative to $AlCl_3$, and they recorded spectra of Ti(III) in each mixture. As the mole percentage of $AlCl_3$ was decreased to 66.7%, the two peaks shifted only slightly and the splitting increased to about 5,000 $cm^{-1}$. At 51 mol% $AlCl_3$, the observed spectrum (Figure 7.3b) has very similar shape, molar absorptivity, and peak position to that ascribed by Gruen and McBeth to tetrahedral $TiCl_4^-$ in the neutral melt $CsGaCl_4$.[14] In slightly basic melt (49% $AlCl_3$—51% KCl) the spectrum

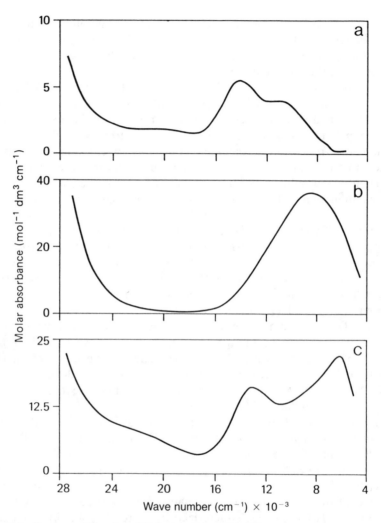

**Figure 7.3.** Spectra of Titanium(III) in (a) 100 mol% $AlCl_3$ at 500 K, (b) 51% $AlCl_3$—49% KCl at 700 K, and (c) 49% $AlCl_3$—51% KCl at 750 K.

observed by Sørlie and Øye and shown in Figure 7.3c is very similar in all features to that observed by Gruen and McBeth in the strongly basic eutectic LiCl—KCl,[14] and ascribed by them to octahedral $TiCl_6^{3-}$ in this medium of high $Cl^-$ availability. While this spectrum is somewhat reminiscent of the two-peak spectrum observed by Sørlie and Øye and shown in Figure 7.3a for $TiCl_3$ in 100% $AlCl_3$, both peak positions are at much lower energies, shifted by about 2,000 and 4,500 $cm^{-1}$, respectively. By analogy with the work in LiCl—KCl, it seems reasonable to postulate that, in 49% $AlCl_3$—51% KCl, Ti(III) is present as the octahedral anion $TiCl_6^{3-}$ and, again by comparison with Ti(III) in neutral $CsGaCl_4$, it is likely to be present as the tetrahedral anion $TiCl_4^-$ in 51% $AlCl_3$—49% KCl. Another possibility in this very weakly acidic medium is that tetrahedral coordination is achieved in the species $Ti(AlCl_4)_2$, each $AlCl_4^-$ ligand, the dominant species in the melt, being bidentate. Obviously as the availability of free chloride is decreased, that is, as we move into more acidic chloroaluminates, anionic speciation of Ti(III) becomes less likely. However, the environment in 100% $AlCl_3$ is octahedral. Whatever $Cl^-$ is potentially available will have formed $Al_2Cl_7^-$, and, therefore, purely on the grounds of ionic equilibria, $TiCl_6^{3-}$ is very unlikely to be formed. Additionally, the Ti(III) spectra in acidic and basic chloroaluminate media, although each strongly suggestive of an octahedral environment, are sufficiently different to suggest that speciation is different in acidic and basic media. It is postulated here that Ti(III) in 100% $AlCl_3$ is $Ti^{3+}$, possibly weakly solvated by three bidentate $Al_2Cl_6$ molecules.

Larsen and co-workers[10] used the acidity of aluminium trihalides to produce small crops of high-purity $ZrCl_3$, $ZrBr_3$, and $ZrI_3$ which were crystallized from $AlCl_3$, $AlBr_3$, and $AlI_3$ as the appropriate zirconium tetrahalides were reduced in the molten aluminium trihalides by metallic Al and Zr, the latter a conproportionation reaction favored by acidity of the medium. They reported that, for the chloride and bromide systems, there was a "competing disproportionation reaction" yielding Zr(II) compounds. Overall, this can be viewed as an equilibrium from the conproportionation of Zr and $ZrX_4$ to give $ZrX_2$. They found that the rate of reduction of the hafnium tetrahalides was slower than that of the corresponding zirconium compounds.

Electrochemical reduction of Zr(IV)[20] in a weakly acidic chloroaluminate ($AlCl_3$:NaCl, about 51:49) at about 200°C produced, not surprisingly, insoluble $ZrCl_3$. This is consistent with all reported solubilities of binary chlorides in neutral or slightly acidic chloroaluminates, whether at high temperature or at ambient temperature. In somewhat more acidic melts ($AlCl_3$:NaCl, 60:40), described by the authors as "very acidic," cyclic voltammograms were interpreted as showing reduction to Zr(III) in solution which disproportionated to Zr(IV) and Zr(II), rapidly above 250°C but slowly at temperatures below 140°C.

## 7.1.5 Disproportionation in Chloroaluminate Melts

As just discussed, there have been proposals,[10,20] reasonably soundly based, that Zr(III) disproportionates in $AlCl_3$ and acidic chloroaluminates while it is stable in nearly neutral melts. Fung and Mamantov[15] demonstrated that Ti(II) was stable in acidic melts, but the work of Sørlie and Øye[18] on $TiCl_2$ in chloroaluminates of varying acidity and basicity appears to be the soundest and most comprehensive in demonstrating disproportionation in chloroaluminates of a $d$-transition element. In section 7.2.4 there is a detailed reappraisal of earlier work reported on the acid-base dependence of the stability in chloroaluminates of the $f$-transition element uranium in oxidation state (III).

Sørlie and Øye[18] dissolved TiCl$_2$ in pure AlCl$_3$ at 500 K and reproduced the earlier spectrum recorded by Øye and Gruen.[7] (See Figure 7.2b.) This spectrum, although shifted to lower energies, is similar to that observed for Ti(FH)$_6^{2+}$ in SbF$_5$—HF$_4$ (Figure 7.2a). Sørlie and Øye's spectrum has peaks at 14,600 and 7,500 cm$^{-1}$ and is shown in Figure 7.4a. They then recorded spectra of TiCl$_2$ in a range of chloroaluminate melts made progressively less acidic by increasing the KCl:AlCl$_3$ ratio until the molar ratio was 51:49; that is, the melt was weakly basic.

In melts containing 82, 75, and 66.6 mol% AlCl$_3$ a new high-energy band appeared, the three peak positions now being 21,550, 14,250 and 7,250 cm$^{-1}$, with the two lower-energy peaks having relative intensities and molar absorbances very similar to those of the corresponding peaks in 100% AlCl$_3$. This is shown for 66.7% AlCl$_3$—33.3% KCl in Figure 7.4b, a spectrum very similar to those in 82 and 75% AlCl$_3$. It is proposed that Ti$^{2+}$, octahedrally coordinated in pure AlCl$_3$ by Al$_2$Cl$_6$ molecules, probably bidentate, is still essentially octahedrally coordinated, but that there is considerable distortion of the environment, perhaps as a result of inclusion of one Al$_2$Cl$_7^-$ or AlCl$_4^-$ in the coordination sphere. Another possibility is that the weakly solvated TiCl$^+$ cation can be formed and that the new high-energy band results from a charge-transfer process. At 51% AlCl$_3$, a very weakly acidic medium, the two peaks at about 14,000 and 7,000 cm$^{-1}$, which are

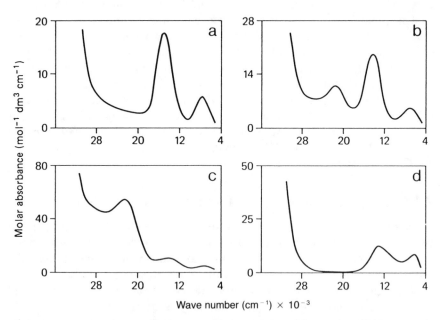

**Figure 7.4.** Spectra of Titanium Dichloride in (a) 100 mol% AlCl$_3$ at 500 K, (b) 67% AlCl$_3$—33% KCl at 600 K, (c) 51% AlCl$_3$—49% KCl at 700 K, and (d) 49% AlCl$_3$—51% KCl at 750 K.

characteristic of $Ti^{2+}$, were still present but weak by comparison with the 22,000 $cm^{-1}$ peak, as seen in Figure 7.4c.

As the melt mixture became basic (at 51 mol% KCl), the spectrum for the system to which $TiCl_2$ had been added changed dramatically to that for Ti(III) in a basic melt. The spectrum in Figure 7.4d is very similar in peak positions and molar absorptivities to that observed by Sørlie and Øye for $TiCl_3$ in 49% $AlCl_3$—51% KCl,[18] and to that recorded by Gruen and McBeth for $TiCl_3$ in the strongly basic eutectic melt LiCl—KCl.[14] Sørlie and Øye report that metallic titanium deposits on the walls of the spectrometric cell under these conditions. These observations are consistent with an idealized disproportionation reaction:

$$3Ti^{2+} + 12Cl^- \rightarrow Ti + 2TiCl_6^{3-} \qquad\qquad 7.1$$

The comprehensive study by Sørlie and Øye provides a significant example of selective stabilization of low oxidation states of transition metals in acidic media. Although, conventionally, Ti(III) is regarded as a low oxidation state for that element, it must be remembered that $Ti(OH_2)_6^{3+}$ is stable in acidic aqueous solution. Strictly, within the stated format of this book, its behavior in nonaqueous media should have been presented in chapter 6. However Ti(III) and Ti(II) in molten chloroaluminates present an interesting comparison in oxidation-state stabilities. Ti(III) and Ti(II) are both stable in acidic chloroaluminates and Ti(III) remains so in basic media. However, Ti(II) disproportionates in the presence of significant amounts of $Cl^-$ (equation 7.1).

## 7.2 Cations of f-Transition Metals in Low Oxidation States

Because of its technological importance, uranium chemistry has been widely and comprehensively studied, especially in the solid-state area where large numbers of phase studies and of crystallographic determinations have been reported. To a great extent solution studies—aqueous and nonaqueous— have been directed toward uranium speciation and reactions of the element in high oxidation states, IV and especially VI. There is a good body of molten salt work on U(IV) in molten halides, some of it alluded to in section 6.1.2.3. In the mid-1980s some elegant work was done on U(III) and U(IV) in room-temperature chloroaluminates, which will be discussed in section 7.2.4. Earlier work on U(III) and U(IV) in high-temperature melts, conducted in the early 1960s, needs reinterpretation, as shown in section 7.2.5.

Although widely considered on experimental grounds to be very difficult to use as a solution medium, anhydrous HF is per se essentially a very simple solvent, as will be discussed in chapter 9. Studies in HF solution chemistry, particularly for U(III), have led to an ability to reinterpret much work in more complex solvents. Therefore, as in chapter 6 and the earlier parts of this chapter, the solution chemistry in HF of $f$-transition elements in low oxidation states will be presented before that in the other nonaqueous systems dealt with in this book.

For the lanthanides, low oxidation states are easy to define. For discrete cations, the only departure from the normal oxidation state (III) is the formation of dipositive cations of three of the lanthanide group. Values of E° for the $Ln^{3+}/Ln^{2+}$ redox couple in aqueous solution are $-0.35$, $-1.05$, and $-1.55$ V for the Eu, Yb, and Sm systems, respectively.[21] Solid compounds of Eu(II), for example, $EuSO_4$, are easily prepared and $Eu^{2+}$ is stable in acidic aqueous solution. Yb(II) and Sm(II) solid compounds can be prepared, for example, the di-iodides, but $Yb^{2+}$ and $Sm^{2+}$ reduce $H_2O$ although solutions of some of their compounds are stable in some nonaqueous media, as cited in the paper which deals with their generation in HF.[22] (See section 7.2.1.)

U(III) reduces aqueous solutions and is included in this chapter. There is no known solution chemistry in oxidation state (III) for thorium and protactinium. For actinides heavier than uranium, oxidation state (III) is stable in aqueous solution. So, the relevant Np(III) and Pu(III) nonaqueous solution chemistry was presented in chapter 6.

## 7.2.1 Solutions in Hydrogen Fluoride

Metallic Eu, Yb, and Sm reacted directly with anhydrous HF to give solutions containing the solvated cations $Eu^{2+}$, $Yb^{2+}$, and $Sm^{2+}$, which were remarkably stable in solution.[22] Even $Sm^{2+}$, the most easily oxidized of the three cations, showed the spectral characteristics of the ion after 5 days at room temperature. The lifetime of $Sm^{2+}$ in oxygen-free acidified aqueous solution is of the order of minutes. $Eu^{2+}$ was stable indefinitely in HF.

For the reasons given in the early part of section 7.1.1, the lanthanide metals used for solution preparation were in "massive" crystalline form. Surface oxide was removed by three or four washings with HF solution. Then the addition of quadruply distilled HF gave colorless solutions of $Eu^{2+}$ and $Yb^{2+}$ and yellow-orange $Sm^{2+}$ solution according to the equation

$$Ln + 2HF \rightarrow Ln^{2+}(solv) + 2F^-(solv) + H_2 \qquad 7.2$$

where Ln = Sm, Eu, Yb.

It is of interest to note that, whereas preparation of HF solutions containing dipositive $d$-transition metal cations required the use of appropriate Lewis acids (section 7.1.1), the lanthanide metals dissolved in "neat" HF and were stable in solution in HF made basic by the increase in $F^-$ ion concentration, indicated from equation 7.2. This should not be surprising. Ionic radii of $Ln^{2+}$ ions are comparable with those of $Ca^{2+}$ and $Sr^{2+}$ and the difluorides of these two alkaline earth elements are appreciably soluble in neat HF. Lanthanide trifluorides are virtually insoluble in neat HF. So it was possible to study very "clean" lanthanide(II) spectra in solutions free of Ln(III) species. Attempts to produce Tm(II) solutions by interaction of the metal with HF were unsuccessful,[22] although it will be shown in section 7.2.4 that electrochemical reduction of Tm(III) in acidic chloroaluminate melts gave Tm(II) at a potential very close to that for reduction of Al(111) in the melt.

The lanthanide(II)—HF UV-visible spectra were compared[22] with their aqueous counterparts. $Yb^{2+}$ and $Sm^{2+}$, which are oxidized by $H_2O$, can be maintained in oxidation state(II) under conditions of continuous electrolysis in the spectral cell. The spectra in HF showed superior resolution to those in other media, but comparison of aqueous and HF spectra indicated the presence of solvated $Ln^{2+}$ cations in HF. As was found for spectra of the tripositive lanthanide cations in HF, peaks resulting from *f-f* transitions were shifted to slightly higher energies than in $H_2O$. The energies of broad bands associated with *4f-5d* transitions are very sensitive to the ligand field and exhibited much bigger shifts to higher energies.

Exposure of the three HF solutions containing $Ln^{2+}$ cations to the gaseous, weak Lewis acid $GeF_4$ resulted in precipitation of $LnGeF_6$ in each case. Details of characterization of $SmGeF_6$, the solid least likely of the three to be a pure lanthanide(II) compound, have been reported.[23] This provides a new route to relatively easy synthesis of solid lanthanide(II) compounds at or below room temperature and will be discussed again within a broader context in section 10.2.5.1.

Uranium(III) compounds reduce water to hydrogen, being oxidized to uranium(IV) species. The spectrum of $U^{3+}_{aq}$ was recorded[24] in a solution which contained $UO_2^{2+}$ initially and in which $U^{3+}$ was being produced by continuous cathodic reduction within the spectral cell. When the reducing potential was no longer being applied, $U^{3+}_{aq}$ would have reduced $H_2O$ and been oxidized to U(IV).

The first stable solution containing $U^{3+}$ in a simple protonic solvent was produced by treating "massive" metallic U, from which surface oxide had been removed by preliminary treatment with acidic HF, with anhydrous HF saturated with $BF_3$, which is a weak but nonoxidizing Lewis acid of the HF solvent system.[25] Protons, represented somewhat ideally as $H_2F^+$, were reduced by U to $H_2$. The clear lilac-colored solution resulting from this reaction was shown to contain $U^{3+}$ by comparison of the UV-visible spectrum of the solution with that for the aqueous solution produced cathodically as earlier. These two spectra are shown in Figure 7.5a and b. The very similar spectrum in Figure 7.5c is that for U(III) in a room-temperature melt and will be discussed in detail in section 7.2.4. The similarity of the aqueous and HF spectra for U(III) suggests very strongly that the HF solution contains solvated $U(FH)_x^{3+}$, the analogue of $U(OH_2)_x^{3+}$ in water. The relative intensities of the major bands at 800–900 nm are reversed in the HF spectra, compared with those in water, but the peak positions are very similar, being shifted 20–30 nm to higher energy relative to those in water, as is common

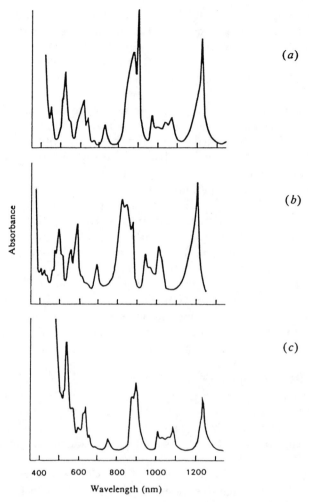

**Figure 7.5.** Spectra of Uranium(III) in (a) 1 M $DClO_4$ (b) $BF_3$—HF, and (c) $AlCl_3$—BuPyCl, 2:1.

for HF spectra of actinides. A detailed comparison of peak positions for U(III) in HF and in water was made in the original paper reporting the HF spectra.[25]

As noted above, the HF solution for which the U(III) spectrum is reproduced (Figure 7.5b) is weakly acidic. $BF_3$ was used as the Lewis acid in the original experiment because it is nonoxidizing and would, if it produced a solution, be least likely to oxidize U(III) to U(IV), totally or in part. The stronger Lewis acids $AsF_5$ and $SbF_5$ would give a more acidic environment but could oxidize U(III). An experiment was conducted[25] in which a small amount of metallic U was introduced into the spectral cell in order to act as a reductant when U(III)—AsF$_5$—HF was introduced to the cell. The spectrum for this solution was virtually identical with that in HF—BF$_3$ with peaks shifted by another 2–5 nm to shorter wavelengths.[25] These results suggest the possibility of some degree of fluoro complexation of U$^{3+}$, possibly as UF$^{2+}$, in HF—BF$_3$, which is less acidic, that is, more basic, than HF—AsF$_5$. The shifts for Np$^{3+}$ between the two HF media were less than for U$^{3+}$,[25] raising the possibility that Np$^{3+}$ is less subject to fluorocomplexation in HF than U$^{3+}$.

## 7.2.2  Disproportionation of Uranium(III) in Hydrogen Fluoride

During 1984 experiments were conducted in Melbourne to investigate the possibility of shifts or broadening of peaks in the spectra of U$^{3+}$ in HF solutions, initially acidified with BF$_3$, as the base F$^-$ was added gradually. Evidence for species in solution such as UF$^{2+}$ and UF$_2^+$ was being sought. Instead, disproportionation of U(III) into U(IV) and U(0) was observed.[26] The solution, initially the characteristic lilac color of U$^{3+}$, became colorless, indicating no uranium species in solution, and a precipitate of green UF$_4$ mixed with black specks of metallic uranium was observed. In one particularly favorable reaction a bronze-colored mirror of metallic uranium formed

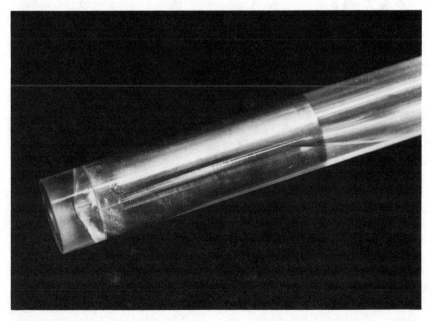

**Figure 7.6.** Mirror of Metallic Uranium on the Walls of a "Synthetic Sapphire" Reaction Tube.

on the walls of the "synthetic sapphire" reaction tube in addition to the black metal in the precipitate. This mirror formation can be seen in Figure 7.6. The reaction is a simple base-induced disproportionation:

$$4U^{3+} + 12F^- \rightarrow U + 3UF_4 \qquad\qquad 7.3$$

## 7.2.3 Dissolution and Disproportionation in Other Superacids

The very small amount of work that has been done on efforts to stabilize low oxidation states in protonic superacids other than HF has highlighted the special efficacy of HF in this regard. $H_2SO_4$ would not be expected to be a suitable medium for generation of low-oxidation-state cations because of the availability of the strong oxidant $SO_3$, particularly in oleums. For a similar reason, $HSO_3F$ has not proved encouraging. $HSO_3F$ dissociates fairly readily into $SO_3$ and HF and so, although intrinsically more acidic than $CF_3SO_3H$ as a medium and, on these grounds, it is more likely to stabilize low oxidation states, its oxidant strength is greater than that of $CF_3SO_3H$.

In studying the Cr(0)—Cr(II)—Cr(III) system in $CF_3SO_3H$ acidified with $B(OSO_2CF_3)_3$ (see section 7.1.3), Adrien[13] showed that $CF_3SO_3H$—$B(OSO_2CF_3)_3$ with a value of about $-18$ for its Hammett Acidity Function ($H_0$) is less effective in stabilizing Cr(II) in solution than $BF_3$—$HF^4$ (see section 7.1.1) for which $H_0$ would be about $-17$. This is consistent with the warning given in section 4.2.4 against a simplistic application of numerical values of $H_0$ in comparing stability and disproportionation reactions in different protonic superacids.

Following the experimental approach adopted by the Melbourne group for possible reaction between HF and the lanthanide metals, Adrien[13] found no reaction between the two metals Yb and Sm and the neat acid $CF_3SO_3H$, whereas the metals had interacted directly with neat HF to give solvated $Yb^{2+}$ and $Sm^{2+}$. Increasing the acidity of $CF_3SO_3H$ with each of the two Lewis acids $TaF_5$ and $B(OSO_2CF_3)_3$ led to reaction of metallic Yb to give yellow solutions of solvated $Yb^{2+}$. The spectra were similar to those for the $Yb(OH_2)_x^{2+}$. For the ligand $CF_3SO_3H$ with oxygen donor atoms, spectral peak positions were at lower energies and closer to those for the aquocation spectra than for the HF-solvated species.

Reaction between metallic Sm and $CF_3SO_3H$ solutions acidified with $TaF_5$ and with $B(OSO_2CF_3)_3$ gave, not blood-red $Sm^{2+}$, but yellow $Sm^{3+}$. In summary, neat $CF_3SO_3H$ cannot produce stable solutions containing cations of Yb or Sm, but $CF_3SO_3H$ acidified with $TaF_5$ or $B(OSO_2CF_3)_3$ is sufficiently acidic to produce stable solutions of solvated $Yb^{2+}$ but not $Sm^{2+}$.

Adrien has also studied U(III) and U(IV) spectra in $CF_3SO_3H$ and $HSO_3F$.[13] He found that when suitably cleaned metallic uranium was dissolved in a solution of $CF_3SO_3H$ which was 0.4 M in $B(OSO_2CF_3)_3$, for which the Hammett Acidity Function ($H_0$) is about $-18$, the spectrum was virtually identical with those for $U_{solv}^{3+}$ in acidic media as in Figure 7.5. However, when the metal was treated with a $CF_3SO_3H$ solution 0.4 M in $TaF_5$ the spectrum was very similar to that for $U_{solv}^{4+}$ in aqueous $HClO_4$ (Figure 6.9a A). He has shown that 0.4 M $TaF_5$—$CF_3SO_3H$ is about 2 $H_0$ units less acidic than 0.4 M $B(OSO_2CF_3)_3$—$CF_3SO_3H$, indicating that the $TaF_5$—$CF_3SO_3H$ solvent system, while sufficiently acidic to generate concentrations of $U^{4+}$ suitable for spectroscopic investigation, was not sufficiently acidic to stabilize the lower oxidation state. $U^{3+}$ formed initially would interact with the small concentration of the base $CF_3SO_3^-$ in $TaF_5$—$CF_3SO_3H$ to disproportionate to $U^{4+}$ and U, the latter being reoxidized through U(III) to U(IV).

When Adrien treated metallic U with the very acidic solvent 1 M $SbF_5$—$HSO_3F$ ($H_0 = -18.9$) the spectrum observed was not that for $U^{3+}$ as expected on the grounds of acidity of the medium but the one ascribed to $UF_2^{2+}$ (Figure 6.9b). It was realized that $HSO_3F$ is particularly susceptible to dissociation into HF and $SO_3$. If this were to have occurred, U(IV) in equilibrium with U(III) would form $UF_2^{2+}$ and that reaction would drive the U(IV)—U(III) equilibrium in favor of formation of $UF_2^{2+}$. This interpretation was supported by treating metallic uranium with $CF_3SO_3H$ which was 0.4 M in $B(OSO_2CF_3)_3$ and which contained deliberately added free HF. Whereas in the earlier experiment 0.4 M $B(OSO_2CF_3)_3$—$CF_3SO_3H$ had generated $U_{solv}^{3+}$, the spectrum showed that when HF was present, the preferred stable species was $UF_2^{2+}$. This can be rationalized by proposing that U(III) which would be formed because of the intrinsic acidity of the system would disproportionate into U(0) and $UF_2^{2+}$ because of the stability of the $UF_2^{2+}$ species and that the U(0) resulting from disproportionation would be oxidized in turn to U(III) and U(IV).

## 7.2.4  Solutions in Chloroaluminate Melts

In high-temperature acidic chloroaluminates ($AlCl_3$:NaCl, 60:40 at 175°C) Gilbert and co-workers[27] produced stable solutions of Eu(II), Yb(II), and Sm(II) as a result of one-electron, reversible voltammetric reduction of the three lanthanides in oxidation state III.

Later, members of the same research group conducted electrochemical and spectroscopic studies for lanthanide(III)—lanthanide(II) systems at 40°C in chloroaluminates based on 1-n-butylpyridinium chloride (BuPyCl) ranging from acidic ($AlCl_3$:BuPyCl, $> 60$: $< 40$), through more weakly acidic ($AlCl_3$, 60 to 50.1 mol%), to neutral and basic. In separate papers they reported electrochemical and spectroscopic behavior of Yb(III)—Yb(II) and Sm(III)—Sm(II)[28] and of Eu(III)—Eu(II) and Tm(III)—Tm(II).[29] Reversible one-electron reduction occurred in acidic melts and values of $E_{1/2}$ (relative to an Al electrode in 57 mol% $AlCl_3$) were $+1.85$, $+1.23$, $+0.66$, and $+0.02$ V for Eu(III—II), Yb(III—II), and Tm(III—II), respectively. It is of interest that differences in these $E_{1/2}$ values reflect quite closely the differences in experimentally determined and calculated values of $E°$ for the same systems in aqueous solution, namely, $-0.35$, $-1.05$, $-1.55$, and $-2.3$ V.[21]

Gilbert and his colleagues report[28] that, as acidic media are replaced by those approaching neutrality and ultimately are basic, values of $E_{1/2}$ for any particular redox couple shift to more cathodic potentials. Like most molten salt chemists, they report that, near neutrality, lanthanide chlorides are sparingly soluble in the chloroaluminates, with lanthanide trichlorides less soluble than the dichlorides, as is reasonable considering the radii and charges of the "cations" in each case. They postulate the existence of $Sm^{3+}$ cations in acidic melts and ascribe to charge transfer in $SmCl^{2+}$ a strong absorption band at 283 nm that is observed at $pCl \simeq 11.1$ when the mole fraction of $AlCl_3$ is well below 60%. They say that "the various forms of Sm(III) and Sm(II) can be represented as solvated $Sm^{3+}$, $Sm^{2+}$ and $SmCl^{2+}$, $Sm^{2+}$ at pCl higher and lower than 11.5, respectively." They postulate more extensive chlorocomplexation of $Yb^{3+}$ and $Yb^{2+}$ than for $Sm^{3+}$ and $Sm^{2+}$, which is consistent with ionic radius trends resulting from the lanthanide contraction.

Species such as $SmCl_6^{3-}$ are proposed as existing in basic melts. It is stated that "the electron transfer becomes irreversible, probably due to the formation of hexachlorocomplexes." They report further "that in basic melt, the Ln(II) precipitates dissolve but a blue coloration appears immediately in the melt" and they then say that "this color is characteristic of the butyl viologen monocation monoradical. The Ln(II) is then reducing the melt and this is further evidence of the difference in structure of lanthanides between the acidic and basic compositions." In describing the formation of the color in Gilbert's work, Osteryoung[30] has written, "thus the divalent lanthanides in the basic melt are strong reducing agents." In section 7.2.5, an alternative explanation, based on disproportionation of Ln(II), is offered to explain reduction of the $BuPy^+$ cation.

There have been several reported studies of uranium(III) in molten salts—usually in alkali chlorides or fluorides, that is, in basic media. In 1963 Morrey and co-workers[31,32] prepared solutions of uranium(III) in a series of chloroaluminate melts ranging from acidic ($AlCl_3$:KCl, 2:1) to basic ($AlCl_3$:KCl, 0.4:1) and purported to demonstrate enhanced *direct* reduction of U(III) by metallic aluminium in near-neutral melts. This work has been totally reinterpreted[33] in terms of disproportionation of U(III) to U(IV) and U(0) in near-neutral melts and reduction of U(IV) by aluminum, as will be shown in section 7.2.5. The reason for giving a preliminary description of Morrey's work here is that he stated incorrectly that the U(III) species across the whole acidity-basicity scale were anionic chlorouranate(III) species. It is of particular significance that Morrey and co-workers reported that "on moving to the $AlCl_3$-rich region the absorption spectrum, typical of $UCl_6^{3-}$ is no longer observed."[31]

It is proposed here that in basic melts chloroanions of the type $[UCl_{3+n}]^{n-}$ would predominate and that, in acidic chloroaluminates, the uranium(III) would be present as solvated cations—a proposition that does not exclude the possibility of solvated $UCl^{2+}$ and $UCl_2^+$. This is supported by recent Belgian potentiometric and spectroscopic studies of uranium(III) and uranium(IV) in an acidic low-temperature (40°C) melt, $AlCl_3$—N-butylpyridinium chloride.[34] The spectrum obtained in the acidic melt ($AlCl_3$:BuPyCl, 2:1) is shown as Figure 7.5c and its similarity to spectra for $U^{3+}$ in $H_2O$ and in HF strongly suggests the existence of cationic uranium(III) in this melt. The peak positions in the HF spectrum are shifted significantly to shorter wavelengths than in $H_2O$, while in the melt spectrum the peak positions are shifted to slightly longer wavelengths. Earlier Polish work[35] reported dissimilar spectra for U(III) in either an acidic melt ($AlCl_3$:NaCl, 60%:40%) at 325°C or in a basic melt ($AlCl_3$:NaCl, 49.5%:50.5%) at 395°C, the data indicating different speciation in the different melts. The resolution of the spectrum in the acidic high-temperature melt was much poorer than for those in both $H_2O$ and HF and in the low-temperature acidic melt; however, the positions of the major peaks are almost identical with those in $H_2O$ and in the low-temperature melt.

In the Polish work,[35] the spectrum for U(III) in basic chloroaluminate at 395°C was recorded only within the visible region, 400–700 nm. However, in earlier work, Gruen and McBeth[36] recorded the spectrum from 360 to 1,260 nm in the more basic medium, the LiCl—KCl eutectic, at 400°C (Figure 7.7). They showed the same broad intense absorption as in the basic chloroaluminate[35] for the region 400–700 nm, with values of the molar absorption coefficient in the range 600–700, much greater than the typical values of 40–70 for the same spectrum in the wavelength range 700–1,200 nm, and for actinide spectra generally in acidic media. Gruen and McBeth cite Jørgensen's work on U(III) spectra in concentrated HCl solutions[37] and say that the intense broad absorption in the melt between 400 and 700 nm "may be taken as evidence for the existence of U(III) chlorocomplexes in the melt."

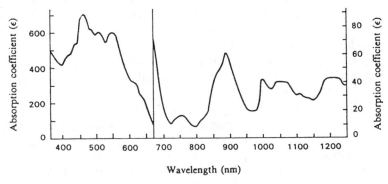

**Figure 7.7.** Uranium(III) in Basic Melt (LiCl–KCl Eutectic) at 400°C (Ref. 36).

Another major difference between U(III) spectra in acidic and basic melts should be noted. While, in both types of media, there is quite strong absorption in the region 800–900 nm, the spectra in acidic media (Figure 7.5) are characterized by a sharp intense peak somewhat above 1,200 nm. No such feature is evident in the spectrum from LiCl—KCl (Figure 7.7).

It is significant that spectra for U(III) in basic fluoride melts also show intense broad absorption in the region 300–500 nm, with broad, less-intense bands beyond 800 nm and with no sharp absorption above 1,200 nm;[38] this suggests that U(III) is present in basic fluoride melts as anionic complexes.

It becomes easy, then, to agree with most authors that, in basic melt media, uranium(III) exists as anionic halocomplexes. The consistent differences between spectra both in basic chloride and in basic fluoride from those in acidic chlorides, whether near ambient temperature or at elevated temperature, as well as the similarity of the spectra in acidic melts to those in HF and acidified water, lend strong support to the proposal that U(III) exists in acidic melts as solvated cations, even though the cations may be $UCl^{2+}$ or $UCl_2^+$, depending on pCl.

While the experimental spectra in low-temperature and high-temperature acidic melts support the proposition that U(III) is cationic in acidic melts, it should be noted that in both reports[34,35] the authors put forward the idea, however tentatively, that $U^{3+}$ is coordinated in the melts directly, and solely, by $Al_2Cl_7^-$ or $AlCl_4^-$.

## 7.2.5  Disproportionation in Chloroaluminates

There is some evidence for disproportionation of $f$-transition-element cations in low oxidation states in room-temperature chloroaluminate melts as the acidity of those melts is reduced to neutrality and the melts are then made progressively more basic. This process of disproportionation as the basicity of the medium is increased has been demonstrated clearly and directly for U(III) in HF (section 7.2.2) and, by strong inference, for U(III) in $CF_3SO_3H$ (section 7.2.3).

Members of the two Belgian research groups who have carried out extensive electrochemical and spectroscopic investigation of $f$-transition metals in the room-temperature melt system, $AlCl_3$—BuPyCl, report behavior of solutes which is consistent with disproportion in near-neutral melts. Gilbert's group at Liege observed,[28] as stated in section 7.2.4, that in neutral melts the dichlorides $SmCl_2$ and $YbCl_2$ precipitate and that in basic melt the Ln(II) precipitates redissolve and that a blue coloration is observed that they ascribe to reduction of the $BuPy^+$ cation by Ln(II) anions.

The group at Leuven[39] has reported similar reduction of the $BuPy^+$ cation for U(III) solutions in $AlCl_3$—BuPyCl melts originally acidic and taken through neutrality to basicity by addition of solid BuPyCl. They observed precipitation of $UCl_3$ just before neutrality ($AlCl_3$:BuPyCl, 1:1) with dissolution of that precipitate on addition of more chloride to provide a deep-blue basic solution, which provided the combined spectra of $UCl_6^{2-}$ and of the radical cation which had been shown by previous electrochemical reduction[40] to be a reduction product of the $BuPy^+$ cation. They had previously reported a mixed precipitate of $UCl_4$ and $UCl_3$ when solid BuPyCl was added to a slightly acidic solution of U(III).[34]

A reasonable explanation of the Liege[28] and Leuven[34,39] observations is that when $Cl^-$ was added to $LnCl_2$ or $UCl_3$ suspended in nearly neutral chloroaluminates, the redissolution process in a medium that was not necessarily uniform would have led, in part at least, to disproportionation of Ln(II) to Ln(0) and Ln(III) and of U(III) to U(0) and U(IV) in the presence of a slight excess of the base $Cl^-$. This would account for the formation of anions $LnCl_6^{3-}$ and $UCl_6^{4-}$ in the final basic melts. More importantly, disproportionation would provide the highly reducing species metallic Ln(0) and U(0) in an extremely fine state of subdivision—ideal reductants for the cation $BuPy^+$.

In section 7.2.4 there was allusion to a series of experiments carried out by Morrey and co-workers.[31] They prepared $UCl_3$ in situ in sealed tubes containing small amounts of uranium and very much larger comparable volumes of metallic aluminium and of chloroaluminate melts ranging from acidic ($AlCl_3$:KCl, 2:1) through neutral (1:1) to basic (0.4:1). After prolonged heating and subsequent quenching, the immiscible metallic phase now containing Al(0) and some U(0) was analyzed for uranium in each case. The results were interpreted in terms of supposed *direct* reduction of U(III) by aluminium. Formation of metallic uranium was dramatically greater in the region of the neutral melt (i.e., $AlCl_3$:KCl, 1:1), as shown in Figure 7.8, in which the variation in the ratio of concentration of uranium in the metallic phase to that in the salt phase ($[U]_m$:$[U]_s$) is plotted against melt acidity and basicity, expressed as the ratio $AlCl_3$:KCl. Figure 7.8 shows that there was little uranium in the metallic phase when the nonmetallic chloroaluminate phase was acidic or basic.

In a second paper[32] they interpreted their observations as resulting from *direct* reduction of U(III) by aluminium, proposing that the extent of such supposedly direct reduction could be interpreted in terms of relative concentrations in the different melts of $Cl^-$, $AlCl_4^-$, $Al_2Cl_7^-$, and $UCl_6^{3-}$, the last species being postulated as existing across the whole acidity-basicity range, despite their spectroscopic observations reported in section 7.2.4.

It is difficult to see how uranium(III) species could be generated in situ in the various melts if U(III) can be reduced by elemental aluminium. It seems easier to postulate, as has been done in section 7.2.4, that in acidic melts U(III) is cationic and that it is anionic in species such as $[UCl_{3+n}]^{n-}$ in basic melts, and to postulate further that both types of U(III) species are resistant to reduction by aluminium. Gruen and McBeth had demonstrated experimentally, before the publication of Morrey's work, that chlorouranate(III) is not reduced in melts by aluminium, and that chlorouranate(IV) is reduced to the uranium(III) anion.[36]

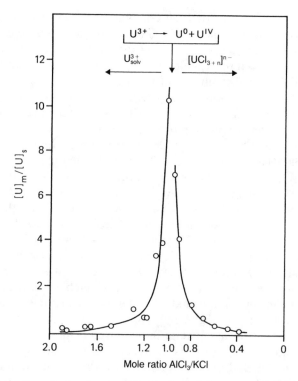

**Figure 7.8.** Acid-Base Dependence of Formation of Metallic Uranium in Chloroaluminate Melts.

The key postulation made here is that in near-neutral chloroaluminate, U(III) disproportionates to U(0) and U(IV), with the U(IV) probably present as a chlorouranate(IV) species. This interpretation would suggest that the high yield of metallic uranium from neutral melts is at least as much a result of disproportionation as it is of direct reduction. The aluminium could reduce U(IV) formed through disproportionation to U(III) and cause a cyclic formation of metallic uranium through further disproportionation. Progressive dissolution of metallic uranium in the immiscible liquid Al layer would "drive" the disproportionation reaction occurring in the nearly neutral chloroaluminate melt. This explanation is consistent with the observed disproportionation of $Ti^{2+}$ and $U^{3+}$ in HF solution (sections 7.1.2 and 7.2.2) and with the disproportionation of $Ti^{2+}$ in melts described in section 7.1.5.

## 7.3 Summary

Chapter 7 has been devoted to a description of monatomic cations of transition metals in oxidation states that are so low that they cannot be generated or have very low stability in aqueous solution. They require, for their generation and stability, nonaqueous media of enhanced acidity—frequently extremely high. It is not the numerical value of the oxidation states that is of interest. Obviously, there are many cations in oxidation state (I) that are stable in aqueous solution, for example, cations of the alkali metals and of Tl and Ag, although it should be noted that, of these, $Ag^+$ is the only monatomic unipositive transition metal cation that has significant concentration and stability in water. Hg(I) is stable in acidic aqueous solutions as the diatomic cation $Hg_2^{2+}$ (section 5.1), but it disproportionates to Hg(0) and Hg(II) on addition of aqueous base.

The oxidation states described in chapter 7 have been those which, for transition-metal compounds, would reduce water to hydrogen. There does not appear to be a substantiated case of a monatomic aquocation of a transition metal in a *low* oxidation state disproportionating in water—a general property of low-oxidation-state transition-metal cations in nonaqueous weakly acidic to neutral media, as has been seen in this chapter. It is interesting to speculate on whether $Cr(OH_2)_6^{2+}$, stable in acidic water, would disproportionate on addition of base if air were to be excluded rigorously. Disproportionation in aqueous solution occurs for compounds of $d$- and $f$-transition elements in *high* oxidation states, but here the species involved are not simple solvated monatomic cations or in low oxidation states. A well-studied example is $UO_2^+$.

There is a close parallel between stability of low-oxidation-state cations in water and ease of preparation in nonaqueous acidic solvents. This is easiest to illustrate for solutions in HF—the solvent system for which many cationic species have been generated over a wide range of acidity conditions. The large dipositive lanthanide cations are generated[22] in basic HF—reaction of metallic lanthanide with HF reduces $H_2F^+$ to $H_2$ displacing the HF self-equilibrium in the direction of excess $F^-$. Such solutions would have Hammett Acidity Function values about $-10$ or $-11$. $Cr^{2+}$,[4] stable in $H_2O$, and $V^{2+}$[4] and $U^{3+}$,[25] which both reduce water relatively slowly, can all be generated at $H_0$ values of about $-16$ or $-17$ in HF acidified with the weak Lewis acid $BF_3$. There is no stable $Zr^{3+}$ in water and $TiCl_2$ is reported to react explosively when added to water. HF-solvated $Zr^{3+}$[9] and $Ti^{2+}$[4] are generated in HF acidified by the very strong Lewis acid $SbF_5$. $H_0$ values for such solutions would be about $-20$ to $-22$.

Another consequence of the more extensive study of transition-metal cations in HF than in other acidic nonaqueous media is that, for this solvent system, there are better grounds for postulating simple solvation of cations than for other nonaqueous media. As reported in section 7.2.1, HF solutions of $U^{3+}$ were prepared using the Lewis acids $BF_3$, $AsF_5$ and $SbF_5$.[25] There was not sufficient difference between the spectra for these solutions to suggest that $BF_4^-$, $AsF_6^-$, or $SbF_6^-$ anions were involved in the coordination sphere. The only observed differences were slight shifting and slight broadening of peaks in the HF—$BF_3$ solution spectra, suggesting that, in the less acidic HF—BF$_3$ medium, some degree of fluorocomplexation to solvated UF$^{2+}$ or UF$_2^+$ might occur.

At this early stage of the development of study of transition-metal cations in acidic nonaqueous solutions, it is too early to make many correlations with the general body of transition-metal chemistry. It is now well established[41] that for binary transition-metal halides as well as for many other compounds of these elements, low oxidation states are much easier to stabilize for the first-row member of a group of the periodic classification than for the second- and third-row members; and the converse holds—higher oxidation states become easier to generate and stabilize as one goes down a group. Additionally, third-row elements show somewhat higher stable oxidation states than second-row elements. Thus the lowest-oxidation-state binary fluorides for the subgroup VI elements are CrF$_2$, MoF$_3$, and WF$_4$ and for subgroup VII, the highest binary fluorides are MnF$_4$, TcF$_6$, and ReF$_7$.

Consistent with these trends are the oxidation states of subgroup IV cations generated in HF and in chloroaluminates. Ti$^{2+}$,[4] Zr$_3^+$,[9] and Hf$^{3+}$,[9] have been studied in HF. TiCl$_2$ and TiCl$_3$ are both stable even in weakly acidic chloroaluminates.[18] Zr(III) is reported[20] to be the electrochemical reduction product of Zr(IV) in weakly acidic chloroaluminate (AlCl$_3$:NaCl, 51:49). In more acidic chloroaluminate (60:40) Zr(III) is formed and disproportionates to give Zr(II), slowly below 140°C but rapidly above 250°C.[20] Chemical reduction of HfCl$_4$ in AlCl$_3$ is reported to be slower than for ZrCl$_4$ in AlCl$_3$.[10]

Finally, in summary, it should be stressed that the discussion of stability of oxidation states in this chapter—particularly insofar as it affects disproportionation—relates only to *cations*. Sørlie and Øye report[18] only slight instability of Ti(II) in the LiCl—KCl eutectic—a basic medium where Ti(II) would be *anionic*, TiCl$_4^{2-}$ or TiCl$_6^{4-}$, and not cationic. Therefore it would not be subject to the base-induced disproportionation proposed throughout this chapter for cationic species.

# References

1. C.G. Barraclough, R.W. Cockman, T.A. O'Donnell, *Inorg. Chem.*, *16*, 673 (1977).

2. R. De Waele, L. Heerman, W. D'Olieslager, *J. Electroanal. Chem.*, *142*, 137 (1982).

3. J.P. Schoebrechts, B. Gilbert, *Inorg. Chem.*, *24*, 2105 (1985).

4. C.G. Barraclough, R.W. Cockman, T.A. O'Donnell, W.S.J. Schofield, *Inorg. Chem.*, *21*, 2519 (1982).

5. C.G. Barraclough, J. Besida, P.G. Davies, T.A. O'Donnell, *J. Fluorine Chem.*, *38*, 405 (1988).

6. C.K. Jørgensen, *Adv. Chem. Phys.*, *5*, 33 (1963).

7. H.A. Øye, D.M. Gruen, *Inorg. Chem.*, *3*, 836 (1964).

8. D.H. Brown, A. Hunter, W.E. Smith, *J. Chem. Soc., Dalton Trans.*, 79 (1979).

9. K. Male, B.Sc. (Honors) Report, University of Melbourne (1979).

10. E.M. Larsen, J.W. Moyer, F. Gil-Arnao, M.J. Camp, *Inorg. Chem.*, *13*, 574 (1974).

11. J. Besida, T.A. O'Donnell, unpublished work.

12. H. Willner, F. Mistry, G. Hwang, F.G. Herring, M.S.R. Cader, F. Aubke, *J. Fluorine Chem.*, *52*, 12 (1991).

13. R. Adrien, Ph.D. Thesis, University of Melbourne (1992).

14. D.M. Gruen, R.L. McBeth, *Pure Appl. Chem.*, *6*, 23 (1963).

15. K.W. Fung, G. Mamantov, *J. Electroanal. Chem.*, *35*, 27 (1972).

16. R.T. Carlin, R.A. Osteryoung, J.S. Wilkes, J. Rovang, *Inorg. Chem.*, *29*, 3003 (1990).

17. D.M. Ferry, G.S. Picard, B.L. Trémillon, *J. Electrochem. Soc.*, *135*, 1443 (1988).

18. M. Sørlie, H.A. Øye, *Inorg. Chem.*, *20*, 1384 (1981).

19. A.B.P. Lever, *Inorganic Electronic Spectroscopy*, 2nd ed. Elsevier, Amsterdam, 1984, pp. 378–382.

20. B. Gilbert, G. Mamantov, K.W. Fung, *Inorg. Chem.*, *14*, 1802 (1975).

21. L.R. Morss in *Standard Potentials in Aqueous Solution*, A.J. Bard, R. Parsons, J. Jordan, eds. IUPAC Publication, Marcel Dekker, New York, 1985, pp. 587–629.

22. C.G. Barraclough, R.W. Cockman, T.A. O'Donnell, *Inorg. Chem.*, *30*, 340 (1991).

23. C.G. Barraclough, R.W. Cockman, T.A. O'Donnell, *Inorg. Chem.*, *30*, 343 (1991).

24. D. Cohen, W.T. Carnall, *J. Phys. Chem.*, *64*, 1933 (1960).

25. M. Baluka, N. Edelstein, T.A. O'Donnell, *Inorg. Chem.*, *20*, 3279 (1981).

26. L. Dawkins, B.Sc. (Honors) Report, University of Melbourne, 1984.

27. B. Gilbert, V. Demarteau, G. Duyckaerts, *J. Electroanal. Chem.*, *89*, 123 (1978).

28. J.P. Schoebrechts, B.P. Gilbert, G. Duyckaerts, *J. Electroanal. Chem.*, *145*, 127 (1983).

29. J.P. Schoebrechts, B.P. Gilbert, G. Duyckaerts, *J. Electroanal. Chem.*, *145*, 139 (1983).

30. R.A. Osteryoung in *Molten Salt Chemistry*, G. Mamantov, R. Marassi, eds. D. Reidel, 1987, p. 346.

31. R.H. Moore, J.R. Morrey, E.E. Voiland, *J. Phys. Chem.*, *67*, 744 (1963).

32. J.R. Morrey, R.H. Moore, *J. Phys. Chem.*, *67*, 748 (1963).

33. T.A. O'Donnell, *Chem. Soc. Rev.*, *16*, 1 (1987).

34. R. De Waele, L. Heerman, W. D'Olieslager, *J. Less-Common Metals*, *122*, 319 (1986).

35. S. Poturaj-Gutniak, *Nukleonika*, *14*, 269 (1969).

36. D.M. Gruen, R.L. McBeth, *J. Inorg. Nucl. Chem.*, *9*, 290 (1959).

37. C.K. Jorgensen, *Acta Chem. Scand.*, *10*, 1503 (1956).

38. J.P. Young, *Inorg. Chem.*, *6*, 1486 (1967).

39. L. Heerman, R. De Waele, W. D'Olieslager, *J. Electroanal. Chem.*, *193*, 289 (1985).

40. D.A. Habboush, R.A. Osteryoung, *J. Electrochem. Soc.*, *127*, 2167 (1980).

41. T.A. O'Donnell in *Comprehensive Inorganic Chemistry*, Vol. 2. R.S. Nyholm, H.J. Eleméus, J.C. Bailar, A.G. Trotman-Dickenson, eds. Pergamon Press, Oxford, 1973, pp. 1073–1080.

# CHAPTER 8

# Polyatomic Anions in Basic Media

The principal aim of this book has been to demonstrate the general nature of the dependence of the stability of cationic species—often in unusually low or fractional oxidation states—on the level of the acidity of the media in which they can be generated and, where possible, to relate this to wider areas of general inorganic chemistry—even, as in this chapter briefly, to stability of anionic species in strongly basic solutions.

Chapter 1 was used to show the general framework within which acidity or basicity of an individual solvent is increased or decreased, regardless of whether the solvent is water or a protonic or a nonprotonic nonaqueous solvent, whether the solvent is more acidic or more basic than water, or whether the solvent is a liquid only at temperatures well outside the ambient experimental temperature range, as is the case for molten salts. Chapter 2 dealt with experimental methods used to determine levels of acidity in protonic solvents, particularly superacids, and with procedures for indicating acidity-dependent speciation of solutes in superacids, while chapter 3 covered the details of the interaction between Lewis acids and the superacids that increase further the acidity levels of the pure superacids themselves.

The main thrust of chapters 4–7 was to use the material of chapters 1–3 to outline the evidence for formation in superacidic media of polyatomic cations of main-group nonmetals and metals in fractional oxidation states and of monatomic cations of *d*- and *f*-transition elements, often in abnormally low oxidation states, and to relate the stabilities *in solution* of all of these cations to differing levels of acidity in the superacidic media.

Chapter 8 is offered to demonstrate a direct corollary to the stabilization of unusual cations in highly acidic media, namely, that polyatomic anions— the classic Zintl ions of groups IV and V and the polyanions of group VI of the periodic classification—are generated in highly basic media, or to be more specific, in the absence of even weakly acidic species. This is the obverse of the position for unusual cationic species which are conveniently generated in highly acidic media, but, in the final analysis, they disproportionate in the presence of even weakly basic species.

J.D. Corbett, one of the major research contributors to the field, has recently published an excellent review entitled "Polyatomic Zintl Anions of the Post-Transition Elements,"[1] which will be used extensively below. This chapter is not intended to be an exhaustive or comprehensive review of the field. In a strict sense, the subject of Zintl anion stability in basic media lies outside the scope of this book, but examples will be offered as representative of the work that has gone into generation of these ions in basic media and their isolation from such media or as "naked" ions. Only those polyatomic anions will be considered which are homopolyatomic or which are very closely related heteropolyatomic anions. These are all ions in which the component elements have fractional formal negative charge. Obviously, heteropolyatomic anions, such as $BrF_4^-$, $TeF_5^-$, and $SbF_6^-$ or similar halo- or oxoanions, will not be discussed. As argued in chapter 4 for corresponding cations, these are anions in which electronegative ligands are coordinated to a central element of high formal positive charge.

Not surprisingly, a very large amount of the research in the field of Zintl ion chemistry has been directed toward elucidation of the often-interesting structures and bonding of the ions in solution and in the solid state. In this chapter, there will be only incidental allusion to structures of these ions. Interest will be focussed principally on generation and stabilization of these ions in *strongly basic solution* and, to a less extent, their isolation from such solutions.

## 8.1 Generation and Isolation of Polyatomic Anions

Historically the earliest Zintl ion studies were conducted in anhydrous ammonia. Subsequently the more basic solvent ethylenediamine provided an effective medium for their generation and, even more importantly, for their isolation as stoichiometric solids, the structures of which could be determined by X-ray crystallography. Other basic solvents such as dimethylformamide have been used to a smaller extent. Polychalcogenide anions, which require less basic media for their generation than the cluster anions of the elements of main groups IV and V, have been isolated from very basic aqueous solutions. In chapters 4 and 5 it was shown that quite a reasonable amount of research had been conducted in molten salts on the existence of polyatomic cations of metallic and nonmetallic elements. There has been surprisingly little thorough investigation of chain or cluster anions *in melts* as solvents.

## 8.1.1 Preparation in Basic Solutions

Corbett[1] has reported the background research of Joannis, late in the nineteenth century, and of Kraus, after the turn of the century, which preceded the very elegant investigations of Zintl and co-workers, who first identified these anionic clusters definitively and after whom the field was named. They were postulated as being formed when alloys of Na and a heavier element such as Sn, Pb, Sb, Bi, Se, or Te were dissolved in liquid $NH_3$. There would have been mutual oxidation and reduction of Na and the heavier element.

Zintl performed potentiometric titrations, adding $PbI_2$ dissolved in $NH_3$ to solutions of Na in $NH_3$. $PbI_2$ was initially reduced to Pb, which was then reduced further to $Pb_7^{4-}$ and $Pb_9^{4-}$, as indicated by "steps" in the titration curve that is reproduced as Figure 1 of Corbett's review.[1] Chemical analysis was used to confirm the nature of the polyanions. Zintl postulated the existence of an extensive range of polyanions in addition to $Pb_7^{4-}$ and $Pb_9^{4-}$. These included $Sn_9^{4-}$ and a range of anions $A_n^{3-}$ for A = As, Sb, and Bi and $n$ = 3, 5, and 7 (except for $Sb_5^{3-}$) and the chalcogens Se and Te as dinegative anions, containing 2, 3, and 4 chalcogen atoms as well as $Se_5^{2-}$ and $Se_6^{2-}$. These and their colors in $NH_3$ solution are listed in Corbett's Table I.

Corbett reports two early observations which have significant implications concerning the acid-base dependence of the stability of complex cations and anions, as spelt out for cations in earlier chapters and for anions in this one. He writes that "the $Pb_9^{4-}$ species produced electrolytically was observed to be unstable when a $Et_4NI$ electrolyte was used," presumably as a supporting electrolyte. This is not surprising when $Et_4N^+$ is recognized as an acid of the $NH_3$ solvent system. Here we have the obverse of the situation where polyatomic cations disproportionate in the presence of bases of suitable strength. Corbett also states that "although $Sn_9^{4-}$ solutions could not be readily achieved by electrolysis, this and the lead analogue were also obtained in a disproportionation reaction following Bergstrom, $viz.$

$$11Sn + 6KNH_2 \xrightarrow{NH_3} K_4Sn_9 + 2KSn(NH_2)"$$

This can be interpreted as being due to greater stability of the polyanions in $NH_3$ made more basic with $NH_2^-$ than in straight $NH_3$.

Zintl and co-workers were unsuccessful in isolating solids containing the polyanions directly from $NH_3$ solutions. Solids which initially appeared to contain the polyanions and the cation $Na(NH_3)_n^+$ decomposed when attempts were made to pump off excess $NH_3$. This will be discussed further in section 8.1.2.

There appears to have been a gap of about four decades between Zintl's pioneering work in this area and subsequent solution studies on Zintl anions carried out by Rudolph and co-workers, when the usually preferred solvent became ethylenediamine (to be referred to as "en" throughout this chapter) rather than $NH_3$ and the investigative technique used was multinuclear NMR spectroscopy. They dissolved Na—Sn, Na—Sn—Pb, Na—Sn—Ge, Na—Sn—Te, and Na—Sn—Tl alloys in en and recorded solution spectra showing the presence in solution of the Zintl ions $Sn_9^{4-}$,[2] $(Sn_{9-x}Pb_x)^{4-}$ for $x$ = 0–9,[2] $Sn_4^{2-}$,[3] $(Sn_{9-x}Ge_x)^{4-}$ for $x$ = 0–9,[3] $SnTe_4^{4-}$,[3] and $TlSn_8^{5-}$.[3] Reduction of $Na^+$, present as a solution 0.1 M in NaI in en, at an electrode which was the 1:1 alloy Pb—Sn, was shown by NMR spectroscopy to produce all species in the series $(Sn_{9-x}Pb_x)^{4-}$, with a predominance of the species $Sn_9^{4-}$ and $Pb_9^{4-}$.[4]

An octadentate macrocyclic ligand, $N(C_2H_4OC_2H_4OC_2H_4)_3N$, referred to generally as 2,2,2-crypt or more simply as crypt, has proved extremely useful in isolating from $NH_3$ or en crystalline solids containing Zintl anions for structural analysis, as will be discussed in some detail in section 8.1.2. Its presence in $NH_3$ or en solutions has been found to facilitate the dissolution of those alloys which are used as precursors for Zintl ions.

Zintl had originally reported the existence of $As_3^{3-}$, $As_5^{3-}$, and $As_7^{3-}$ as a result of electrochemical studies in $NH_3$. In 1985 a French group[5] dissolved K—As alloys in an approximate 1:2 stoichiometry in en containing excess of crypt. EXAFS spectra of a solution containing K and As corresponding to a $K_3As_5$ stoichiometry indicated the presence in solution of an equilibrium mixture of $As_4^{2-}$ and $As_6^{4-}$.

Many authors do not classify the polyatomic anions of the chalcogens as Zintl ions, although Zintl himself did report their formation in solution, as mentioned earlier. Almost exclusively, the chalcogen anions are chain structures, rather than the rings and cages exhibited by anions of groups IV and V and the mixed anions containing group III elements. Schrobilgen and colleagues at McMaster University[6] provided a link between the two types of anions by using multinuclear NMR spectroscopy to determine a wide range of structures in solution of the Zintl anions $HgCh^{2-}$, $CdCh_2^{2-}$, $SnCh_3^{2-}$, $TlCh_3^{3-}$, $SnCh_4^{4-}$, and $Tl_2Ch_2^{2-}$ where $Ch$ = Se and/or Te. They extracted the ions with $NH_3$ or en in the presence of 2,2,2-crypt from alloys such as $KM(Te/Se)$ and $NaM(Te/Se)$ where M = Hg, Cd, Tl, or Sn.

Later they carried out a similar study[7] in which they restricted their investigations to anions of chalcogen elements themselves, characterizing by NMR spectroscopy the homo- and heteropolychalcogenide anions $Ch^{2-}$, $HCh^-$, $Ch_2^{2-}$, $Ch_3^{2-}$, and $Ch_4^{2-}$ where $Ch$ = Se and/or Te. They generated the homopolyatomic anions in $NH_3$ or en, with or without crypt, by equilibrating, in the basic solvents, 2 mol of Na or K with n mol of Ch to produce $2 M^+$ and $Ch_n^{2-}$. For the heteropolychalcogenide anions they reacted $K_2Ch'$ with $n$ mol of Ch to yield $2 K^+$ and $Ch'Ch_n^{2-}$.

Earlier, Sharp and Koehler[8] had used UV-visible spectroscopy to characterize the ions $Se_3^{2-}$, $Se_4^{2-}$, and $Se_6^{2-}$ in liquid $NH_3$, after dissolving stoichiometric amounts of Na and Se in $NH_3$. They found $Na_2Se_5$ to be an equimolar mixture of $Na_2Se_4$ and $Na_2Se_6$, the latter being the highest polyselenide observed. They reported that $Na_2Se_2$ could be prepared only in equilibrium with $Na_2Se_3$ and $Na_2Se$ and proposed that $Se_2^{2-}$ disproportionates to $Se^{2-}$ and $Se_3^{2-}$ in $NH_3$. It is interesting to speculate whether $Se_2^{2-}$ might be stable in a more basic medium, for example, in $NH_3$ made basic with $NH_2^-$ or in en alone or with its anion.

This latter type of speculation is supported by recent work on the stability of polysulfides in solution in liquid $NH_3$ and shows that there would probably be advantage to further studies of polysulfides in $NH_3$ made more basic with $NH_2^-$ or in the corresponding en systems. Lelieur and colleagues[9] generated ammonium polysulfides in $NH_3$ and used UV-visible and Raman spectroscopy to study polysulfide speciation. They reported identification of the polysulfides $S_n^{2-}$ for $2 \le n \le 6$ with the radical anion $S_3^-$ always observed in equilibrium with $S_6^{2-}$ for n > 1. It is highly significant within the context of the general discussion in this chapter that they observed selective disproportionation in the case of the more highly reduced anions. They reported that "for $n \leq 4$, the disproportionation of polysulfides is systematically higher for $(NH_4)_2S_n$ solutions than for $Li_2S_n$ solutions." Disproportionation was particularly marked in the Raman observations where the necessary use of concentrated solutions (1 M) would lead to high concentrations of $NH_4^+$. This is consistent with lower stability of the polysulfides in $NH_3$ containing $NH_4^+$, the acid of the solvent system, than in the presence of $Li^+$, which is an ion of vanishingly small acidity in $NH_3$ by comparison with $NH_4^+$.

In other reports this group generated $Li_2S_n$ in situ in $NH_3$ by taking appropriate stoichiometric proportions of Li and S to form the individual polysulfides of Li and S and recording Raman and UV-visible spectra. They showed[10,11] that $S_6^{2-}$ which is in equilibrium with $S_3^-$ is the least reduced polysulfide in $NH_3$ and that $S_4^{2-}$ and $S_2^{2-}$ also exist in $NH_3$. They report that the existence of $S_5^{2-}$ is questionable and that of $S_3^{2-}$ is also unlikely.

There was a comment early in this chapter that there appears to have been very little investigation of Zintl-type anions in molten salts. In a 1968 study described by the authors as exploratory, Corbett and co-workers[12] carried out phase studies on solutions of Au, Tl, Sn, Pb, Sb, and Bi in the molten eutectic NaCl—NaI (37:63) and reported significant dissolution of the heavy metals only in the cases of Sb and Bi. They suggested the possibility of formation of Zintl phases in this nonacidic melt but had no firm evidence. They commented that thermal stabilities of Zintl ions in melts "must be limited" and that unfavorable entropy effects for complex anions would be expected to be substantial. They commented that "the intermetallics of interest readily reduce the useful solvent $NaAlCl_4$." This would eliminate use of a melt for which acidity and basicity can be nicely controlled. The less-reducing blue radical anion $S_3^-$, previously characterized by Raman spectroscopy in dimethylformamide,[13] has been identified by resonance Raman spectroscopy in the basic melt, 55 mol% CsCl—45 mol% $AlCl_3$.[14]

## 8.1.2 Isolation of Solids From Solution

In the preceding section, it was reported briefly that potentiometric and spectroscopic investigations had been used to provide evidence for the existence of polyatomic anions in strongly basic nonaqueous solvents. A natural extension would be to attempt to isolate from the solutions crystalline solids which could be subjected to X-ray diffraction structural analysis in order to obtain further, and perhaps more definitive, information on anion speciation.

Zintl and co-workers attempted such isolation from $NH_3$ solutions. Evaporation of $NH_3$ at low temperature appeared to give initially nonmetallic solids such as $[Na(NH_3)_n^+]_4Pb_9^{4-}$ and $[Na(NH_3)_n^+]_3Sb_7^{3-}$, but attempts to pump off excess $NH_3$ gave, ultimately, gray solids with a metallic appearance. It is likely that decomposition of $[Na(NH_3)_n^+]_4Pb_9^{4-}$ gave the alloy $NaPb_{2.25}$ and $NH_3$, the weakly coordinated $Na^+$ losing its monodentate $NH_3$ ligands, while $[Na(NH_3)_n^+]_3Sb_7^{3-}$ yielded the alloy $NaSb$ and elemental $Sb$. In the absence of strong complexing of the cations, electrons were delocalized from the anions to the cations, a mutual oxidation-reduction which leads to the formation of the intermetallic phases and the heavy element.

Kummers' group[15] was the first to isolate stoichiometric solids in which $Na^+$ ions were coordinated by molecules of the solvent ethylenediamine $H_2NCH_2.CH_2.NH_2$, abbreviated throughout this chapter as en. Ethylenediamine has been widely used in inorganic synthetic and structural chemistry as an effective bidentate ligand. Alloys of $Na$ with $Sn$, $Ge$, and $Sb$, when treated with en, yielded stable solutions of Zintl ions from which $[Na_4.7en]Sn_9$, $[Na_4.5en]Ge_9$, and $[Na_3.4en]Sb_7$ were isolated, with the first compound sufficiently crystalline to allow structural determination of the anion.

Recognizing the need to sequester the cation as effectively as possible in order to isolate solids containing the highly reducing Zintl anions, Corbett[1] turned to the macrocyclic ligand crypt, referred to in section 8.1.1, which had been used by Dye to isolate the sodide anion in the solid compound $[Na^+.crypt]Na^-$.[16] Crypt, for which the abbreviation 2,2,2-crypt is also used, has the formula $N(C_2H_4OC_2H_4OC_2H_4)_3N$ and, conceptually, is an extension of the branched-chain compound $H_4EDTA$ which is widely used to provide a powerful hexadentate ligand, wrapping itself around cations. Crypt is a bicyclic ligand containing a three-dimensional intramolecular cavity lined with six oxygen and two nitrogen donor atoms, making it a very powerful octadentate ligand.

Corbett and his colleagues produced several papers describing the dissolution in en of crypt and an alloy which is a precursor of a Zintl ion and then isolating compounds such as $[K^+.crypt]_2Pb_5^{2-}$, starting from the alloy $KPb_{2.5}$. Corbett[1] has reviewed much of this work in which homopolyatomic anions such as $Sn_9^{4-}$, $Sn_5^{2-}$, $Ge_9^{4-}$, $Ge_9^{2-}$, $Pb_5^{2-}$, $As_{11}^{3-}$, $Sb_7^{3-}$, $Sb_4^{2-}$, $Bi_4^{2-}$, and $Te_3^{2-}$ and the heteropolyatomic anions $Sn_2Bi_2^{2-}$, $Pb_2Sb_2^{2-}$, $Tl_2Te_2^{2-}$, $TlSn_8^{5-}$, $As_2Se_6^{3-}$, and $As_2Te_6^{3-}$ were prepared and isolated using this general procedure and anionic structures were then determined.

Birchall and colleagues[17] isolated several Zintl anions from en as solids containing the cryptated sodium cation and recorded $^{119}Sn$ Mössbauer spectra, which they interpreted in terms of known structures for the anions $Sn_5^{2-}$, $Sn_9^{4-}$, $TlSn_8^{5-}$, $Sn_2Bi_2^{2-}$, $SnTe_4^{4-}$, and $[SnSe_{3-x}Te_x]_z^{2z-}$.

The classic Zintl ion $Sn_9^{4-}$ has been reported[18] as having been isolated "without the use of cryptate ligands." The authors said that treatment of a dimethylformamide (DMF) solution of $Me_4N.BPh_4$ with an alloy of nominal composition $K_4Sn_9$ instantly gave a precipitate of $(Me_4N)_4Sn_9$. It is highly significant that this solid decomposes at 25°C to give a nearly quantitative yield of $Me_3N$ and lower yields of various methylated stannanes, principally $Me_6Sn_2$. Obviously the cation is subject to reduction by the anion in this compound. Sequestered cations give stable compounds.

Some representative examples from the recent literature suggest that polychalcogenide anions can be generated under conditions that are less basic than for the elements of earlier groups of the periodic classification and that they are stable when the nature of the countercation is such that it would be reduced by the classic Zintl anions. Thus the bicyclic polyselenide $Se_{10}^{2-}$ was prepared [19] from DMF with the countercation $[Ph_3PNPPh_3]^+$. The compounds $[Ph_4P^+]_4$ $[In_2Se_{21}^{4-}]$,[20] $[Et_4N^+]_4[Hg_7Se_{10}^{4-}]$,[21] and $[Et_4N^+]_{4n}[Hg_7Se_9]^{4n-}$, which is based on an extended anionic framework,[21] were all isolated from DMF.

A further indication of the reduced basicity of the medium required for generation and isolation of polychalcogenide anions relative to that required for the ring- and cagelike Zintl ions of earlier groups is the report[18] that the compounds $(Bu_4N)_2Ch_n$ (where Ch = Te, $n = 5$; Ch = Se, $n = 6$; Ch = S, $n = 6$) can be obtained by the *aqueous* extraction of binary alkali metal/chalcogen alloys in the presence of $Bu_4N.Br$. A statement like this can be somewhat misleading. These polychalcogenide-chain anions are, in fact, generated in extremely basic aqueous solution because, experimentally, the chalcogen is introduced into a small volume of water as an alloy with Na or K and subsequent redox reactions yield caustic solutions. However, the general principle holds. Generation of other Zintl ions requires strongly basic *nonaqueous* solvents. Another important point to arise from this work is that when the compounds $(Bu_4N)_2Ch_n$ were dissolved in acetone for spectroscopic investigation, some homolytic bond cleavage occurred. It was observed[18] that, "on the basis of visible absorption and Raman spectroscopies, the extent of dissociation [of the anions] apparently decreases for the heavier elements with $S_9^{2-} >> Se_6^{2-}, > Te_5^{2-}$."

## 8.1.3 In Association With Naked Cations

Earlier in the book, in sections 4.1.1.3 and 4.2.3.2, for example, there was a presentation of the isolation of polyatomic nonmetal cations as "naked" cations. In that context elemental iodine and tellurium were shown to have been oxidized to the appropriate cations in liquid $SO_2$ by and in the presence of strong Lewis acids such as $SbF_5$ and $AsF_5$ and the polyatomic cations were then isolated with the extremely weakly basic $Sb_2F_{11}^-$ and $AsF_6^-$ as counteranions.

The situation in this section is very different, involving formation of polyatomic anions by direct element-to-element redox interaction, and does not warrant extensive treatment. The possibility of producing polyatomic anions in this way is indicated for the sake of completeness, but the ions are not generated in solution and therefore no acid-base–dependent interactions are involved.

Many simple discrete polyatomic anions have been synthesized by direct element-to-element reaction, as exemplified by the synthesis of $P_7^{4-}$, $As_7^{4-}$, and the intermediate mixed anions. In a review,[22] "Homoatomic Bonding of Main Group Elements," von Schnering described thermal interaction between potassium and phosphorus to form $K_3P_7$. He described the formation of the corresponding $As_7^{3-}$ anion[23] in $Ba_3As_{14}$ formed by reaction of Ba and As in ratios 1:4 to 1:5 at temperatures between 1,000 and 1,100 K and subsequently reported[24] mixed crystals $Rb_3[P_{7-x}As_x]$ as being synthesized from the elements in a stoichiometric ratio in sealed Ta ampules at 900 K. Not surprisingly, these anions as well as $P_7^{4-}$ and $As_7^{4-}$ are soluble and stable in en. This allowed $^{31}P$ NMR investigation of "scrambling" of atoms in those anions. More recently, infinitely polymeric anions in chain and sheet form have been isolated using a general procedure that Corbett has developed to characterize metal–metal bonded sheets and ions in compounds like ZrCl and the formally "lower" halides of scandium and lanthanides.[25] Typically, reacting elements, and compounds where appropriate, are sealed into capsules, frequently made from tantalum or silica, heated for very long periods, and cooled slowly. Using this general approach, Corbett and others have reported what they describe as Zintl phases with layered or chain structures, examples occurring in the compounds $KSi_3As_3$,[26] $K_2SiAs_2$,[27] and $K_3Ga_3As_4$.[28] These types of infinitely extended anions are so different in nature from the discrete polyanions presented in sections 8.1.1 and 8.1.2 that they will not be discussed further here.

## 8.2 Summary

It can be seen that the classic Zintl ions of metallic elements require strongly basic conditions for their formation in solution and highly complexed cations for their isolation. $Sn_9^{4-}$, with a low charge-to-metal ratio, has been isolated from basic nonaqueous solution with the relatively simple counterion $(CH_3)_4N^+$.[18] Obviously, the polychalcogenide anions can be formed in basic nonaqueous media, but they can also be isolated with a suitable cation from extremely basic aqueous solution. The natural extension of this is the well-known formation of polyhalide anions, for example, $I_3^-$, in nonbasic aqueous solution without any restriction on the countercation.

Thus, polyhalide anions require less "forcing" conditions for their synthesis and stability than do the chalcogen anions and, in turn the anions of elements of earlier groups. Within groups, the same trends apply as for polyatomic cations; that is, the greater the atomic number, the easier an ion of any given formula is to generate and the stabler it is when formed. It is almost trivial to state that $I_3^-$ is formed more easily than $Br_3^-$ and $Cl_3^-$. Similar stability trends for the chalcogen anions have been reported,[18] as indicated at the end of section 8.1.2.

It is difficult to sort out relative stabilities of anions of the metallic and semimetallic elements. There are no studies involving control of acidity or basicity in $NH_3$, en, or other basic solvents that correspond with the work summarized in Tables 4.2 and 4.3 on dependence of halogen and chalcogen polycations on acidity and basicity levels in superacidic media. There do not appear to have been any systematic attempts to investigate anionic stability as a function of addition of $NH_2^-$ or of $NH_4^+$ to $NH_3$ or of addition of corresponding basic or acidic species to ethylenediamine or other amines or amides. There is the very strong suggestion in section 8.1.1 that the presence of $Et_4N^+$ in $NH_3$ destabilized $Pb_9^{4-}$ and that $NH_2^-$ favored the formation of $Sn_9^{4-}$. Obviously, deliberate control of acidity and basicity levels in the basic solvents needed for Zintl ion generation would provide much evidence on relative stabilities of the ions.

It was noted in section 1.1.2.1 that a well-known property of liquid $NH_3$ was its ability to sustain solvated electrons as stable species. Three other groups of examples of the dependence of unusual low oxidation states in anionic species on extreme basicity of the medium are worth very brief mention here. The first group are the alkalide anions. Dye, who first isolated $Na^-$ in a solid compound, has reviewed the spectroscopic and other evidence for the existence of alkali metal ions in solution in ammonia and amines.[29] He crystallized $[Na.crypt^+]Na^-$ from ethylamine and characterized the compound by structural and other methods.

Ionic ozonides have been prepared in solution in liquid $NH_3$.[30] $KO_3$ was extracted with $NH_3$ from the products of a gas-solid reaction between $KO_2$ and $O_3$. Mixing of this solution with a solution of $(CH_3)_4N^+O_2^-$ in $NH_3$ led to precipitation of $KO_2$. Slow evaporation of $NH_3$ gave crystalline $(CH_3)_4N^+O_3^-$ in nearly quantitative yield.

Very recently, Beck has reviewed the preparation of highly reduced transition-metal carbonyls,[31] with the metals in the lowest-known formal oxidation states. Compounds containing anions such as $Cr(CO)_4^{4-}$, $V(CO)_4^{5-}$, and $Zr(CO)_4^{6-}$ were prepared by reducing carbonyls or carbonylmetallates, containing the transition metals in less-negative formal oxidation states, with metallic Na or alkali metal naphthalenides in liquid ammonia. Beck reports that "$V(CO)_5^{3-}$ reacts with $NH_4Cl$ in liquid ammonia with $H_2$ elimination ($H^+$ as oxidizing agent) and formation of $[V(CO)_5NH_3]^-$"; that is, increase in acidity of the solvent destabilized $V(CO)_5^{3-}$.

It is acknowledged that this chapter, which has dealt nonexhaustively with "unusual" anions, is far from a complete treatment. As far as possible it has been restricted to polyatomic anions generated in strongly basic solutions or isolated from such solutions. It has not included anything on structure or bonding in polyatomic anions. These matters are dealt with in reviews that are readily available.[1,22] In one of those[1] Corbett makes the point that several Zintl ions are isoelectronic and often isostructural with some of the polyatomic cations presented in chapters 4 and 5. Thus $Bi_5^{3+}$ can be compared with $Pb_5^{2-}$ and $Sn_5^{2-}$, $Te_4^{2+}$ with $Sb_4^{2-}$ and $Bi_4^{2-}$, and $Bi_9^{5+}$ with $Sn_9^{4-}$ and $Ge_9^{4-}$.

This chapter has been included to stress the quite-generally-well-known point that basic conditions favor the formation of Zintl anions; as stated earlier, this is effectively a corollary of the proposition which forms the main basis of this book, namely, that the "unusual" cations presented in earlier chapters must be generated in the virtual absence of basic species; otherwise they disproportionate. As was shown in those chapters, a convenient way of ensuring these conditions is to carry out syntheses of the cations in superacidic media.

# References

1. J.D. Corbett, *Chem. Rev.*, *85*, 383 (1985).

2. R.W. Rudoph, W.L. Wilson, F. Parker, R.C. Taylor, D.C. Young, *J. Am. Chem. Soc.*, *100*, 4629 (1978).

3. R.W. Rudolph, W.L. Wilson, *J. Am. Chem. Soc.*, *103*, 2480 (1981).

4. B.S. Pons, D.J. Santure, R.C. Taylor, R.W. Rudolph, *Electrochimica Acta*, *26*, 365 (1981).

5. J. Roziére, A. Seigneurin, C. Belin, A. Michalowicz, *Inorg. Chem.*, *24*, 3710 (1985).

6. R.C. Burns, L.A. Devereux, P. Granger, G.J. Schrobilgen, *Inorg. Chem.*, *24*, 2615 (1985).

7. M. Björgvinsson, G.J. Schrobilgen, *Inorg. Chem.*, *30*, 2540 (1991).

8. K.W. Sharp, W.H. Koehler, *Inorg. Chem.*, *16*, 2258 (1977).

9. P. Dubois, J.P. Lelieur, G. Lepoutre, *Inorg. Chem.*, *27*, 1883 (1988).

10. P. Dubois, J.P. Lelieur, G. Lepoutre, *Inorg. Chem.*, *27*, 73 (1988).

11. V. Pinon, J.P. Lelieur, *Inorg. Chem.*, *30*, 2260 (1991).

12. M. Okada, R.A. Guidotti, J.D. Corbett, *Inorg. Chem.*, *7*, 2118 (1968).

13. R.J.H. Clark, D.G. Cobbold, *Inorg. Chem.*, *17*, 3169 (1978).

14. R.W. Berg, N.J. Bjerrum, G.N. Papatheodorou, S. Von Winbush, *Inorg. Nucl. Chem. Lett.*, *16*, 201 (1980).

15. L. Diehl, K. Khodadadeh, D.D. Kummer, J. Strähle, *Chem. Ber.*, *109*, 3404 (1976).

16. J.D. Dye, *Angew. Chem., Int. Ed. Engl.*, *18*, 587 (1979).

17. T. Birchall, R.C. Burns, L.A. Devereux, G.J. Schrobilgen, *Inorg. Chem.*, *24*, 890 (1985).

18. R.G. Teller, L.J. Krause, R.C. Haushalter, *Inorg. Chem.*, *22*, 1809 (1983).

19. D. Fenske, G. Kräuter, K. Dehnicke, *Angew. Chem. Int. Ed. Engl.*, *29*, 390 (1990).

20. M.G. Kanatzidis, S. Dhingra, *Inorg. Chem.*, *28*, 2024 (1989).

21. K.-W. Kim, M.G. Kanatzidis, *Inorg. Chem.*, *30*, 1966 (1991).

22. H.G. von Schnering, *Angew. Chem., Int. Ed. Engl.*, *20*, 33 (1981).

23. W. Schmettow, H.G. von Schnering, *Angew. Chem. Int. Ed. Engl.*, *16*, 857 (1977).

24. W. Hönle, H.G. von Schnering, *Angew. Chem. Int. Ed. Engl.*, *25*, 352 (1986).

25. J.D. Corbett, *Acc. Chem. Res.*, *14*, 239 (1981).

26. W.-M. Hurng, J.D. Corbett, S.-L. Wang, R.A. Jacobson, *Inorg. Chem.*, *26*, 2392 (1987).

27. W.-M. Hurng, E.S. Peterson, J.D. Corbett, *Inorg. Chem.*, *28*, 4177 (1989).

28. T.L.T. Birdwhistell, E..D. Stevens, C.J. O'Connor, *Inorg. Chem.*, *29*, 3892 (1990).

29. J.L. Dye in *Progress in Inorganic Chemistry*, Vol. 32. S.J. Lippard, ed. Wiley-Interscience, 1984, pp. 327–441.

30. W. Hesse, M. Jansen, *Angew. Chem. Int. Ed. Engl.*, *27*, 1341 (1988).

31. W. Beck, *Angew. Chem. Int. Ed. Engl.*, *30*, 168 (1991).

# CHAPTER 9

# Principles Underlying Stabilization of Cationic Species in Acidic Media

Experimental approaches to inorganic synthesis have often been constrained by conventional ideas based on the limits imposed by hydrolysis occurring in the narrow window of acidity and basicity available in the common solvent water, or by the redox limits of that solvent. Attempts to avoid those limitations frequently have gone further than seeking "inert," noninteracting solvents such as halogenated hydrocarbons and ketones or, at worst, coordinating solvents such as DMF, DMSO, or MeCN.

Venturing into strongly acidic or basic media provides the opportunity to prepare unusual cations or anions, the stability of which depends on the availability—or, more specifically, the nonavailability—of basic or acidic species in the solvents. Significant concentrations of these latter species in the solvents can lead to decomposition, often through disproportionation, of the desired products. This chapter will be used to show in summary form that, to a first approximation, stabilization of particular cations (or anions) depends on the level of acidity (or basicity) of the solvent, or more specifically, on the low level of basic (or acidic) species in the media.

## 9.1  General Dependence of Cation Stability on Acidity of Media

In the introductory section of chapter 6 there is a reasonably detailed account of the complexity of speciation of cations in aqueous solution, the extent of complexity depending on the pH of the solution and the presence of complexing species. However there is general acceptance of the proposition that, in strongly acidic aqueous solution, transition metal compounds, e.g. those of Cr(III), are present as aquocations, e.g. $Cr(OH)_6^{3+}$, whereas in basic solution they are anionic, e.g. $Cr(OH)_4^-$, and in neutral solution they exist as neutral sparingly soluble compounds, idealized as $Cr(OH)_3$ in Figure 9.1 The insoluble Cr(III) species is very complex, being an indefinite assembly of Cr(III), O, OH, and $H_2O$ units. Depending on the acidity of the solution, many cationic species such as $Cr(OH)^{2+}$, $Cr_2(OH)_4^{2+}$, and $Cr_3(OH)_4^{5+}$

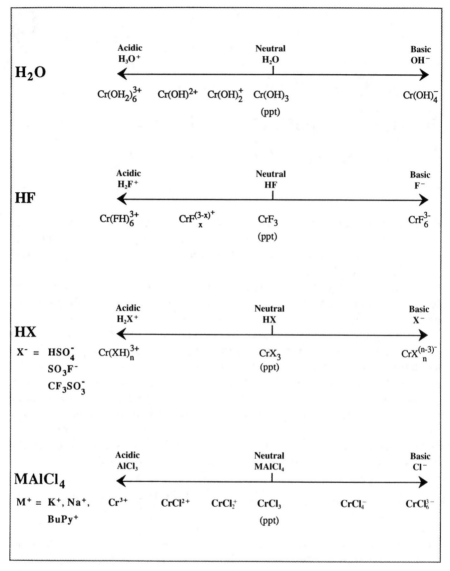

**Figure 9.1.** Generalized Scheme for Acid-Base Dependence of Formation of Cations, Binary Precipitates, and Complex Anions in Different Media.

form as the solution approaches neutrality, but, regardless of complexity, these species are cationic, as long as the solution is acidic. The number of positive charges per metal atom is reduced and the charge is attenuated over progressively larger entities as neutrality is approached.

Although there is very much less information available on the dependence of speciation on acidity and basicity in superacids and in chloroaluminate melts than in aqueous solution, the general structure seems to be the same in all media. Transition metal fluorides in oxidation states (II) and (III) are insoluble in neutral HF, as are the corresponding bisulfates, fluorosulfates, and triflates in the neutral parent acids. Much evidence was put forward in chapters 6 and 7 to show that binary, neutral transition-metal chlorides are insoluble in high-temperature and room-temperature melts when the media are close to the neutral point.

The number of attempted observations on the amphoteric nature of the insoluble fluorides, bisulfates, fluorosulfates, and triflates is minimal. Unpublished work from the University of Melbourne has shown that $CrF_6^{3-}$ can be observed in basic HF[1a] and that the solubilities of $Co(HSO_4)_2$ and $Co(CF_3SO_3)_2$ in $H_2SO_4$ and $CF_3SO_3H$ are enhanced by addition of the bases $HSO_4^-$ and $CF_3SO_3^-$, respectively.[1b] There is a strong and compelling body of spectroscopic and electrochemical evidence that transition metals exist as anionic tetra- or hexachlorometallates in basic chloride melts. Much of this was presented in chapters 6 and 7.

Strong spectroscopic evidence was also put forward in chapters 6 and 7 to show that di- and tripositive cations of $d$- and $f$-transition elements exist in acidified HF, $H_2SO_4$, $HSO_3F$, and $CF_3SO_3H$, with the proposal that the cations were coordinated in solution by the parent solvent molecules in each case. It is not easy to suggest definite models for the overall cationic species of these elements in acidic melts, other than to say that the similarity of spectra suggests the existence of cations, as such, in acidified water in protonic superacids and in acidic chloroaluminates. The components of the latter media, namely, $Al_2Cl_6$, $Al_2Cl_7^-$, and $AlCl_4^-$, are very weakly coordinating species. There is a difference of nearly 2 V between the redox potentials for Fe(III)/Fe(II) in acidic and basic room-temperature chloroaluminates, suggesting very strongly, when spectra are also considered, that the entities are anionic in basic melts and cationic under acidic conditions.

Figure 9.1 is offered to indicate the overall generality of the nature of speciation in acidic, neutral, and basic media, regardless of whether the medium in each case is water, HF, another protonic superacid, or a high-temperature chloroaluminate based on NaCl or KCl or a room-temperature melt based on butylpyridinium chloride (BuPyCl). The figure reflects the fact that varying levels of information are available for data on solution speciation in different solvents.

## 9.2 Experimentally Important Differences Between Solvents

Section 9.1 was used to indicate that, for a wide range of reaction media, there is a high degree of generality in the overall nature of solute speciation—cations in acidic media, binary precipitates under nearly neutral conditions and anions when basic. However, there must be and are significant chemical differences between solvents. Some of these will be discussed in chapter 10, insofar as they affect choice of experimental conditions in undertaking synthesis of inorganic species in superacids or melts. One such difference will be presented here because, in the section below, stability of cationic species in unusually low oxidation states will be related to the degree of acidity of the medium, and possible oxidation or reduction of or by the medium then becomes important.

By comparison with other reaction media being discussed here, water has a rather small usable potential range of about 2 V. It can be reduced to hydrogen or oxidized to oxygen over a potential range which is pH dependent, but is of this order. In contrast, the usable potential range for HF is much larger, 4–4.5 V. The implications for synthesis of unusual inorganic species are very great. As shown in the next section, cations in very low oxidation states like $Ti^{2+}$ and $Sm^{2+}$ which would reduce water are stable in HF. High-oxidation-state anions like $MnF_6^{2-}$[1c] and $NiF_6^{2-}$ have been generated in HF, and addition of the strong Lewis acid $AsF_5$ to HF solutions of $NiF_6^{2-}$ and $AgF_4^-$ has produced precipitates of the strongly oxidizing binary compounds $NiF_4$ and $AgF_3$ by competitive displacement of these compounds by the stronger Lewis acid.[3]

The other protonic superacids discussed here are all capable of supporting cations in low fractional oxidation states without being reduced; for example, $I_2^+$ and $S_8^{2+}$ are stable in oleums and acidified $HSO_3F$ and $CF_3SO_3H$. It may be that they cannot be made sufficiently acidic to prevent ions like $Ti^{2+}$ from disproportionating, but they are not readily reduced. The principal disadvantage, from a redox point of view, of the sulfur-containing superacids is that they are all oxidants, to a greater or less degree, under solution chemistry conditions. $SO_3$ in oleums is a very strong oxidant. It occurs, in small amount, in equilibrium with HF in $HSO_3F$ and so $HSO_3F$ is itself a reasonable oxidant. $CF_3SO_3H$ is not a strongly oxidizing solvent but Carre and Devynck report that S(VI) in $CF_3SO_3H$ can be reduced[4] and electropositive metals like Ti have been found to reduce $CF_3SO_3H$ to the $S_5^+$ radical cation.[5]

The high-temperature chloroaluminates $KCl—AlCl_3$ and $NaCl—AlCl_3$ have long usable potential ranges terminating in oxidation of chloride and reduction of aluminium(III). Titanium(II) is stable in acidic chloroaluminates. Room-temperature chloroaluminates are limited by oxidation of the organic cation at the positive limit, with aluminium deposition being the negative limit.

There will be greater discussion in chapter 10 of the grounds for choice of superacidic media for synthesis of inorganic compounds. Apart from physical properties, which affect ease of manipulation of the solvents, relative self-ionizations of the superacids will be discussed in connection with buffering and other control of acidity levels.

## 9.3 The Essential Role of Acidity Level in Stabilizing Cations in Superacidic Media

### 9.3.1 "Normal" Oxidation States

In the introduction to this chapter there was a general statement of the fact that binary fluorides, bisulfates, triflates, and fluorosulfates of transition metals in "normal" oxidation states are sparingly soluble in the "neat" parent superacids; however, they can be brought into solution as solvated cations by enhancing the acidity of the media using appropriate Lewis acids, the counteranion usually being a compound of the Lewis acidic used and the base of the solvent.

Dissolution of these sparingly soluble compounds in protonic superacids appears to depend on crystal energetics in similar fashion to the established patterns for solubility in water and other common solvents—formal cationic charge and radius can be related to solubilities. By far, the most information that is currently available applies to solubility in HF, and examples in this section will be restricted to that solvent system.

A generalized representation of the dependence of solubility of binary transition-metal fluorides in HF of varying acidity was given in equations 6.1 and 6.2. Some supportive detail of those generalizations can now be provided after the presentation in chapters 6 and 7 of solution chemistry of transition metal cations in HF. Thus fluorides of dipositive Ni and Co are soluble in HF—$BF_3$ whereas the fluoride of the tripositive first-row transition element Cr requires the much stronger Lewis acid $SbF_5$ for dissolution.

However, for the much larger tripositive cation of lanthanides and actinides, for example, $Pr^{3+}$, $Nd^{3+}$, $U^{3+}$, and $Np^{3+}$, weakly acidic HF—$BF_3$ provides working solutions. It is not realistic to talk of tetrapositive cations of first-row transition metals. The tetrafluorides of those elements dissolve essentially as molecular entities, but solutions containing the tetrapositive actinide cations $U^{4+}$ and $Np^{4+}$ are generated quite easily in HF—$AsF_5$. The large dipositive lanthanide cations $Eu^{2+}$, $Yb^{2+}$, and $Sm^{2+}$ are soluble even in basic HF.

In summary, for dissolution of fluorides in HF, the fluoride ion concentration in equilibrium with $d$- and $f$-transition-metal fluorides must be reduced progressively with increase in cationic charge and with decrease in cationic radius.

### 9.3.2 "Unusual" Oxidation States

A second aspect of acidity control, separate from that described immediately above, is that enhanced acidity of the medium, whether of HF, other superacids, or chloroaluminates, is necessary in the preparation of solutions containing $d$- and $f$-transition in unusually low oxidation states and in the generation of polyatomic cations of main-group elements exhibiting fractional formal oxidation state.

Oxidation by the protons of acidified HF of first-row $d$-transition elements in the metallic state provides a nice example.[6] $Cr^{2+}$, stable in acidified aqueous solution, and $V^{2+}$, marginally stable in that medium, could both be produced in HF acidified with the weak Lewis acid $BF_3$. $Ti^{2+}$, which is reported to reduce water explosively, required the very strong Lewis acid $SbF_5$ at a concentration of about 3 M in HF to form a stable solution. Even in the presence of $SbF_5$ at concentrations less than 2 M, $Ti^{2+}$ is converted to Ti(III) by a disproportionation reaction involving the very small amount of $F^-$ present in HF of this acidity. $AsF_5$, intermediate in acidity between $SbF_5$ and $BF_3$, produced Ti(III) solutions when interacting with metallic Ti in HF. As stated in section 9.3.1, the larger dipositive lanthanide cations were stable in "neat" HF.

There has been very much less investigation of the dependence of stability of lower oxidation states of transition metals on acidity for other media than for HF. What is available follows similar trends. U(III) can be generated in $CF_3SO_3H$ and Cr(II) is in equilibrium with Cr(III) within that solvent.[5] Ti(II) is stable in acidic high-temperature chloroaluminates but not in the neutral medium,[7] and $U^{3+}$ is stable in acidic room-temperature melts.[8]

Much evidence was put forward in chapters 4 and 5 to show that polyatomic cations of the halogens and the chalcogens and of some main-group metals require very high acidity for their generation and stability in solution; and, even if they are generated under naked conditions, they decompose by disproportionation by interacting with small amounts of basic species present adventitiously or otherwise. Many examples were given of the need for high levels of acidity for stabilization of these polyatomic cations in the four protonic superacids considered in this book and in chloroaluminates, the melt system considered here.

Taking polyiodide cations as an example, Table 4.1 shows the threshold of acidity of HF below which these cations do not form because the fluoride ion levels, although low, are sufficiently high to cause disproportionation. As the acidity is increased the cations $I_5^+$, $I_3^+$, and $I_2^+$ are formed in turn—the higher the acidity, the lower (for nonmetal cations) the charge-to-element ratio, that is, the lower the formal fractional oxidation state of the element. Table 4.2 and its footnotes show exactly similar trends for iodine cations in other superacids and in melts, for example, $I_3^+$ and $I_5^+$ are stable in neutral chloroaluminates but $I_2^+$, stable in acidic melts, cannot be observed in the neutral medium.

Although bromine and chlorine cations, in turn, are much more difficult to stabilize than their iodine counterparts, and fewer experimental observations are available for determining trends, the same general pattern for these ions can be seen in Table 4.2 as is shown for the iodine cations.

Semiquantitative comparisons of stabilities of polychalcogen cations are limited in Table 4.3 to behavior in oleums and $H_2SO_4$ itself and in $HSO_3F$.

The cations of general formula $Ch_4^{2+}$ afford a nice basis for comparisons. Greater stability of tellurium cations than of selenium and sulfur, in turn, is shown by the fact that $Te_4^{2+}$, $Se_4^{2+}$, and $S_4^{2+}$ can be generated, respectively, in basic $H_2SO_4$, in 100% $H_2SO_4$, and in oleums containing not less than 40% $SO_3$. The cations of decreasing fractional oxidation state, $S_4^{2+}$, $S_8^{2+}$, and $S_{19}^{2+}$, require $H_2SO_4$-based media containing not less that 40, 15, and 5% $SO_3$, respectively.

For polyatomic cations of other main-group metals, much less information on other dependence of cation stability on acidity levels of media is available than for the halogen and chalcogen cations. Isolation of the metal cations has, in the main, been directed toward structural analysis of the cations. As can be seen from chapter 5, there has been little study of them *in solution*. Almost invariably they have been isolated from melts or as "naked" cations from weakly basic solvents such as $SO_2$ or $AsF_3$, selected because they would not be expected to undergo acid-base interaction with the desired cations.

As reported in section 5.3, Corbett demonstrated that increased acidity of melt media favored formation of $Cd_2^{2+}$. He showed spectroscopically that $Cd_2^{2+}$ production in mixed $AlCl_3$—$CdCl_2$ melts was at a maximum when the melt was acidic ($AlCl_3$:$CdCl_2$, 3:1) and that, when NaCl—$AlCl_3$ melts contained 60% or more of NaCl, $Cd_2^{2+}$ decomposed to $Cd^{2+}$, presumably by disproportionation.

When a range of polyatomic cations of a particular metallic element can be generated in melts, the trend is the reverse of that observed for polycations of nonmetals. Increase in acidity of the melt favors formation of cations with progressively *lower* charge-to-metal ratios, that is, progressively *lower* formal fractional oxidation states. As reported in section 5.4, $Bi^+$ and $Bi_9^{5+}$ were isolated together from a neutral melt, whereas growth of pure crystals for the structure determination of $Bi_5^{3+}$ required a small excess of about 10 mol% $AlCl_3$ over the neutral composition and $Bi_8^{2+}$ required a 33 mol% excess of $AlCl_3$ over neutrality.

# 9.4 Base-Induced Disproportionation of Cations

The major aim of this book is to establish that a similar set of principles applies to generation and stabilization of unusual cations in solution, regardless of whether the medium is a protonic superacid or an acidic melt. The chemical nature or the temperature domain of the reaction medium has relatively little effect on the redox equilibria involved. The primary requisite for preparation of cations in such solution media is a high level of acidity, which implies a low level of available basic species. This latter is the fundamental requirement. Even if unusual cations are synthesized, not in solution, but under "naked" conditions—either by direct interaction of the elements of the final compound or in an inert, noninteracting solvent—availability of species of sufficient basicity can lead to decomposition of the cationic species by disproportionation.

## 9.4.1 Disproportionation of Monatomic Cations

The simplest case of disproportionation to consider is that of a transition-metal cation in a very low oxidation state, for example, $U^{3+}$ in HF. It was reported in sections 7.2.1 and 7.2.2 that reaction of metallic U with HF acidified with $BF_3$ produces a stable solution containing $U^{3+}$ with $BF_4^-$ as the counteranion. Addition of NaF in HF to that solution causes disproportionation of U(III) to U(0) and U(IV), the latter being present as the very insoluble $UF_4$.

Similar disproportionation of uranium(III) has been observed in other protonic superacids and in chloroaluminates. It was reported in section 7.2.3 that Adrien[5] treated metallic U with $HSO_3F$ acidified with the strong Lewis acid $SbF_5$, expecting to obtain a strongly acidic solution containing $U^{3+}$. Instead, the solution spectrum was that for the U(IV) species $UF_2^{2+}$. The origin of this reaction was demonstrated by generating $U^{3+}$ in triflic acid acidified with $B(OSO_2CF_3)_3$ and then adding HF when the spectrum changed from that for $U^{3+}$ to that for $UF_2^{2+}$. This showed that the small equilibrium amount of HF present in $HSO_3F$ as a result of decomposition to HF and $SO_3$ was sufficient to progressively form $UF_2^{2+}$ and "drive" the disproportionation of U(III) to U(0) and U(IV), the latter present as $UF_2^{2+}$. The very finely divided metallic uranium produced in the disproportionation reaction would be in an ideal physical state for reoxidation to $U^{3+}$, initially, which would then disproportionate. This process of disproportionation and reoxidation is now believed to be very common in chemistry in acidic nonaqueous solutions and in melts and is highlighted in section 9.5.

The uranium(III) oxidation state has proved to be very nicely poised for the illustration of acid-base effects in stabilization and disproportionation of transition-metal cations. There is strong spectroscopic evidence for the generation of a stable $U^{3+}$ in acidic room-temperature chloroaluminates[8] as shown in section 7.2.4. In earlier work in high-temperature melts $U^{3+}$ had been produced in $AlCl_3$-rich melts, but not recognized as such. The real significance in the context of this book of the U(III)—U(IV) work in high-temperature melts lies in the postulated disproportionation of U(III) to U(0) and U(IV) in nearly neutral melts and ensuing cyclic reoxidation and disproportionation. This has been presented in section 7.2.4 and will be referred to again in sections 9.5 and 10.1.4.

Preparation of stable solutions of Ti(II) in HF required uses of a narrow "window" of acidity as previously illustrated in section 7.1.1. $Ti(FH)_6^{2+}$ was found to be stable only in HF solutions containing $SbF_5$ at concentrations between 2 and 3 molal. It is believed that there is sufficient free molecular $SbF_5$—a strong oxidant—at concentrations above 3 molal to oxidize Ti(II). $SbF_6^-$, the dominant species in more dilute $SbF_5$—HF, is a much weaker oxidant than $SbF_5$. However, in more dilute $SbF_5$—HF solutions there is sufficient fluoride ion available to cause disproportionation of Ti(II). Besida[1a] made small, gradual additions of NaF in HF to solutions of $Ti(FH)_6^{2+}$ and showed the progressive formation of Ti(III) and Ti(IV) (section 7.1.2).

Ti(II) was shown to be stable in the acidic medium, molten $AlCl_3$, the spectrum suggesting the presence of $Ti^{2+}$ (section 7.1.4). Although there appeared to be some change in speciation as the melt medium was made more basic by the addition of KCl, oxidation state II was retained by the titanium until the melt acidity approached neutrality, when there was spectroscopic evidence for the formation of Ti(III) as $TiCl_6^{3-}$ and direct physical evidence for deposition of metallic Ti—disproportionation had occurred (section 7.1.5).

## 9.4.2  Disproportionation of Polyatomic Cations

There was a considerable amount of detailed discussion in chapter 4 about base-induced disproportionation of polyatomic cations of nonmetallic elements. Such cations can be synthesized in highly acidic solutions or under "naked" conditions. The requirement for their stability is that the solutions in which they are generated must contain such small concentrations of basic species such that they will not disproportionate or such that, under "naked" conditions, the counteranions or any other species "seen" by the polyatomic cations during preparation or isolation are so weakly basic that disproportionation does not occur.

Disproportionation occurs when a highly electrophilic polyatomic nonmetal cation interacts with a nucleophilic base to form, in the first instance, a polyatomic cation of lower charge *per* nonmetal atom (lower formal fractional oxidation state) and a compound, essentially covalent, formed between the base and the nonmetal of the cation with that nonmetal now in a higher oxidation state than in the parent cation. With greater availability of base, the polyatomic cation first formed will disproportionate to a lower-charge polyatomic cation and ultimately to the free element, from which the polyatomic cations were formed.

The system that has had the most detailed quantitative study of base-induced disproportionation of polyatomic cations in solution is that involving equilibria in HF of controlled acidity and basicity between the cations $I_2^+$, $I_3^+$, and $I_5^+$ with the ultimate disproportionation products $I_2$ and $IF_5$. This experimental work was outlined in Table 4.1, discussed in detail in section 4.1.1.1, and summarized in equation 4.6a–d of section 4.1.1.4. It was seen that $I_2^+$, stable in HF of the highest acidity, in which availability of the base $F^-$ was minimal, disproportionated on availability of slightly larger amounts of $F^-$ to $I_3^+$ and $IF_5$ and to $I_5^+$ and $IF_5$, in turn, and ultimately to the final disproportionation products $I_2$ and $IF_5$. The latter, in HF of sufficient basicity, could ultimately become $IF_6^-$, but that is incidental to the disproportionation process.

Similar disproportionations, based on smaller bodies of experimental evidence, were observed for polyiodine cations in $CF_3SO_3H$, as described at the end of section 4.1.1.1, and in chloroaluminates (section 4.1.1.2). Equations representing the disproportionation reactions were given as equations 4.7a–c and 4.4.

Throughout chapter 4, evidence was put forward for base-induced disproportionation of the polyatomic cations of bromine, chlorine, sulfur, selenium, and tellurium. Because the polyiodine cations have been studied so much more exhaustively than those other nonmetal cations, formal treatment of disproportionation equilibria has been restricted here to the iodine system, but close perusal of chapter 4 shows that the same principles apply in the cases of all polyatomic cations of nonmetals.

Despite the foregoing paragraph, some brief mention should be made of the polyatomic cations of metallic elements which were presented in chapter 5. In simple cases, disproportionation of polymetallic cations is the direct parallel of that for the nonmetallic cations. For example, equation 5.2 summarizes the reversible disproportionation of $Cd_2^{2+}$ to Cd(0) and Cd(II).

The difference between polyatomic cations of nonmetals and of metals that should be recognized is that, with decreasing basicity of the medium, nonmetal cations progress to a higher fractional oxidation state whereas polymetallic cations exhibit progressively lower fractional oxidation states. The ultimate disproportionation products will appear to be similar in chemical character, that is, the element itself and a compound of the base and the element in an oxidation state higher than in the polyatomic cation—$I_2^+$ will give $I_2$ and $IF_5$ (or $IX_3$ or $IX$, depending on the medium HX) and $Bi_8^{2+}$ will give Bi and $BiX_3$. A nonmetal polyatomic cation such as $I_2^+$, on progressive disproportionation, will give cations of *lower* fractional charge and a compound such as $IF_5$. However, disproportionations involving metallic cations of higher fractional charge will produce cations of low charge and the elemental metal. We saw at the end of section 9.3 above that melts had to be made progressively more acidic with $AlCl_3$ to generate $Bi^+$, $Bi_9^{5+}$, $Bi_5^{3+}$, and $Bi_8^{2+}$ in turn. The obverse of this is that reaction of $Bi_8^{2+}$ with $Cl^-$ will give $Bi_5^{3+}$ and $Bi_5^{3+}$ with $Cl^-$ will give $Bi_9^{5+}$ and Bi. Metallic Bi will be produced in each disproportionation step until the final disproportionation products are $BiCl_3$ and Bi.

It should cause little surprise then that the literature describing attempts to produce polymetallic cations in melts is sprinkled with observations that finely divided metal was observed in the reaction mixtures very frequently. Now that it is known that ideal growth conditions for compounds containing the $Bi_5^{3+}$ cation require slightly acidic melts, it is reasonable to note that when Friedman and Corbett[9] crystallized $Bi^+Bi_9^{5+}(HfCl_6^{2-})_3$ from a neutral melt they speculated that their compound might have resulted from disproportionation of $Bi_5^{3+}$ in the hypothetical compound $(Bi_5^{3+})_2(HfCl_6^{2-})_3$ to $Bi^+$ and $Bi_9^{5+}$. Burns et al.[10] found that when they tried to generate $Bi_5^{3+}$ with counteranions that were not very weakly basic, they obtained white compounds of Bi(III)—likely disproportionation products.

## 9.5  Cyclic Disproportionation and Reoxidation

When polyatomic cations of nonmetals disproportionate, the products are a cation of lower charge-to-element ratio and a higher oxidation-state product, not subject to redox reactions with the medium. For example, $I_2^+$ disproportionates, in turn, to $I_3^+$ and $IF_5$, $I_5^+$ and $IF_5$, and finally to $I_2$ and $IF_5$ (equations 4.6a–d).

However, the addition of a small amount of base to a solution in a superacid of a monatomic cation of a $d$- or $f$-transition metal in low oxidation state or of a polyatomic cation of a main-group metal results in formation of the appropriate metal, as outlined in sections 9.4.1 and 9.4.2. The metal so produced will be very finely divided, in a form that is highly reactive, and if the acidity is sufficiently great, it will be oxidized by the solvent, and the lowest-oxidation-state cation will form transiently. This will then disproportionate. This reaction cycle will continue as long as the solvent is sufficiently acidic until all the metal which is produced progressively has been consumed, leaving only the high-oxidation-state disproportionation product.

Cyclic disproportionation can occur without deliberate or adventitious reduction of the acidity of the medium containing the metal cations in situations where the initial high-oxidation-state product is effectively removed from solution.

Again the uranium(III)—uranium(IV) system provides a series of examples of disproportionations governed by the solution acidity and solute speciation. The fundamental disproportionation is an equilibrium represented most simply as

$$4U(III) \rightleftharpoons U(0) + 3U(IV) \qquad\qquad 9.1$$

In the original disproportionation encountered in HF (section 7.2.2)[1d], this equilibrium was driven to the right by both the addition of excess fluoride and the insolubility of $UF_4$:

$$4U^{3+} + 12F^- \rightarrow U + 3UF_4 \qquad\qquad 9.2$$

In that experiment[1d] metallic uranium was observed with $UF_4$ as a disproportionation product because the resulting solution was too basic for oxidation of metallic U by $H_2F^+$ to occur.

However, the disproportionation equilibrium can be driven to the right by factors other than reduction of acidity of the medium. As reported in sections 7.2.3 and 9.4.1, Adrien[5] found that $U^{3+}$, stable in $CF_3SO_3H$, which was acidified with $B(OSO_2CF_3)_3$, changed to a solution of $UF_2^{2+}$ when HF, itself acidic, was added to $U^{3+}$ in the acidified $CF_3SO_3H$. This allowed rationalization of the observation that treatment of metallic uranium by $HSO_3F$—$SbF_5$, a medium of acidity far in excess of that required to stabilize

$U^{3+}$, gave instead U(IV) as soluble $UF_2^{2+}$. Spontaneous partial decomposition of $HSO_3F$ to $SO_3$ and HF provided available fluoride to drive the disproportionation equilibrium according to an equation which represents the overall reactions in both $CF_3SO_3H$ with added HF and in $HSO_3F$;

$$4U^{3+} + 6F^- \rightarrow U + 3UF_2^{2+} \qquad\qquad 9.3$$

It has already been reported (section 7.2.1) that the weak acid $BF_3$ in HF provides sufficient $H_2F^+$ to oxidize U to $U^{3+}$ with liberation of $H_2$. It was shown in Figure 2.5 that $TaF_5$ is considerably stronger as a Lewis acid in HF than $BF_3$. Consequently it was expected that the nonoxidizing acid $TaF_5$ would be even more effective than $BF_3$ in producing solutions of $U^{3+}$ in acidified HF. However, Dawkins[1d] found that treatment of metallic uranium with HF—$TaF_5$ produced a pale green precipitate of uranium(IV) fluorotantalate with a colorless supernatant HF solution showing no evidence for the lilac-colored $U^{3+}$ cation. Both the nonoxidizing properties of $TaF_5$ and its ability to produce very insoluble precipitates from HF based on the $Ta_2F_{11}^-$ anion will be discussed in section 10.2. The best interpretation of Dawkins's observations appears to be that the extreme insolubility of the uranium(IV) fluorotantalate pushes the general disproportionation equilibrium shown in equation 9.1 virtually to completion.

In all of these disproportionation reactions of U(III) in acidic media, initial displacement of the fundamental disproportionation equilibrium in favor of formation of U(IV) products leads to formation of finely divided U that is then reoxidized by protons of the solvent, setting up a cycle of disproportionation and reoxidation which leads effectively to complete conversion of $U^{3+}$ to uranium(IV) products even in acidic media.

An investigation of uranium(III) in molten salt media of varying acidity and basicity has already been presented in section 7.2.4. It will be presented again in section 10.1.4 as an illustration that cyclic disproportionation and reoxidation provides a better rationalization of the experimental observations than that offered by the original workers. Also in chapter 10 other examples will be given where better interpretations than those previously offered will be based on disproportionation of solutes and reoxidation by species other than protons from the solvent.

## 9.6 Cataclysmic Irreversible Disproportionation

In many of the examples cited in section 9.4 disproportionation was shown to be a reversible process. The best evidence for this is provided by the detailed study presented in section 4.1.1.1 on the acid-base dependence of the stability of polyatomic cations in HF, where these cations were generated by oxidizing excess $I_2$ with $F_2$ and their base-induced disproportionation was followed spectroscopically. The reversibility of cation disproportionation was demonstrated by conproportionation reactions at various acidities of $I_2$ and $IF_5$—the ultimate disproportionation products of the cations. Although less detailed, reactions of Cd, $Cd^{2+}$, and $Cd_2^{2+}$ in chloroaluminates of varying acidities showed similar reversibility (section 5.2).

In these cases of reversibility of disproportionation of complex cations, the bases involved were those of the solvent systems, $F^-$ in the case of HF and $Cl^-$ in the case of chloroaluminates. For the sake of completeness, it is worth recording, however briefly, that if disproportionation is caused by bases very much stronger than those of the appropriate solvent systems in which the cations are generated, irreversible drastic decomposition will occur.

The role of water is of vast importance here because it is a common, but unwanted, impurity in many syntheses of unusual cations. It is obvious that cations such as those of the halogens and chalcogens would undergo massive hydrolysis if compounds containing them were to be added to water. This would be an extreme case of base-induced disproportionation. Polyiodine cations, for example, would be converted to oxoanions of iodine and elemental iodine which then might or might not react further with the aqueous solution, depending on its composition.

Hydrolytic disproportionation of these types of ions can occur in synthetic and other experimental procedures even when supposedly nonaqueous conditions are being observed. For example, Kemmitt and co-workers[11] reported that $I_2$ reacts in $IF_5$ with pentafluorides of P, As, Sb, Nb, and Ta to give deep-blue solutions with spectra characteristic of $I_2^+$ but that attempts to remove the last traces of solvent led to decomposition of the blue species. They reported that the solids are highly reactive and immediately hydrolyzed by water. This effect has frequently been observed during attempted syntheses of moisture-sensitive species in HF. Traces of $H_2O$ in the original HF solvent remain as involatile protonated water ($H_3O^+$) in residues as the HF is removed by distillation, become highly concentrated, and are available for hydrolysis.

Other examples of massive disproportionation of polyatomic cations have been given in sections 5.1 and 5.2. $Hg_2^{2+}$, stable as the nitrate in aqueous solution that is probably somewhat acidic, disproportionates to Hg and compounds of Hg(II) when strong base in the form of $NH_3$ or $OH^-$ is added to the solution. $Hg_3^{2+}$ is reported as disproportionating rapidly and completely to Hg and Hg(II) in the presence of water and other basic substances. Corbett reported that $Cd_2(AlCl_4)_2$ gave a dark precipitate of metallic Cd on contact with basic solvents such as water, dioxane, ammonia, ethanol, and ethers.

While hydrolysis and solvolysis generally are an accepted part of the overall fabric of inorganic chemical reactions, the importance of the examples presented in this section has been that in each case the *element* has been observed as a product of interaction of the cation with water or other base, that is, disproportionation has occurred, but with bases much stronger than the bases of superacid solvent systems or chloroaluminates—the situation described in sections 9.4 and 9.5.

## 9.7  Acidity Level as Principal Determinant of Cationic Oxidation States

It was stated relatively early in chapter 4 that there was general agreement in the extant chemical literature that formation and stability of cations in unusual oxidation states, particularly of polyatomic cations, depends on stoichiometry of the redox reactants and on the degree of acidity of the media in which the cations are generated—or more particularly on the limited availability of basic species which could lead to disproportionation of the cations. The experimental data that have been presented in the intervening chapters allow us to look critically at these two stability criteria.

The first criterion is important only if one of the reactants is a very strong oxidant or reductant. Few, if any, cations have been generated in highly acidic media using very strong reductants, whereas very strong oxidants have been used frequently as indicated by examples in the next paragraph. Of course, synthesis of cations by conproportionation reactions involves oxidation-reduction; but a distinction is made later in this section between the relatively mild redox properties of disproportionation products and those of reagents conventionally regarded as very strong oxidants or reductants. So we need look here only at the effect of very strong oxidants.

Typical examples of use of very strong oxidants to generate unusual cations would be oxidation of a halogen by $F_2$ in HF or oxidation of a chalcogen by $S_2O_6F_2$ in $HSO_3F$ to produce polyatomic cations of nonmetals. Polyatomic metal or nonmetal cations could be generated in acidic chloroaluminates by oxidation of elements by $Cl_2$, or metals could be oxidized to low-oxidation-state monatomic cations in protonic superacids or acidic chloroaluminates by $F_2$ or $Cl_2$. In each of these cases it would be essential not to exceed the oxidant stoichiometry—it may be advantageous to have a substoichiometric amount of oxidant.

Even if acidity of the medium is appropriate for the generation of a particular cation, use of excess oxidant will lead to formation of cations of different formal charge from that sought. A large excess of oxidant would lead to products which might not be cationic at all. For example, oxidation of $I_2$ in HF with a large excess of $F_2$ would give products like $IF_5$ or $IF_7$ depending on the amount of $F_2$ available. An attempt to generate $Cd_2^{2+}$ or $Cd_3^{2+}$ by oxidizing Cd with $Cl_2$ in an acidic chloroaluminate would lead to formation of $CdCl_2$ if a very large excess of $Cl_2$ were to be used.

As indicated in chapter 4, Gillespie was highly successful in using *stoichiometric* amounts of $S_2O_6F_2$ in $HSO_3F$ to oxidize elemental halogens and chalcogens to polyatomic cations in media that were sufficiently acidic to provide stability of the cations after formation. There will be further discussion of this work in terms of media acidity at the end of this section.

Providing the use of very strong oxidants is avoided, the level of acidity of the medium becomes the principal—perhaps sole—determinant of the speciation of cations generated. This has been demonstrated to be the case in conproportionation reactions—that is, when the oxidants and reductants used for generation of particular cations are the ultimate disproportionation products of those cations. Regardless of whether the ratio of oxidant to reductant is 10:1 or 1:10, generation of individual polyiodine cations in HF from $IF_5$ and $I_2$ or in $CF_3SO_3H$ from $ICl_3$ and $I_2$ depends only on the level of acidity of the medium, as was shown in section 4.1.1.1. It was also shown in a limited study in that section that if compounds comparable in oxidant strengths to disproportionation products are used to produce cations, acidity levels dictate cation speciation. Polyatomic cations were produced by conproportionation of $IO_3^-$ and $I_2$ and polysulfur cations from $S_8$ and either $SO_3^{2-}$ or $SO_4^{2-}$.[12]

An uncritical approach to the implications of experimental acidity levels in the pioneering work of Gillespie and his colleagues on generation of polyatomic cations of halogens and chalcogens in $HSO_3F$ might lead to some doubts about the dependence of the stability of specific cations on the acidity of the medium. Besida and O'Donnell[12] showed that, providing the redox reagents used to generate iodine cations in HF were the disproportionation products $I_2$ and $IF_5$ or compounds of comparable redox strength, acidity level of the medium determined whether cations formed at all or whether the predominant species were $I_5^+$, $I_3^+$, or $I_2^+$ as the acidity of the medium was increased. This was the case regardless of whether oxidant or reductant was in excess.

Gillespie's group used fixed, appropriate stoichiometries of $I_2$ and $S_2O_6F_2$ to synthesize $I_5^+$, $I_3^+$, and $I_2^+$ in $HSO_3F$.[13] They did find that $I_2^+$ was not particularly stable in "neat" $HSO_3F$. They needed to increase the acidity of the medium to obtain reliable spectra for $I_2^+$. But $I_5^+$ and $I_3^+$ were both stable indefinitely in $HSO_3F$. At first sign, this may appear to indicate that acidity level is not of great importance in obtaining stable solutions of these cations.

The real situation, however, is that "neat" $HSO_3F$ is the medium of sufficient acidity for stabilization of $I_3^+$, but not for $I_2^+$. $I_5^+$, on the other hand, could have been generated at lower acidity; however, having been generated in "neat" $HSO_3F$, $I_5^+$ remained as a stable entity in solution. More oxidant would have been needed to convert it to $I_3^+$. The stoichiometric amount of $S_2O_6F_2$ had been used to produce $I_5^+$, and there was not present in solution any disproportionation product or other mild oxidant that could have driven the equilibrium reaction represented as equation 4.7b to the left. Similar considerations apply to Gillespie's preparation of a range of chalcogen cations of different fractional oxidation state in $HSO_3F$ (section 4.2.1.1).

Of course, acidity-dependent conproportionation-disproportionation equilibria can be affected by extraneous factors, as indicated by the major disturbances of the U(0)—U(III)—U(IV) equilibria that were described in section 9.5.

## 9.8 Generalized Scheme for Stability of Cations in Acidic Media

Figure 9.1 provides a summary view of the general proposition that in neutral media, regardless of the nature of the solvent, transition metals form insoluble neutral binary compounds with the base of the solvent. Depending on the extent to which these insoluble neutral compounds are amphoteric, species in basic media are anionic—they will certainly not be cationic. As the acidity of the medium is increased from neutrality cationic species will be formed. At sufficiently high levels of acidity, these cationic species can be considered as simple solvated cations. At intermediate acidity, the cations, while still solvated, may be bonded to one or more anions which are the base of the solvent. Species such as $[MX_m(HX)_p]^{(n-m)+}$ will form at the expense of the simple solvated cation $[M(HX)_{m+p}]^{n+}$.

Transition-metal compounds have been selected for presentation in Figure 9.1 and in this discussion because they differ from members of the alkali and alkaline earth-metal groups which exist exclusively as cations over the whole acidity-basicity range. Their hydroxides are not sparingly soluble in neutral solution, nor are they amphoteric, and so they do not provide complex anionic species in basic solutions. There are no discrete neutral binary compounds such as hydroxides, fluorides, or bisulfates, to which one or more ions that are the bases of the solvents concerned can coordinate to form anionic complexes.

The cations of most interest in the presentation of this book are not the alkali, alkaline earth, or transition-element cations in normal oxidation states but transition-metal cations in unusually low oxidation states or polyatomic cations of nonmetallic elements or of metallic elements of the posttransition-metal groups—principally the polyatomic cations of groups IV–VII of the periodic classification. All of these latter groups of cations require, for their generation and stability in solution, high levels of solvent acidity, that is, low levels of basicity. Even when these cations are synthesized as solids under "naked" conditions, any basic species present during synthesis or as the counteranion must be extremely weak—the higher the acidity needed to generate the cations in solution, the weaker must be the bases present under conditions of "naked" synthesis.

There has been a considerable body of evidence put forward in earlier chapters to relate cation speciation to level of acidity in the medium. Thus for polyatomic cations of the halogens and chalcogens $X_n^+$ and $Ch_m^{2+}$ (or $Te_6^{4+}$ in one instance), the higher the level of acidity the higher the charge-to-element ratio, or formal fractional oxidation state. This was amply illustrated in Tables 4.1, 4.2, and 4.3. For metallic cations, whether monatomic, for example, $Ti^{n+}$, or the polyatomic cations of elements like Hg, Cd, Bi, and so on, lower charge-to-element ratios (formal oxidation states) were stabilized as acidity was increased.

For all of these cations, disproportionation occurred as the acidity of the medium was decreased, that is, as the availability of the base was increased to the level where it would interact with the cation under consideration. For all the cations, the ultimate disproportionation products were the element itself and a compound, covalent or insoluble, of the base of the solvent and of the element in a higher oxidation state than in the disproportionating cation. Typical examples of metallic cations would be $Ti^{2+}$, disproportionating through $Ti^{3+}$ and Ti to a Ti(IV) compound and Ti, or $Bi_8^{2+}$ disproportionating in a melt through Bi and cations of intermediate formal charge to ultimate disproportionation products Bi and $BiCl_3$. Polyatomic cations of nonmetals disproportionate through intermediate cations and a covalent compound to a mixture of that compound and the element. Thus $I_2^+$ would disproportionate through $I_3^+$ and $I_5^+$ and compounds such as $IF_5$ or $I(OSO_2F)_3$ ultimately to elemental iodine and the high-oxidation-state covalent compound. These sequences were detailed in equations 4.6a–d and 4.7a–c.

*In overall outline these stabilization and disproportionation processes are quite general and do not depend in any fundamental manner on the chemical nature or the temperature domain in which they occur.* The overall pattern is the same regardless of whether the reaction medium is one of the protonic superacids or a high-temperature chloroaluminate melt based on NaCl or KCl or a low-temperature chloroaluminate based on organic chlorides such as 1-butylpyridinium chloride or 1-methyl-3-ethylimidazolium chloride.

There will, of course, be minor differences in detail between the stabilization and disproportionation processes in different solvents. For example, the predominant covalent disproportionation produce for iodine cations is $IF_5$ in HF systems and $IX_3$ where the acid HX is $H_2SO_4$, $HSO_3F$, or $CF_3SO_3H$.

Finally, there should be a warning against any idea that stability of cations in unusual oxidation states can be related in an absolute sense to levels of acidity in solvents as measured by Hammett Acidity Functions. Such relationships are valid *within* any one solvent system, but there are small but significant differences in the acidity levels at which a particular cation is stabilized *in different solvents*.

As shown in Table 4.2, $I_3^+$ is stable in $H_2SO_4$ at a value of $-11.9$ for $H_0$, whereas a value of $-13.8$ is observed for $HSO_3F$, and $I_3^+$ is stable in HF to an acidity level of $-16$. Table 4.3 shows that the cations $S_4^{2+}$, $S_8^{2+}$, and $S_{19}^{2+}$ require higher absolute acidity levels for stabilization in $HSO_3F$-based media than in oleums. It should be noted that it is only for the $I_n^+$—HF system (Table 4.1) that a detailed study of the dependence of cation stability on precisely fixed acidity levels has been made. The levels quoted in Tables 4.2 and 4.3 are taken from individual papers on the synthesis and reactions of the cations concerned and do not necessarily define the acidity levels with great precision, but the trends in Tables 4.2 and 4.3 support the general proposition given above that there are small but significant differences in acidity levels for cation stabilization in chemically different media.

The reason for these differences probably lies in differing electronegativity or nucleophilicity of the bases which interact with the elements of the polyatomic cations to form covalent disproportionation products. On this basis the bond strengths of iodine fluorides would be expected to be greater than those of iodine fluorosulfates and bisulfates in turn. Therefore in a generalized disproportionation reaction scheme, iodine fluorides would be expected to form more readily than fluorosulfates and bisulfates. The consequence of this is that, in order to prevent disproportionation of a particular cation in solvents based on $H_2SO_4$, $HSO_3F$, and HF, progressively lower concentrations of base must be available in each solvent, that is, progressively higher acidities must be maintained.

Although the chemical nature of the ultimate disproportionation products of the chalcogen cations is not as well known as for the halogen cations, similar considerations would be expected to be applicable in rationalizing the acidity-dependence of the stabilities of the cations of Table 4.3.

# References

1. Unpublished observations, University of Melbourne (a) J. Besida, (b) P.B. Smith, (c) R.W. Cockman, (d) L. Dawkins.

2. L. Stein, J.M. Neil, G.R. Alms, *Inorg. Chem., 8*, 2472 (1969).

3. B. Zemva, K. Lutar, A. Jesih, W.J. Casteel, Jr., N. Bartlett, *Chem. Commun.*, 346 (1989).

4. B. Carre, J. Devynck, *Anal. Chim. Acta, 159*, 149 (1984).

5. R. Adrien, Ph.D. Thesis, University of Melbourne (1992).

6. C.G. Barraclough, R.W. Cockman, T.A. O'Donnell, W.S.J. Schofield, *Inorg. Chem., 21*, 2519 (1982).

7. M. Sørlie, H.A. Øye, *Inorg. Chem., 20*, 1384 (1981).

8. R. De Waele, L. Heerman, W. D'Olieslager, *J. Less-Common Metals, 122*, 319 (1986).

9. R.M. Friedman, J.D. Corbett, *Inorg. Chem., 12*, 1134 (1973).

10. R.C. Burns, R.J. Gillespie, W.-C. Luk, *Inorg. Chem., 17*, 3596 (1978).

11. R.D.W. Kemmitt, M. Murray, V.M. McRae, R.D. Peacock, M.C.R. Symons, T.A. O'Donnell, *J. Chem. Soc. (A)*, 862 (1968).

12. J. Besida, T.A. O'Donnell, *Inorg. Chem., 28*, 1669 (1989).

13. R.J. Gillespie, M.J. Morton, *Quart. Rev., 25*, 553 (1971).

# CHAPTER 10

# Earlier Reactions Revisited: Future Synthetic Strategies

The experimental data presented in chapters 4–7 on synthesis and stability of polyatomic and monatomic cations in media of very low basicity were used to provide a summary in chapter 9 of the acid-base principles that govern the stabilization of such cations. The principles are now to be used in two contexts in chapter 10.

First, there will be presented some previously published reactions for which better interpretations of the experimental observations are now available as a result of the generalizations set out in chapter 9. The reactions cited will be representative examples chosen to illustrate different interdependent principles and will not constitute any attempt at an exhaustive list. References used in this chapter will be to earlier sections of the book rather than to original literature sources because the additional experimental data necessary to support the arguments presented below are available in those sections.

Second, there will be a brief account, based on reassessment of earlier reactions and on other principles established throughout this book, of suggestions for improvements in the experimental approach to synthesis of new and existing monatomic and polyatomic cations in acidic media.

## 10.1 Earlier Reactions Revisited

Some examples are to be presented below of work on generation of cations that has been limited in its potential scope because of the experimental procedures that were adopted or for which the interpretations of the observations given at the time are now believed to be presented better in terms of equilibria governed by the acid-base principles developed in this book.

## 10.1.1 Nonspecificity of Reagents

In much reported work, confusion about interpretation of observed reactions can occur because a single compound has been used as a Lewis acid, an oxidant, and a precipitant (or the source of the counteranion used to isolate the generated cation). A particularly representative example of a compound with such a multiple role is $SbF_5$, which, as a very strong Lewis acid, has been used in many syntheses to decrease the basicity of media and to act as an oxidant. In the latter role, it produces $SbF_3$ on reduction. Additionally it can accept $F^-$ to form $SbF_6^-$ or $Sb_2F_{11}^-$ as a counterion. In this context clean syntheses have frequently been made difficult or impossible by coprecipitation of the $SbF_3$ or by mixed-oxidation-state anion formation between $SbF_5$, $SbF_3$, and $F^-$. Reactions that lead directly to solid residues yield contaminated products because of the very low volatility of $SbF_3$.

Edwards's exellent crystallographic work that led to the structure of the compound $Br_2^+ Sb_3F_{16}^-$ gives a nice example of the preparative difficulties in using a single compound as acid, oxidant, and precipitant and of a resolution to it. It was shown in section 4.1.2.1 that McRae had identified $Br_3^+$, characterized by its UV-visible spectrum, in the products resulting from the reaction between $Br_2$ and $SbF_5$ shown in equation 4.8. Because of contamination of the compound $Br_3^+ Sb_3F_{16}^-$ by involatile $SbF_3$ no single compound could be isolated from the reaction mixture for structural analysis. Edwards and co-workers used the much stronger oxidant $BrF_5$ instead of $SbF_5$ in the reaction with $Br_2$ and $SbF_5$ shown in equation 4.9 and obtained a more highly oxidized cation and a clean sample free of $SbF_3$ for structural analysis of the $Br_2^+$ compound. It should be noted that the Edwards group found that they had to adhere very strictly to the stoichiometry of equation 4.9. If even small excesses of $BrF_5$ were used, the very strong oxidant caused oxidation of $Br_2$ beyond the oxidation state 0.5.

There are many examples in the literature where the presence of $SbF_3$, remaining in the reaction mixture after reduction of $SbF_5$, led to products that could not be characterized adequately. Occasionally, as in the isolation of $(Te_2Se_2)(Sb_3F_{14})(SbF_6)$ after oxidation of equimolar Te and Se with $SbF_5$ in $SO_2$ (see Table 4.4), the residual $SbF_3$ formed a discrete anion and characterization of the cation was possible.

Although a weaker Lewis acid than $SbF_5$, $AsF_5$ provides fewer problems as an oxidant. The reduction product $AsF_3$ is very volatile and is very weakly basic. It does not exhibit fluoride-donor capability toward even $SbF_5$. Therefore it does not interact to any observable extent with cationic species. The Gillespie and the Passmore groups have used it very effectively as an oxidant in the successful isolation of many solids containing unusual cations, for example, $Cl_3^+ AsF_6^-$ (section 4.1.3), $I_3^+ AsF_6^-$ (section 4.1.1.3), polyselenium (section 4.2.2.2), and polytellurium (section 4.2.3.2) cations and several heteropolyatomic cations of chalcogen elements (Table 4.4). The reduction product $AsF_3$ is sufficiently weakly basic to have been used as the solvent medium for the successful synthesis of many cationic species, for example, $Cd_3^{2+}$ and $Cd_4^{2+}$ (section 5.2) and several heterocations in Table 4.4.

The dual role of $SO_3$ in acting as an oxidant of Te in oleum and in simultaneously increasing the acidity of the oleum are presented in section 10.1.4.

In the chemistry of molten chloroaluminates, $AlCl_3$ frequently plays a dual role, about which there is much confusion. It both determines the acidity of the melt and provides the basis for an anion for isolation of the cation under investigation. The limitations of its use as an acid and as the precursor to a counter anion are discussed in section 10.1.2.

## 10.1.2 Necessity for Control of Acidity Levels

Dependence of the stability of unusual cationic species on the level of acidity of the media is now becoming fairly well recognized for synthetic reactions in protonic superacids, if only as the result of the collection of a large body of reported syntheses in a variety of superacids, each studied over a wide range of acidity and basicity, as in Tables 4.2 and 4.3. However, as will be illustrated later in this section, there has been little exploitation of the possibility of synthesis of new polyatomic cationic species in chloroaluminate melts by deliberately enhancing the acidity of the media.

The detailed study by Besida and O'Donnell on the levels of acidity in HF at which $I_5^+$, $I_3^+$, and $I_2^+$ are stable (section 4.1.1.1) underpins observations such as those of Gillespie whose work reported in the same section showed that if he took the appropriate stoichiometric reacting proportions of the oxidant $S_2O_6F_2$ with $I_2$, he generated $I_5^+$ and $I_3^+$, which were both stable indefinitely in $HSO_3F$, made very slightly basic by the formation of $SO_3F^-$, the reduction product of $S_2O_6F_2$, and the base of the solvent system. The fact that each should be stable in a medium of the same acidity after being prepared in this way has been dealt with in section 9.7. However, Gillespie found that the more highly oxidized cation $I_2^+$ was unstable in this medium, disproportionating to $I_3^+$ and $I(SO_3F)_3$. It was necessary to increase the acidity of the medium by addition of $SbF_5$ to obtain stable solutions of $I_2^+$.

There are many examples in chloroaluminate chemistry to show that investigators failed to exploit the possibility of obtaining new polycations by increasing the acidity. In the earliest reports on formation of $I_3^+$ and $I_5^+$ in melts, $I_2$, $ICl$, and $AlCl_3$ were melted in the correct stoichiometric proportions to produce the compounds $I_3AlCl_4$ and $I_5AlCl_4$ (section 4.1.1.2). The melts so produced were neutral—they contained neither excess $AlCl_3$ nor excess $Cl^-$. Later work by Mamantov is described in that section, in which Mamantov gave strong Raman spectroscopic evidence for the existence of $I_2^+$ in the acidic melt 63 mol% $AlCl_3$—37 mol% $NaCl$; but there does not appear to be any report of isolation of a solid chloroaluminate of $I_2^+$ from such a melt.

It was shown in section 4.2.4 that polysulfur cations require higher acidities for stabilization in protonic superacids than the corresponding cations of selenium and tellurium. The limitation of working in neutral melts is illustrated by the fact that no polysulfur cations have been isolated from chloroaluminate melts (section 4.2.1.2) whereas $Se_8(AlCl_4)_2$ and $Se_4(AlCl_4)_2$ (section 4.2.2.3) and $Te_4(AlCl_4)_2$ and $Te_4(Al_2Cl_7)_4$ (section 4.2.3.3) were isolated from neutral melts. Enhanced acidity would be required in quests for compounds of $Te_6^{4+}$ and the sulfur cations from melts.

After many earlier inconclusive investigations of the nature of species in Cd—CdCl$_2$ melts, Corbett showed that reduction of CdCl$_2$ by Cd reached a maximum in acidic chloroaluminate medium, when the ratio AlCl$_3$:CdCl$_2$ in the melt was 3:1 or greater (section 5.2), and he isolated Cd$_2$ (AlCl$_4$)$_2$ from such melts. It will be proposed in section 10.2.2.3 that Cd$_3^{2+}$ and Cd$_4^{2+}$, which have been formed by oxidation of Cd by AsF$_5$ in AsF$_3$, might be able to be identified or isolated from melts that are sufficiently acidic.

Section 5.4 shows that Corbett was able to isolate Bi$_5$(AlCl$_4$)$_3$ from neutral NaAlCl$_4$ but that Bi$_8$(AlCl$_4$)$_2$ obtained from this medium was not pure and needed additional AlCl$_3$ to make it acidic. Later work from the Krebs group, reported in that same section, demonstrates that, in order to obtain "good" crystals for crystallographic purposes, the melt for Bi$_5$(AlCl$_4$)$_3$ isolation needed to contain an excess of about 10 mol% AlCl$_3$ over the neutral composition and that the more highly reduced cation in Bi$_8$(AlCl$_4$)$_2$ needed a 33 mol% excess of AlCl$_3$.

## 10.1.3 Competitive Acid-Base Equilibria

The synthetic methods as reported for preparation of pure samples of [I$_2^+$][Sb$_2$F$_{11}^-$] (section 4.1.1.3) and [Br$_2^+$][Sb$_3$F$_{16}^-$] (section 4.1.2.1) give interesting examples of the way in which base strength of the medium can determine the speciation of the product of reaction. Aubke and co-workers reacted I$_2$ and Br$_2$ directly with S$_2$O$_6$F$_2$ in the precise stoichiometric proportions to produce I$_2^+$SO$_3$F$^-$ and Br$_2^+$SO$_3$F$^-$. Instead, they observed the expected disproportionation products for the cations, ISO$_3$F and I$_3^+$SO$_3$F$^-$, in the first instance and Br$_2$ and BrSO$_3$F in the second. It is consistent with the relative stabilities of I$_2^+$ and Br$_2^+$ in media that are not extremely highly acidic that, at that acidity level, namely, in the presence of SO$_3$F$^-$, I$_2^+$ would disproportionate to the lower-oxidation-state cation I$_3^+$ whereas Br$_2^+$ would disproportionate all the way through to the element.

When they treated each set of reaction products with a vast excess of SbF$_5$ and removed all volatile products, the residues were I$_2^+$Sb$_2$F$_{11}^-$ and Br$_2^+$Sb$_3$F$_{16}^-$. The authors reported this second stage in each synthesis as occurring because of solvolysis of SO$_3$F$^-$ to a fluoroantimonate anion. This solvolysis is probably quite incidental. The main driving force for the acid-base equilibria involving I$_2^+$, Br$_2^+$, their disproportionation products, SO$_3$F$^-$, and fluoroantimate anions, as set out in equations 4.5 and 4.10, appears to be that I$_2^+$ and Br$_2^+$ would be stable in the presence of the very weakly basic anions Sb$_2$F$_{11}^-$ and Sb$_3$F$_{16}^-$ but would disproportionate in the presence of the less-basic anion SO$_3$F$^-$. In terms of the reported syntheses, although oxidant-reductant stoichiometry favored formation of I$_2^+$ and Br$_2^+$, these cations disproportionated in the presence of SO$_3$F$^-$; but, in the far-less-basic environment of first SbF$_5$ and then the fluoroantimonates, conproportionation of the initial products occurred.

A similar basis to that outlined earlier for stability of cations appears to hold for the reported stability of polybismuth cations in a range of attempted syntheses. As reported in section 5.4, Burns et al. studied the possible oxidation of metallic Bi to polyatomic cations by $PF_5$, $AsF_5$, $SbF_5$, $SbCl_5$, $HSO_3F$, and $HSO_3Cl$ in the solvent $SO_2$ and reported, "Interestingly, the cations $Bi_8^{2+}$ and $Bi_5^{3+}$ were only observed when the counteranions were derived from the group 5 pentafluorides. It is somewhat surprising that these cations, or perhaps others, were not formed as $SbCl_6^-$, $SO_3F^-$ and $SO_3Cl^-$ salts." Indeed, they reported that oxidation with $SbCl_5$, $HSO_3F$, and $HSO_3Cl$ led to white crystalline products, admixed with metal. It seems that the anions $SbF_6^-$, $AsF_6^-$, and $PF_6^-$ were sufficiently weakly basic—the last-named marginally so—to stabilize polybismuth cations, but that in the presence of the more basic anions $SbCl_6^-$, $SO_3F^-$, and $SO_3Cl^-$ disproportionation of any potential cations to Bi and white Bi(III) compounds occurred. A cycle of reoxidation of Bi and disproportionation could lead ultimately to formation only of white compounds of Bi(III), depending on experimental conditions as discussed in section 10.1.4.

## 10.1.4 Disproportionation and Reoxidation

There have been many examples throughout this book, some of them touched on in section 9.5, where the ultimate course of a reaction can often be explained in terms of intermediate steps involving disproportionation to an element and its subsequent oxidation, this process occurring cyclically. One such example could also have been dealt with in section 10.1.1. It involves the dual role of $SO_3$ in oleums in acting both as an oxidant and as a Lewis acid to enhance the acidity of the medium.

In section 4.2.3.1 work by the Gillespie group on generation of $Te_6^{4+}$ was cited. They report that Te is oxidized to $Te_4^{2+}$ by $SO_3$ in weak oleums, $SO_2$ being formed, and that $Te_4^{2+}$ is "oxidized" to $Te_6^{4+}$ by increasing the $SO_3$ content of oleums. The role of $SO_3$ in increasing the acidity of the medium and stabilizing $Te_6^{4+}$ is probably much more important than its role as an oxidant—even in relatively dilute oleums there should be sufficient $SO_3$ to act solely as an oxidant for the amount of Te present. They say that $Te_4^{2+}$ and $Te_6^{4+}$ can exist in 100% $H_2SO_4$ but that a precipitate of $TeO_2$ is formed on prolonged standing, a process that is accelerated by heating. This complex system involving Te, $Te_4^{2+}$, $Te_6^{4+}$, $TeO_2$, $H_2SO_4$, and $SO_3$ can be rationalized in terms of a series of disproportionation equilibria leading ultimately to Te and $TeO_2$ and cyclic reoxidation of Te. $SO_3$, acting both as an oxidant and as a Lewis acid, will oxidize Te through $Te_4^{2+}$ to $Te_6^{4+}$, a sequence that will be favored on acid-base grounds as the $SO_3$ content of the oleum is increased. $SO_3$ is reported as ultimately oxidizing $Te_6^{4+}$ to $TeO_2$ on long standing or heating.

It seems reasonable to postulate that in oleums—media of moderately high acidity—$Te_6^{4+}$ is involved in disproportionation equilibria, governed by acidity levels, with $Te_4^{2+}$ and $TeO_2$ and that $Te_4^{2+}$ so formed can be reoxidized by $SO_3$ to $Te_6^{4+}$ which can yield more $Te_4^{2+}$ and $TeO_2$. It is reported that both $Te_6^{4+}$ and $Te_4^{2+}$ can exist in $H_2SO_4$ but disproportionate. Again it is reasonable to assume that $Te_4^{2+}$ is involved in another disproportionation equilibrium, in very small part at least, with $Te_6^{4+}$ and Te. The elemental Te would be oxidized by $SO_3$ via $Te_4^{2+}$ and $Te_6^{4+}$ disproportionations to $TeO_2$. On long standing or heating, a series of cyclic disproportionations would ultimately favor the formation of $TeO_2$ as the principal product.

Another example from superacid chemistry in which cyclic disproportionation and reoxidation explains the experimental observations has been cited at the end of section 5.4 where it was reported that the $Bi_5^{3+}$ cation was found to be slowly oxidized in 100% $H_2SO_4$, as indicated by loss over several hours of the characteristic color of the cation and formation of a white precipitate containing only Bi(III). $Bi_5^{3+}$ would disproportionate to very finely divided metallic bismuth and a Bi(III) compound after passing through progressively less highly charged polyatomic cations of Bi, which, with decrease in charge, would become progressively more prone to disproportionation. Bi, formed in the first and subsequent disproportionation equilibria, would be continuously oxidized through polyatomic cations until a Bi(III) compound was the only product.

There are examples in molten salt chemistry where interpretations based on cyclic disproportionation and reoxidation appear to give better explanations of the observed data than those presented in the original literature. The reported stability or instability of U(III) in acidic, neutral, and basic chloroaluminates described in section 7.2.5 provides what is probably the most compelling example. Morrey and co-workers prepared several ampules each containing a small amount of metallic uranium and large comparable amounts of metallic aluminium and of chloroaluminates, present after reaction as immiscible phases. Across the series of ampules, the melt acidity was varied from quite strongly acidic ($AlCl_3$:KCl, 2:1) through neutral (1:1) to basic (0.4:1). Initially, the U was oxidized in situ to U(III) by the chloroaluminate. The ampules were heated "at the desired temperature for a time sufficient to ensure attainment of equilibrium" and then quenched. In each case, the resulting solid metallic and salt phases were analyzed for uranium and the ratio of uranium in the metallic phase to that in the chloroaluminate salt phase was plotted against melt acidity-basicity as shown in Figure 7.8. It was observed that the formation and transfer of metallic uranium to the metallic aluminium phase was very much greater for ampules where the melts were nearly neutral than for those where the melts were definitely acidic or basic.

In the original papers the dependence of the proposed reduction of U(III) by Al on melt acidity was interpreted simply in terms of equilibria between the anions of the melt $Al_2Cl_7^-$, $AlCl_4^-$, and $Cl^-$ and the U(III) species, postulated as being anionic chlorouranate(III) across the whole acidity-basicity range. Their explanation is at odds with the previously known fact that U(III) in chloroaluminates is not reduced by Al. Now that there is good evidence for U(III) being cationic as $U^{3+}$ in acidic melts and anionic in basic melts, as shown in Figures 7.5 and 7.7 (sections 7.2.4 and 7.2.5), it seems better to postulate that each of these species is reasonably stable in its host environment, that is, in acidic or in basic melt, but that $U^{3+}$ disproportionates to $U(0)$ and $U(IV)$ in melts that are nearly neutral. $U(IV)$ is known to be reduced by Al to $U(III)$, which would then disproportionate in nearly neutral melt. Cyclic disproportionation and reoxidation in the melt would lead to a large transfer of metallic U to the molten Al layer where it would be protected to a large extent from reoxidation to $U(III)$ by the melt.

Some experimental observations of redox reactions in uranium(III)—melt chemistry are open to a much more speculative reinterpretation based on disproportionation processes than that offered by the original authors. It was reported in section 7.2.5 that Belgian workers observed spectra that were characteristic of $U^{3+}$ in acidic room-temperature melts based on $AlCl_3$—butylpyridinium chloride (BuPyCl). They reported that, as they made the melts more basic by addition of BuPyCl, spectra suggestive of $U(IV)$ species were observed and the melts became strongly blue in color, a change attributed to reduction of the $BuPy^+$ cation to a previously characterized radical species. They suggested that anionic chlorouranate(III) is a sufficiently strong reductant such that the $BuPy^+$ cation of the melt is reduced. An alternative explanation, that would need experimental verification, is that as the acidic melt was made progressively more basic on addition of BuPyCl, $U(III)$ would disproportionate to $U(IV)$ and the very powerful reductant, finely divided metallic U, which would then reduce the $BuPy^+$ cation.

## 10.1.5  Summary

The few examples cited and discussed in the last two sections of this chapter indicate that the concept of base-induced disproportionation of cationic species may be more important and much more widely applicable than has been realized to date.

As presented in section 10.1.3, Aubke's very reliable experimental observations show that, in the presence of $SO_3F^-$, a weak base but one of significant strength, the products of reactions designed to generate $I_2^+$ and $Br_2^+$ were, instead, different. They were the disproportionation products to be expected from those cations. The very weakly basic counteranions, the fluoroantimonates, were required to provide stable solid compounds of $I_2^+$ and $Br_2^+$. While reversibility of disproportionation was not demonstrated, Gillespie's co-workers showed that polybismuth cations were stable in the presence of the very weakly basic fluoroanions derived from group V pentafluorides but disproportionated in the presence of somewhat more basic anions.

Cyclic disproportionation and oxidation of disproportionation products was used in section 10.1.4 to provide more adequate rationalization of $U(0)$—$U(III)$—$U(IV)$ reactions in high-temperature chloroaluminates than those provided by the experimentalists who made the observations. The same proposed processes were used above somewhat more tentatively to discuss U(0)—U(III)—U(IV) reactions in room-temperature melts and the stabilities of polytellurium and polybismuth cations in $H_2SO_4$.

Sometimes quite elaborate explanations have been put forward in discussing the stability of unusual cations when a simple explanation employing the concept of base-induced disproportionation seems to suffice or to be superior.

A case in point has been the attempted rationalization of the stability of $Cd_2^{2+}$ cations in acidic chloride melts at the expense of the related entities Cd and $CdCl_2$. In section 5.2 Corbett is quoted as saying that $Cd_2^{2+}$ is formed as a result of reduction of $CdCl_2$ by Cd in acidic melts because charge-transfer complexes held together by a chloride bridge are destroyed when chloride is replaced by the much-less-basic $AlCl_4^-$ ion. Corbett had given another view, also quoted in section 5.2, to the effect that the relative lattice energies of the cations $Cd_2^{2+}$ and $Cd^{2+}$ in combination with anions $Cl^-$ and $AlCl_4^-$ might lead to an energetic favoring of $Cd_2(AlCl_4)_2$ at the expense of $Cd_2Cl_2$, whereas $CdCl_2$ would be expected to form at the expense of $Cd(AlCl_4)_2$. As stated in section 5.2, it seems to be sufficient to postulate that, regardless of the nature of the counteranion, $Cd_2^{2+}$ forms by conproportionation of $Cd^{2+}$ and Cd in acidic melts and disproportionates to Cd and $CdCl_2$ on interaction with $Cl^-$ in basic melts, as in the simple equilibrium set out as equation 5.2.

## 10.2  Future Synthetic Strategies

To a large extent section 10.1 has been used for a critical examination of previously reported methods of synthesis of cations in low and fractional oxidation states and, in several cases, for modification of the interpretations of observed experimental data, as put forward by earlier authors of research papers. This followed a codifying in section 9.8 of some general principles which had been developed in chapters 4–7 for stabilities of cations in media which were highly acidic—or, more specifically, of very low basicity. Sections 10.1.1, 10.1.2, and 10.1.3 pointed to the need for appropriate control of media acidity and basicity levels and to the need for specificity in selection of reagents in designing synthetic procedures for generation of stable solutions or compounds containing unusual cations.

This section will be used to provide a generalized, more reliable approach to future synthesis of inorganic cationic species than has occurred in much earlier work. The necessity will be demonstrated for selection of the most suitable *combination* of experimental conditions, for example, deliberate choice of the chemical nature of an acidic reaction medium and of the oxidant-reductant stoichiometry, as well as the level of acidity or basicity needed to stabilize the required species in the medium and then, if necessary, to isolate from solution pure solid compounds containing these species.

## 10.2.1  Choice of Reaction Media

It should now be obvious that the most important requirements for synthesis of any particular cation in solution are the selection of a medium of sufficient acidity to sustain the synthetic reaction and the "fine tuning" of the acidity level in that medium by deliberate use of Lewis acids or bases of the solvent system. This will be discussed in section 10.2.2. However, other properties of acidic reaction media are also important in any synthetic strategy. The physical, redox, and self-ionization properties can be influential in the quest for a particular cation.


### 10.2.1.1  Physical Properties of the Medium

All of the protonic superacids are of sufficiently high dielectric constant such that no serious problems such as ion-pairing are encountered, although high levels of self-ionization, as for $H_2SO_4$, can seriously reduce the reliability of conductance measurements, as was shown in section 2.2.

Melting and boiling points, and to a less extent viscosities, affect ease of manipulation of the solvent superacids. Generally, vacuum systems provide the best experimental procedures for protecting media and products from atmospheric and other adventitious moisture, which would lead to impurities in, or total decomposition of, the wanted cations. High-boiling $H_2SO_4$ is very difficult to handle in a vacuum system whereas HF, boiling at 19.51°C, is very easy, providing the components of the vacuum system are themselves fluoride-resistant. (See section 1.1.3.5.) HF, which exerts a vapor pressure close to 1 atm at room temperature, does not present many of the experimental difficulties encountered in handling other very volatile nonaqueous solvents such as HCl and $SO_2$ with much lower boiling points. $HSO_3F$ and $CF_3SO_3H$, having boiling points between those of $H_2SO_4$ and HF, can be distilled reasonably easily in greaseless, glass vacuum systems, but $HSO_3F$ dissociates fairly easily to $SO_3$ and HF.

Reliable cryoscopic techniques have been developed for each of the protonic superacids and the low melting points of HF, $HSO_3F$, and $CF_3SO_3H$ allow observation of NMR spectra with much better resolution than is possible for room-temperature spectra, as was demonstrated for solutions of $SbF_5$ in HF (section 3.4.1).

For obvious reasons, spectroscopic and electrochemical observations are easier to make in room-temperature chloroaluminates, based on organic cations, than in high-temperature melts such as $NaCl—AlCl_3$, and general manipulation is easier. However, there are trade-offs. Room-temperature melts cannot usually be made as acidic as those based on alkali chlorides and ease of oxidation or reduction of the organic cations, relative to that of alkali metal cations, puts greater limits on the usable potential range.

## 10.2.1.2 Redox Properties of Media

The cations presented in chapters 4, 5, and 7 all contain metals or nonmetals in unusually low oxidation states. There was an earlier discussion (section 9.2) of the ability, or lack of it, of the various solvents water, the protonic superacids, and chloroaluminate melts to support these highly reducing species without undergoing reduction themselves. $SO_3$ in oleums was seen as being able to play two roles in cation synthesis. It can increase the acidity of $H_2SO_4$-based media and can act as a strong oxidant.

Neat $HSO_3F$ exists in equilibrium with small concentrations of HF and $SO_3$, its spontaneous decomposition products. Because of this, $HSO_3F$ is a reasonably strongly oxidizing solvent. Added $SO_3$ has been shown to increase the acidity of solutions of $SbF_5$ in $HSO_3F$ by forming anions such as $[F_3Sb(OSO_2F)_3]^-$ and $[F_2Sb(OSO_2F)_4]^-$. Lewis acids such as $SbF_5$ and $AsF_5$ probably act as "sinks" for $SO_3$ in superacidic media based on $HSO_3F$, so that formation of low-oxidation-state cations in acidified $HSO_3F$ is probably favored both by increase in Lewis acid concentration and decrease in $SO_3$ availability.

Strong reductants can reduce S(VI) in $CF_3SO_3H$ to compounds containing S(IV), but $CF_3SO_3F$ appears to be much less strongly oxidizing than $HSO_3F$. For example, U(III) is stable in acidified $CF_3SO_3H$ but not in acidified $HSO_3F$ (section 7.2.3). Of the four protonic superacids discussed in this book, HF is the most resistant to reduction and oxidation—the one with the longest usable potential range. It is able to support reducing cations like $Ti^{2+}$, $Sm^{2+}$, and oxidizing anions like $NiF_6^{2-}$ and $AgF_6^{3-}$ in stable solutions. As suggested at the end of the preceding section, alkali-metal chloroaluminates have much longer usable potential ranges than those chloroaluminate room-temperature liquids based on organic cations.

## 10.2.1.3 Self-Ionization of Solvents

Values of the ionic-product constants for $H_2SO_4$, $CF_3SO_3H$, $HSO_3F$, and HF were discussed in chapter 1, and differences in both the absolute acidities of the four pure solvents and in their different self-ionizations were summarized in Figure 2.4. In that figure, pure HF and $HSO_3F$ are seen to be of virtually identical acidity, with pure $CF_3SO_3H$ significantly less acidic and pure $H_2SO_4$ very much less so.

Self-ionization processes are simple for HF (equation 1.34), $HSO_3F$ (equation 1.29), and $CF_3SO_3H$ (equation 1.28), giving in each case the solvated proton and the base of the system. The ionic product for HF is less than $10^{-12}$ $mol^2$ $kg^{-2}$ with the values of about $10^{-8}$ for $HSO_3F$ and for $CF_3SO_3H$. The differences are indicated by the relative extents of the "vertical" sections of the plots of $H_0$ vs. concentration of base and acid for each solvent system in Figure 2.4. By comparison with these three acids, $H_2SO_4$ exhibits a self-ionization that is both complex and extensive, as indicated by equations 1.23 and 1.24 and by equilibrium constants associated with them, and by the very flat plot of $H_0$ vs. concentration of base and acid for $H_2SO_4$ in Figure 2.4.

In one sense, these properties make $H_2SO_4$ a very good solvent for synthesis of cations. The medium is strongly buffered against change in $H_0$ caused by adventitious impurities or by basic anionic species which are usually produced in cation syntheses. However, the value of $H_0$ for $H_2SO_4$ is not nearly as negative as those for the other three superacids and the extensive self-ionization means that large concentrations of the base $HSO_4^-$ are available to cause disproportionation of desired cations.

In proceeding through the series of superacids $CF_3SO_3H$, $HSO_3F$, and HF, the neat media become progressively more acidic and self-ionization decreases markedly, so that in a fundamental sense, conditions for generation of low-oxidation-state cations become progressively more favorable. The availability from the solvent self-ionization of basic species that can cause disproportionation becomes less.

However, two other factors must be considered. First, the bond strength between fluoride and the element of the cation is likely to be very much stronger than bond strengths between the same element and the bases $CF_3SO_3^-$ and $SO_3F^-$, as discussed toward the end of section 9.8, and dependence of stabilities of individual cations will not depend only on absolute values of $H_0$ for different superacids. Second, the smaller the self-ionization, the more it will be disturbed and displaced by adventitious impurities such as $H_2O$ or by bases produced in the synthetic reaction. It was reported in section 2.4.1 that distilled HF stored in KelF containers will have an experimentally measured value of about $-11$ for $H_0$ because of impurities present, whereas $H_0$ for pure HF is $-15.1$.

This last point emphasizes the necessity for the use of added Lewis bases and acids to a solvent such as HF to fix the $H_0$ value at any required value and to arrange for a reasonable level of buffer capacity in the reaction medium. This will be presented in detail in section 10.2.2.

## 10.2.1.4  Chemical Complexity of Media

Some of the material which would be appropriate here has been foreshadowed in the preceding section. If neutral or acidic $H_2SO_4$ were being considered as a reaction medium, its very complex self-ionization would need to be taken into account, as well as the possible effect of the presence of the strong oxidant $SO_3$. At the other end of the scale, the "nonoxidizing" HF has a small, simple self-ionization. Offsetting these considerations would be the fact that there is a high degree of acid-base buffering in $H_2SO_4$ and very little in HF, so that acidity adjustment and control, as set out in section 10.2.2 below, would be vital in syntheses in HF calling for well-defined acidity levels.

A conscious effort has been made throughout this book to generalize the solution chemistry in highly acidic media by referring regularly to more familiar aqueous systems. The protonic superacids, HF in particular, offer a simplicity of cation speciation which is not always available in aqueous solution chemistry. The actual self-ionization is very similar in HF and $H_2O$ and the base of each solvent can coordinate to greater or less extent, depending on acidity levels, with cations to form species such as $[M(OH)_m(OH_2)_{6-m}]^{n+}$ and $[MF_m(FH)_{6-m}]^{n+}$. There is some polarographic and potentiometric evidence[1a] to suggest that complexation of metal cations by $F^-$ in HF solution is significantly less than that by $OH^-$ in $H_2O$.

However, the larger factor leading to more complex cationic speciation in water than in HF is associated with the ability of water to lose first one proton to form $OH^-$ and then a second to form $O^{2-}$. HF can undergo only a single deprotonation. This means that there are no equivalents in HF solution chemistry of the complex oxohydroxospecies that occur even in acidic $H_2O$ for many heavy metals. An illustration of the differences between the two solvents is the simplicity of the voltammetry of tungsten(VI) and molybdenum(IV) in HF compared with that in $H_2O$.[1b]

There have been several examples presented in chapter 6 where spectra in HF are better resolved and show sharper peaks than their counterparts in $H_2O$. This is best illustrated for spectra of lanthanides and actinides which are based on the sharp peaks resulting from $f$-$f$ transitions, for example, in Figures 6.6, 6.7, and 6.8. Similar considerations should apply to comparisons of cation speciation in water with that in protonic superacids other than HF, but to date there are few experimental data to support this proposition.

Room-temperature chloroaluminates also provide media for "clean" spectra. In this case, the media themselves are very highly ionized, perhaps fully; but the ions derived from the solvent are very weakly coordinating. Seddon[2] has compared the better resolution of spectra for the single species $[CoCl_4]^{2-}$ in basic chloroaluminates with corresponding aqueous spectra, noting that there is evidence that, even in a solution of $CoCl_2$ in 12 M HCl, $[CoCl_4]^{2-}$, $[CoCl_3(OH_2)]^-$, and $[CoCl_2(OH_2)_2]$ coexist in equilibrium. He reports spectra for species like $[Mo_2Cl_8]^{4-}$ in basic chloroaluminates. It would be interesting to see whether corresponding cations like $[Mo_6Cl_8]^{4+}$ exist in acidic chloroaluminates or whether acidic media favor the formation of low-oxidation-state monatomic cations.

## 10.2.2 Adjustment of Acidity Levels

In the introduction to a general section on choice of reaction media (10.2.1), it was stated that the most important requirement for synthesis of a particular cation in a very low or fractional oxidation state would be choice of a medium having sufficient acidity to sustain the synthetic reaction and having the acid-base properties to allow the necessary "fine tuning" of the acidity level in that medium. This section will consider the acid-base and other characteristics of Lewis acids used in acidity-level adjustment.

## 10.2.2.1 *Lewis Acids in Hydrogen Fluoride*

Far more systematic, quantitative work has been done on Lewis acidity in HF than in the other protonic superacids or in acidic melts. Because of this, it was possible for Besida and O'Donnell to carry out the investigation of the dependence of stability of the different polyiodine cations on acidity levels (section 4.1.1.1) which is the most detailed study of its type yet reported. In that work, acidity levels were adjusted to within 0.2 of a Hammett function unit ($H_0$), where required.

As stated earlier in section 10.2.1.3, neat HF, as distilled and stored, will have a value for $H_0$ much lower than that for pure HF, which is $-15.1$. Residual impurities after distillation and as leached from storage vessels cause "best-available" HF to have $H_0$ values of about $-11$ or $-12$. Even the use of a Hammett function indicator and its method of introduction to the "neat" HF would cause measured $H_0$ values to be considerably less than $-15.1$. Such HF could be brought to a value about that for pure HF by using a $NbF_5/NbF_6^-$ buffer or it can be fixed accurately at $H_0$ values between about $-22$ and $-9$. It can be set approximately at levels outside these limits by using large amounts of $SbF_5$ or NaF, but reliable $H_0$ values are not available for these concentrated solutions.

Formal buffering by using a particular Lewis acid fluoride and its conjugate fluoroanion is usually not necessary, as the most likely entities entering the reaction system are impurities such as $H_2O$ or reaction products other than the cations sought. These will usually be bases of the solvent system. So it is necessary only to ensure that the individual Lewis acid (or base) used has sufficient buffer capacity to cope with small amounts of adventitious base in the solution.

In general, solid Lewis acid fluorides, which are easily weighed under drybox conditions, are more convenient for acidity adjustment than gases, although maintenance of a pressure of $BF_3$, $GeF_4$, or $PF_5$ may be desirable to hold the HF solution near neutrality. Of course, a Lewis acid like $BF_3$ might fix the acidity and be a reagent in its own right, as in the conversion of some metals and some fluorides suspended in HF to solvated cations with $BF_4^-$ as the counteranions.

The relative strengths of Lewis acid fluorides have already been presented in section 3.5, but considerations other than absolute strengths of different Lewis acids will also enter into choice of the appropriate acid for different syntheses. Where cations of low oxidation state are being sought, oxidant strength of the Lewis acid will be of enormous significance. $TaF_5$ and $BF_3$ are nonoxidants in an all-fluoride system. Thus $BF_3$ was used to increase the $H_2F^+$ concentration of HF for metallic U to be oxidized to $U^{3+}$ by $H_2F^+$ with evolution of $H_2$ (section 7.2.1). $NbF_5$ is an extremely weak oxidant and would be safe to use in almost any synthetic situation. *Molecular* $AsF_5$ and $SbF_5$ are strong oxidants, but, at moderate concentrations, $SbF_5$ is present in HF solution essentially as $SbF_6^-$, a very much weaker oxidant, and the strongly reducing $Ti^{2+}$ could be generated in solutions up to 3 m in $SbF_5$ (section 7.1.1). Beyond that concentration, free $SbF_5$ oxidized $Ti^{2+}$. At quite low concentrations, for example, 0.1 molal, $AsF_5$ is in HF solution at room temperature as $AsF_{11}^-$, $AsF_6^-$, and $AsF_5$ in equilibrium (section 3.4.2), a moderately strongly oxidizing solution. It was found that generation of $U^{3+}$ in HF—$AsF_5$ required presence in the systems of excess metallic U. Ti metal treated with HF—$AsF_5$ gave $Ti^{3+}$, not $Ti^{2+}$, and Cr gave Cr(III), not Cr(II).

Solubility in HF of compounds containing fluoroanions of Lewis acid fluorides will also affect choice of a Lewis acid in synthetic strategies. Compounds incorporating the anions $BF_4^-$ and $AsF_6^-$ are generally fairly soluble in HF—a property that may militate against their successful crystallization from HF or their recovery after distillation of excess HF. Compounds based on the usually observed anion $Ta_2F_{11}^-$ have been found to be extremely insoluble in HF and those containing the anion $Sb_2F_{11}^-$ and $Nb_2F_{11}^-$ are usually of small, intermediate solubility. The dimeric anions are usually observed in solids crystallized or precipitated from HF—$SbF_5$, HF—$NbF_5$, and HF—$TaF_5$ whereas solids from HF—$BF_3$ and HF—$AsF_5$ have monomeric anions.

Heavy metals are well known to bond to oxygen in systems containing the metals, halides, and oxides, particularly when the oxygen is present as water. This can be an advantageous factor. Lewis acids such as $SbF_5$ and $TaF_5$ can act as scavengers for traces of moisture—the most likely unwanted impurity in most HF solution syntheses. However, it also means that oxidetetrafluorides, like those of W and Mo which have received some study as potential Lewis acids in preparation of solutions for spectroscopy because of their low oxidant strength and lack of color, can be labile toward loss of bonded oxygen. A solution containing $WOF_4$ and $SbF_5$ in HF was found to form $WF_6$ gradually.

## 10.2.2.2  Lewis Acids in Other Superacids

As stated earlier, much less comprehensive and detailed work has been done on measurement of strengths and characterization of the products of interaction of Lewis acids with superacids other than HF than has been the case for the HF system. In making a choice of the Lewis acid to be used for cation synthesis in $H_2SO_4$, $HSO_3F$, or $CF_3SO_3H$, for example, similar considerations would apply as in the comparison of Lewis acids to be used in HF as presented in the section immediately above.

Thus, oxidant strength of the Lewis acid would be a vital consideration for the generation of metallic or nonmetallic cations in low or fractional oxidation states in these three acids. There was a detailed discussion in section 10.1.4 of the difficulties introduced into a study of the dependence of the stability of tellurium cations on acidity levels in oleums because of the oxidant strength of $SO_3$, which was available in excess. This could have been avoided by using the nonoxidizing Lewis acid $B(HSO_4)_3$ and stoichiometric amounts, calculated to produce individual cations, of a specific strong oxidant such as $S_2O_8^{2-}$, which will be referred to in section 10.2.3.2.

There is some limited evidence from the Melbourne work on synthesis of metal cations in $H_2SO_4$, $HSO_3F$, and $CF_3SO_3H$ that the formal binary compounds which are bisulfates, fluorosulfates, and triflates of transition metals are themselves sparingly soluble and polymeric, probably because of bridging of cations by multidentate anions. For these polymeric "binary" compounds, addition of Lewis acids—and of the Lewis bases of the solvent systems—has enhanced the transition-metal cation concentrations but has generally produced much more dilute solutions than for the corresponding procedure of adding Lewis acid fluorides or the base fluoride to HF solutions containing a suspended insoluble transition-metal fluoride or the metal itself.

Lewis acids, which are now known to be available for the three solvent systems $H_2SO_4$, $CF_3SO_3H$, and $HSO_3F$, have been presented in some detail in sections 3.1, 3.2, and 3.3. Limitations on the use of some of them in synthetic work are also now known. The complication of $SO_3$ acting both as a Lewis acid and as a strong oxidant in $H_2SO_4$-based systems has been referred to earlier. $B(HSO_4)_3$ appears to deserve more consideration in synthetic work than it has received to date.

$B(OSO_2CF_3)_3$, potentially the safest Lewis acid for use in $CF_3SO_3H$ because it is intrinsically nonoxidizing, is known to decompose slowly in solution at room temperature, the rate of decomposition appearing to increase with increased concentration of the Lewis acid. The only other Lewis acids that are known currently to be effective in $CF_3SO_3H$ are the three pentafluorides $SbF_5$, $TaF_5$, and $NbF_5$, which decrease in acid strength in that order. Use of $SbF_5$ because of its acid strength might have to be balanced against its oxidant strength, of which little is known in $CF_3SO_3H$.

$Ta(SO_3F)_5$, which would be expected to have minimal oxidant strength in $HSO_3F$, would appear to be the ideal Lewis acid of this system. It is comparable in strength with the binary or substituted fluorides of Sb(V) (section 3.3.2). Being a binary compound of Ta(V) and the base of the solvent system, it does not introduce extraneous anionic species when used to increase the acidity of $HSO_3F$.

## 10.2.2.3 Acidity in Chloroaluminates

The only acid to be considered for these systems is $AlCl_3$, whether we are dealing with high-temperature melts or room-temperature media. The main point to be made here is that there has been too little deliberate enhancement of acidity in most chloroaluminate studies. Most of the work has involved generation of polyatomic cations of iodine, tellurium, or cadmium in "neat," that is, neutral, mixtures of the element, its chloride, and $AlCl_3$ to form chloroaluminates of $I_3^+$, $I_5^+$, $Te_4^{2+}$, and $Cd_2^{2+}$. Much earlier work involved a difficult search for the species present when a metal was added to its fused chloride, as was done then for systems like Cd—$CdCl_2$ or Bi—$BiCl_3$ in differing stoichiometries.

As reported in section 5.1, $Hg_3^{2+}$ and $Hg_4^{2+}$ were synthesized by Gillespie and colleagues from $SO_2$ by using the single reagent $AsF_5$ as oxidant for Hg, as a medium of very low basicity, and as the source of the very weakly basic counterion $AsF_6^-$. In the same section, it was shown that Mamantov had isolated $Hg_3(AlCl_4)_2$ after reacting $HgCl_2$ with excess of both Hg and $AlCl_3$, that is, in acidic medium, and then removing Hg and $AlCl_3$ from the reaction medium by distillation. In that study Mamantov adduced voltammetric evidence that $Hg_3^{2+}$ was present in fairly acidic chloroaluminate ($AlCl_3$:NaCl, 65:35). It is possible that $Hg_4^{2+}$ could be generated by a conproportionation reaction between $HgCl_2$ and Hg or by controlled oxidation of Hg by $Cl_2$ (section 10.2.3.3) in very acidic melts, perhaps pure $AlCl_3$. Its isolation would present a separate problem, as discussed in section 10.2.4.3.

Similarly, $Cd_3^{2+}$, or even $Cd_4^{2+}$, which Gillespie and co-workers synthesized as for $Hg_2^{3+}$ and $Hg_4^{2+}$, might be formed in melts, and subsequently isolated deliberately by using the reaction mixtures Cd:$CdCl_2$, 2:1 or 3:1 in very acidic chloroaluminate or by controlled oxidation of Cd with $Cl_2$ or electrochemically using controlled oxidation of a Cd electrode or controlled reduction of $CdCl_2$ in very acidic melts.

Polyatomic cations of Hg, Cd, and Zn are formed with greater difficulty as the atomic number of the element decreases. Kerridge and Tariq gave Raman spectroscopic evidence for formation of some $Zn_2^{2+}$ when Zn and $ZnCl_2$ were reacted together (section 5.3). They reported greater reduction of $Zn^{2+}$ to $Zn_2^{2+}$ by Zn when the chloride acceptor $CeCl_3$ was present, that is, when the reaction mixture was more acidic. It is possible that solids containing $Zn_2^{2+}$ suitable for structural analysis could be obtained from acidic reaction media based on $AlCl_3$ using a suitable counteranion. A search for $Zn_3^{2+}$ in highly acidic chloroaluminates would be worthwhile.

The need to increase the acidity of chloroaluminate media in order to obtain stable polyatomic cations with decreasing fractional oxidation states was demonstrated for polybismuth cations in section 5.4 and 9.3.2. Crystallization of the cation with most reduced charge, $Bi_8^{2+}$, was favored from a chloroaluminate melt containing a 33 mol% excess of the acid $AlCl_3$.

## 10.2.3  Choice of Chemical Oxidants and Reductants

If a very strong oxidant is being used for generation of a particular cation in an acidic medium, two experimental requirements must be met. First, the medium must be sufficiently acidic—or nonbasic—to be able to stabilize the cation. It must not be below a certain threshold of acidity; otherwise disproportionation of the cation will occur. It can, of course, be above that threshold. Then, the strong oxidant, for example, $F_2$ in HF or $Cl_2$ in an acidic chloroaluminate, must be used in the correct stoichiometry to generate the desired cation.

If, on the other hand, a conproportionation reaction is used, that is, there is mutual oxidation-reduction between the formal disproportionation products of the cation under investigation, the nature of the cation or mixture of cations in equilibrium will be as dictated by the final level of acidity of the medium. There must be a sufficient concentration of the Lewis acid or base to buffer against change in acidity of the medium caused by basic entities generated in the reaction.

## 10.2.3.1  Oxidants in Hydrogen Fluoride

The "cleanest" chemical oxidant in HF is elemental $F_2$, its reduction product being $F^-$, the base of the solvent system. After $F_2$ oxidation, the level of $F^-$ in the system will be that dictated by the position of the self-equilibrium of the solvent at the appropriate acidity or basicity. If "neat" HF were to be used, displacement of the self-equilibrium with buildup in $F^-$ concentration could be significant; but, for HF containing a reasonable concentration of a Lewis acid or of the Lewis base $F^-$, the buildup of $F^-$ as the reduction product would be insignificantly small.

Since the strength of the oxidant $F_2$ can override the ability of the acidity level of the solvent to dictate cation speciation, $F_2$ must be added strictly stoichiometrically. When it was used to react with excess $I_2$ in the investigation of the dependence of speciation of polyiodine cations on acidity-basicity levels of HF-based media (section 4.1.1.1), it was metered into the reaction vessel using p-v-T measurements.

In that work it was shown also that when the disproportionation products $I_2$ and $IF_5$ were mixed in any proportions, it was the acidity level of the medium alone which dictated cationic speciation. $IF_5$ was sufficiently mild as an oxidant for acidity to be the determinant of speciation. When an oxidant comparable in strength with $IF_5$, namely, $IO_3^-$, was added to HF containing $I_2$, mutual oxidation-reduction led to the generation of cations the identity of which depended only on the level of acidity of the solutions. Figure 5 of the paper describing that investigation is a flowsheet showing the acidity-basicity dependence of disproportionation and conproportionation of iodine cations. In that same paper, it was reported that polysulfur cations $S_n^{2+}$ were formed in acidic HF when $S_8$ was allowed to interact with $SO_3^{2-}$. and with $SO_4^{2-}$.

The experimental approaches described earlier for generation of polyatomic cations of iodine and sulfur have been shown to have general application to synthesis in HF solutions of polyatomic cations of nonmetals and metals generally, an example being the generation of $Bi_5^{3+}$ in HF—$BF_3$ (section 5.4).

Production in acidified HF of those stable solvated transition metal cations which were described in chapter 6 requires no oxidant—the oxidation state of the cation is that of the HF-insoluble metal fluoride which was then bought into solution by adding an appropriate Lewis acid fluoride. The HF-solvated transition metal cations in chapter 7—those in "low" oxidation states—were produced by oxidizing a metallic transition element with solvated protons from a solution of enhanced acidity.

## 10.2.3.2 Oxidants in Other Superacids

Most of the early work on generation of nonmetal cations was done in oleums of different strengths where $SO_3$ acts both as oxidant and Lewis acid. Given the amount of solute to be oxidized, this means that the oxidant is in vast excess and in the case of the formation of polytellurium cations the ultimate product on long standing was $TeO_2$. (See section 10.1.4.) It was suggested in section 10.2.2.2 that, for syntheses in $H_2SO_4$, acidity levels could be fixed with $B(HSO_4)_3$ as the strong nonoxidizing Lewis acid and that the easily weighed peroxydisulfate ($S_2O_8^{2-}$) could be used to provide strict stoichiometric oxidant control. $S_2O_8^{2-}$ would be reduced in the first instance to $SO_4^-$ which, in the oleum, would be protonated to $HSO_4^-$, the base of the solvent system.

Chapter 4 has many examples of the very elegant experimental approach adopted within the Gillespie group for synthesis in $HSO_3F$ of nonmetal cations. For most of this work, $HSO_3F$ itself was of sufficient acidity to maintain the cation in solution as a stable entity after oxidation of the nonmetallic element with peroxydisulfuryldifluoride, $S_2O_6F_2$, that is, $FO_2S$—O—O—$SO_2F$. The reduction product $SO_3F^-$ is totally compatible with the system, being the base of the solvent. Furthermore, in "neat" $HSO_3F$, Gillespie was able to measure the enhanced conductance due to the $SO_3F^-$ ions generated and use the derived $\gamma$ values to deduce information about species in solution after oxidation. In order to obtain some cations (e.g., $I_2^+$ and $Br_3^+$) as stable entities in $HSO_3F$ after oxidation with $S_2O_6F_2$, the acidity of the medium had to be enhanced with the Lewis acid $SbF_5$ or $SbF_2(SO_3F)_3$.

It is unfortunate that syntheses in $CF_3SO_3H$ cannot follow the same track as those in HF and $HSO_3F$ where the specific strong oxidant is itself reduced to the base of the solvent system. The potential oxidant $S_2O_6(CF_3)_2$, that is, $CF_3$—$O_2S$—O—O—$SO_2CF_3$, is the direct analogue for the $CF_3SO_3H$ solvent system of $S_2O_6F_2$ in $HSO_3F$, but it is thermally unstable in $CF_3SO_3H$ at room temperature. $S_2O_6F_2$ has been used as an effective specific oxidant in $CF_3SO_3H$.

An approach to generation of cations in $HSO_3F$ or $CF_3SO_3H$ that depends on conproportionation of the formal disproportionation products of the cations sought is as valid as for syntheses in HF. The last paragraph of section 4.1.1.1 shows that, when Adrien conproportionated $ICl_3$ and $I_2$ in $CF_3SO_3H$ solutions, the acidity of which was adjusted by addition of Lewis acids, polyiodine cations were formed whose nature depended only on the level of acidity of the medium. Regardless of whether oxidant or reductant was in tenfold excess, acidity level of the medium was shown to be the determinant of speciation.

### 10.2.3.3 Oxidants in Chloroaluminates

Most of the work on synthesis of polyatomic cations of metals and nonmetals in chloroaluminates has involved conproportionation reactions, as seen in chapters 4 and 5 for cations of iodine, tellurium, mercury, cadmium, zinc, and bismuth. Mamantov used specific oxidation of $I_2$ by $Cl_2$ to generate $I_2^+$ in acidic melt (section 4.1.1.2). His approach could have wider application if experimentally convenient. $Cl_2$ could be metered into a reaction system by p-v-T measurements and its reduction product is the base of the medium.

## 10.2.4  Electrochemical Oxidation and Reduction

A point stressed through section 10.2.3 was the desirability of using "clean," specific chemical oxidants. The oxidants $F_2$, $S_2O_8^{2-}$, $S_2O_6F_2$, and $Cl_2$ gave, as their reduction products, the bases of the solvent systems HF, $H_2SO_4$, $HSO_3F$, and chloroaluminates. The ultimate in a clean oxidant or reductant is, of course, the electron.

There was a warning at the end of section 4.2.4 that electrochemical oxidation or reduction as it occurs in cyclic voltammetry, for example, can override certain oxidation states and that, in such a technique, the species being sensed at the working electrode is not necessarily that dictated by the acidity of the bulk of the medium. Obviously, in any medium, if voltammetry or some electrochemical technique is used initially to determine the potential at which a desired species is generated, the system can then be maintained at controlled potential to ensure that only the species being sought is being generated.

Voltammetry and potentiometry have been widely used to obtain information about cations in fractional oxidation states in chloroaluminates, as indicated in several sections of chapter 4. It is this writer's belief that such studies would start from more reliable ground if conproportionation were to be used initially to generate a particular cation at a fixed acidity and that this solution should then be investigated by cyclic voltammetry. Fixed-potential oxidation or reduction, coupled with appropriate acidity control of the medium, could then be used to characterize and isolate a range of cations of the system. The known spectra for $I_3^+$ could be used to define the chloroaluminate acidity at which formation of this cation predominated after conproportionation of $ICl_3$ and $I_2$. This would then seem to provide a firm base for systematic identification, and subsequent isolation, of $I_2^+$ after voltammetric determination of the appropriate potential for oxidative synthesis. Similarly, reduction could yield $I_5^+$ and perhaps even $I_7^+$. Similar strategies could be used for generation of cations of other nonmetals and metals in chloroaluminates.

## 10.2.5 Isolation Procedures

The point has been made several times in this book, especially in section 10.1.1, that the effectiveness of the characterization and isolation of many unusual cationic species could have been improved had more specific reagents been used at different stages of the synthesis and isolation of inorganic compounds from highly acidic media. Too often a compound like $SbF_5$ has been used to provide a medium of very low basicity, as an oxidant and as a source of the counteranion for the cation sought. $SO_3$ has been used in solution studies to fix the level of acidity of $H_2SO_4$, but considering its vast stoichiometric excess in the oleum over the reductant concentration, its action as an oxidant has been largely uncontrollable. Acidity control and oxidant selection have been discussed separately in sections 10.2.2, 10.2.3, and 10.2.4. In this section there will be some examination of the requirements in different media for specific isolation of cations generated in solution.

### 10.2.5.1 Isolation From Hydrogen Fluoride

In section 10.2.2.1 solubilities of fluoroanions of Lewis acid fluorides in HF were discussed in the context of providing reasonable reactant concentrations. The situation in considering isolation of compounds is almost the obverse of that. Another important consideration is whether the solid isolated is to be sufficiently crystalline to allow crystallographic analysis or whether a very insoluble, but possibly noncrystalline, solid is sought.

   In principle, isolation from HF—$BF_3$ or HF—$AsF_5$ of crystalline compounds which include the $BF_4^-$ or the $AsF_6^-$ anion with a transition-metal cation, for example, should be achieved relatively easily by slow removal by distillation of the volatile HF and $BF_3$ or HF and $AsF_5$. However, in practice, real difficulties arise. Volatility of $BF_3$ and lability of the $BF_4^-$ anion are such that, toward the end of distillation, $BF_3$ will be lost from the anion, and the residue tends to be a fluoride or a mixture of the fluoride and tetrafluoroborate of the metal concerned. Many fluoroarsenates have been found to be so soluble in HF that removal of HF and $AsF_5$ leads to a very concentrated solution which suddenly becomes glassy, rather than providing a crop of crystals. As HF—$SbF_5$ solutions become more concentrated, polymerization of $SbF_6^-$ to $Sb_2F_{11}^-$, $Sb_3F_{16}^-$, and $Sb_nF_{5n+1}^-$ occurs (section 3.4.1). The ultimate polymers are extremely viscous and partial pressures of both HF and $SbF_5$ become very small, so that recovery is difficult.

   Experience with $TaF_5$ as a Lewis acid in HF suggests that transition-metal fluorotantalates are very insoluble. This can be an advantage if easily recovered, but noncrystalline, compounds are sought. A similar experience has been found with $GeF_4$, which forms very insoluble hexafluorogermanates with dipositive cations from $Co^{2+}$ to the highly reducing $Sm^{2+}$. Sm-$GeF_6$, isolated from HF, is so insoluble that it does not undergo hydrolytic oxidation for some hours when left in a normal atmosphere. Presumably the $Sm^{2+}$—$GeF_6^{2-}$ interaction in HF is similar to that for $Ba^{2+}$ and $SO_4^{2-}$ in $H_2O$. $SiF_4$ in HF also provides insoluble hexafluorosilicates with dipositive transition metal cations.

### 10.2.5.2 Isolation From Other Superacids

Polycations of iodine and sulfur, for example, were characterized by physicochemical measurements and spectroscopy in superacids such as $HSO_3F$, but were isolated for crystallographic analysis after oxidation of the nonmetallic element with $SbF_5$ or $AsF_5$, as fluoroantimonates or fluoroarsenates, $SO_2$ frequently being used as a reaction medium.

   Little work has been reported on isolation from $H_2SO_4$, $HSO_3F$, or $CF_3SO_3H$ of specific compounds of cations generated in those media. Attempted crystallization of compounds with anions which are the bases of these acids is not likely to be successful because of the ability of the anions to form polymeric solids. (See section 10.2.2.2.) Specific precipitants which are reasonably compatible with the solvents need to be sought. Fluoroanions, particularly if di- or trinegative, might be good candidates for investigations using $HSO_3F$ or $CF_3SO_3H$ as reaction media.

### 10.2.5.3 Isolation From Chloroaluminates

With very few exceptions the compounds containing nonmetallic or metallic cations which have been recovered from chloroaluminate reaction media have been based on $AlCl_4^-$ anions. A notable exception was Corbett's characterization of $Bi^+Bi_9^{5+}(HfCl_6^{2-})_3$, although even this compound was isolated from a neutral stoichiometric reaction mixture of Bi, $BiCl_3$, and $HfCl_4$ (section 5.4). It will be proposed shortly that future synthetic strategies for identifying new cations might include adding anions like $HfCl_6^{2-}$ or their germanium, tin, thorium, or uranium(IV) counterparts to acidic chloroaluminate reaction media.

## 10.2.6  Choice of Reaction Media—Summary

In the various subsections of section 10.2 there have been separate presentations of the individual experimental conditions that should be considered before synthesis of cations in low or fractional oxidation state is to be undertaken. The most appropriate medium would be selected on the basis of the way in which its physical properties affect its ease of manipulation, on the basis of its strength as an oxidant—preferably very small—and on the level of its acidity both in the "neat" state and with that acidity adjusted by use of appropriate Lewis acids or bases. The nature of the redox reaction which would generate the cation must be considered, whether to use a strong specific oxidant in strictly stroichiometric amount, whether to use a conproportionation reaction or electrochemical oxidation or reduction. Finally, if a pure compound which contains the cation is required, a specific precipitant should be sought, particularly if it is not possible or easy to isolate the cation in a compound of which the anion is the base of the solvent system.

A hypothetical example might be drawn from chloroaluminate chemistry where deliberate separate selection of specific reagents has often been lacking. $Cd_3^{2+}$ and $Cd_4^{2+}$ have both been prepared as fluoroarsenates by oxidation with $AsF_5$ in $AsF_3$. The most reduced cation of cadmium recovered from chloroaluminates is $Cd_2^{2+}$, isolated as $Cd_2(AlCl_4)_2$ from neutral melt.

Possible synthesis of $Cd_3^{2+}$ and $Cd_4^{2+}$ in solution in chloroaluminates could involve reaction in the most acidic melt practically possible using conproportionation (Cd:CdCl$_2$, 2:1 or 3:1), or direct oxidation of Cd in the melt with the appropriate stoichiometric amounts of $Cl_2$, or controlled electrochemical reduction of $CdCl_2$ dissolved in the acidic melt. A suitable precipitating anion would then be sought. $HfCl_6^{2-}$ might not be the ideal candidate because of acid-base competition between $HfCl_4$ and $AlCl_3$ and because $HfCl_4$ is reasonably volatile by comparison with $AlCl_3$ and may be lost from the reaction medium during removal of $AlCl_3$ to concentrate the solution. Much-less-volatile tetrachlorides such as those of Th and U could be examined, although mono- and trinegative chloroanions should also be considered. An essential feature of the binary chloride used as a precipitant would be that it is a stronger Lewis acid in chloride systems than $AlCl_3$, so that its choroanion would precipitate with the sought cation while the acidity of the melt is maintained with $AlCl_3$.

## 10.3 Conclusion

The basic propositions put forward in this book have been that, despite differences in the chemical nature and temperature domains of various reaction media, elements will be expected to exist as cations under sufficiently acidic conditions and as anions in basic media. For nonmetals, homopolyatomic cations of increasing charge per atom of the element can be stabilized by increasing the acidity of the medium. For metallic elements, the charge per metal on homopolyatomic cations can be decreased by increasing the acidity. The charge on stable monatomic cations of metallic elements also decreases with increasing acidity.

In all of the preceding, increasing acidity of the medium implies decreasing availability of the base of the solvent system. For all of the cations above—metallic or nonmetallic, polyatomic or monatomic—progressive additions of base lead ultimately to disproportionation of a cationic species to the element itself and to a nonionic or insoluble compound formed between the base and the particular element in an oxidation state higher than that in the cation which disproportionated. For polyatomic cations, disproportionation resulting from the addition of a relatively small amount of base may lead to the formation of a cationic species with a formal oxidation state intermediate between that of the parent cation and the zero oxidation state of the element.

Although not dealt with in great detail, it has been shown that the reverse situation holds for stabilization of unusual anionic species. Homopolyatomic anions of groups VI, V, and IV require progressively more basic media for their stabilization and special attention must be given to the nature of the countercation in isolation of solid compounds.

It seems necessary also to direct attention to the need to postulate, as a normal requirement, the existence of cationic species in the interpretation of observed reaction pathways in acidic systems. There has been far too little acceptance of the existence of solvated cationic species in many acidic nonaqueous systems, particularly in acidic melts. This seems strange when we accept cations, however complex, as the expected species in acidic aqueous solutions and anions as the norm under basic aqueous conditions.

## References

1. T.A. O'Donnell, *J. Fluorine Chem.*, *11*, (a) 477–478, (b) 470–473 (1978).
2. K.R. Seddon in *Molten Salt Chemistry*, G. Mamantov, R. Marassi, eds. D. Reidel, 1987, pp. 370–381.

# Subject Index

---

*See* trifluoromethylsulfuric acid